THE FOUNDATIONS OF MATHEMATICS

THE FOUNDATIONS
OF MATHEMATICS

Second Edition

IAN STEWART AND DAVID TALL

OXFORD
UNIVERSITY PRESS

OXFORD
UNIVERSITY PRESS

Great Clarendon Street, Oxford, OX2 6DP,
United Kingdom

Oxford University Press is a department of the University of Oxford.
It furthers the University's objective of excellence in research, scholarship,
and education by publishing worldwide. Oxford is a registered trade mark of
Oxford University Press in the UK and in certain other countries

First Edition published in 1977
Second Edition published in 2015

Impression: 1
10093715'72

Published in the United States of America by Oxford University Press
198 Madison Avenue, New York, NY 10016, United States of America

British Library Cataloguing in Publication Data
Data available

Library of Congress Control Number: 2014946122

ISBN 978-0-19-870644-1 (hbk.)
ISBN 978-0-19-870643-4 (pbk.)

Printed and bound by
CPI Group (UK) Ltd, Croydon, CR0 4YY

TO
PROFESSOR RICHARD SKEMP

whose theories on the learning of mathematics have been
a constant source of inspiration

PREFACE TO THE SECOND EDITION

The world has moved on since the first edition of this book was written on typewriters in 1976. For a start, the default use of male pronouns is quite rightly frowned upon. Educationally, research has revealed new insights into how individuals learn to think mathematically as they build on their previous experience (see [3]).[1] We have used these insights to add comments that encourage the reader to reflect on their own understanding, thereby making more sense of the subtleties of the formal definitions. We have also added an appendix on *self-explanation* (written by Lara Alcock, Mark Hodds, and Matthew Inglis of the Mathematics Education Centre, Loughborough University) which has been demonstrated to improve long-term performance in making sense of mathematical proof. We thank the authors for their permission to reproduce their advice in this text.

The second edition has much in common with the first, so that teachers familiar with the first edition will find that most of the original content and exercises remain. However, we have taken a significant step forward. The first edition introduced ideas of set theory, logic, and proof and used them to start with three simple axioms for the natural numbers to construct the real numbers as a complete ordered field. We generalised counting to consider infinite sets and introduced infinite cardinal numbers. But we did not generalise the ideas of measuring where units could be subdivided to give an ordered field.

In this edition we redress the balance by introducing a new part IV that retains the chapter on infinite cardinal numbers while adding a new chapter on how the real numbers as a complete ordered field can be extended to a larger ordered field.

This is part of a broader vision of formal mathematics in which certain theorems called *structure theorems* prove that formal structures have natural interpretations that may be interpreted using visual imagination and symbolic manipulation. For instance, we already know that the formal concept of a complete ordered field may be represented visually as points on a number line or symbolically as infinite decimals to perform calculations.

---

[1] Numbers in square brackets refer to entries in the References and Further Reading sections on page 383.

Structure theorems offer a new vision of formal mathematics in which formal defined concepts may be represented in visual and symbolic ways that appeal to our human imagination. This will allow us to picture new ideas and operate with them symbolically to imagine new possibilities. We may then seek to provide formal proof of these possibilities to extend our theory to combine formal, visual, and symbolic modes of operation.

In Part IV, chapter 12 opens with a survey of the broader vision. Chapter 13 introduces group theory, where the formal idea of a group—a set with an operation that satisfies a particular list of axioms—is developed to prove a structure theorem showing that elements of the group operate by permuting the elements of the underlying set. This structure theorem enables us to interpret the formal definition of a group in a natural way using algebraic symbolism and geometric visualisation.

Following chapter 14 on infinite cardinal numbers from the first edition, chapter 15 uses the completeness axiom for the real numbers to prove a simple structure theorem for any ordered field extension K of the real numbers. This shows that K must contain elements k that satisfy $k > r$ for all real numbers r, which we may call 'infinite elements', and these have inverses $h = 1/k$ that satisfy $0 < h < r$ for all positive real numbers r, which may be called 'infinitesimals'. (There are corresponding notions of negative infinite numbers k satisfying $k < r$ for all negative real numbers r.) The structure theorem also proves that any finite element k in K (meaning $a < k < b$ for real numbers a, b) must be of the form $a + h$ where a is a real number and h is zero or an infinitesimal. This allows us to visualise the elements of the larger field K as points on a number line. The clue lies in using the magnification $m : K \to K$ given by $m(x) = (x - a)/h$ which maps a to 0 and $a + h$ to 1, scaling up infinitesimal detail around a to be able to see it at a normal scale.

This possibility often comes as a surprise to mathematicians who have worked only within the real numbers where there are no infinitesimals. However, in the larger ordered field we can now *see* infinitesimal quantities in a larger ordered field as points on an extended number line by magnifying the picture.

This reveals two entirely different ways of generalising number concepts, one generalising counting, the other generalising the full arithmetic of the real numbers. It offers a new vision in which axiomatic systems may be defined to have consistent structures within their own context yet differing systems may be extended to give larger systems with different properties. Why should we be surprised? The system of whole numbers does not have multiplicative inverses, but the field of real numbers does have multiplicative inverses for all non-zero elements. Each extended system has properties that are relevant to its own particular context. This releases us from the limitations of our real-world experience to use our imagination to develop powerful new theories.

The first edition of the book took students from their familiar experience in school mathematics to the more precise mathematical thinking in pure mathematics at university. This second edition allows a further vision of the wider world of mathematical thinking in which formal definitions and proof lead to amazing new ways of defining, proving, visualising, and symbolising mathematics beyond our previous expectations.

Ian Stewart and David Tall
Coventry 2015

PREFACE TO THE FIRST EDITION

This book is intended for readers in transition from school mathematics to the fully-fledged type of thinking used by professional mathematicians. It should prove useful to first-year students in universities and colleges, and to advanced students in school contemplating further study in pure mathematics. It should also be of interest to a wider class of reader with a grounding in elementary mathematics seeking an insight into the foundational ideas and thought processes of mathematics.

The word 'foundations', as used in this book, has a broader meaning than it does in the building trade. Not only do we base our mathematics on these foundations: they make themselves felt at all levels, as a kind of cement which holds the structure together, and out of which it is fabricated. The foundations of mathematics, in this sense, are often presented to students as an extended exercise in mathematical formalism: formal mathematical logic, formal set theory, axiomatic descriptions of number systems, and technical constructions of them; all carried out in an exotic and elaborate symbolism. Sometimes the ideas are presented 'informally' on the grounds that complete formalism is too difficult for the delicate flowering student. This is usually true, but for an entirely different reason.

A purely formal approach, even with a smattering of informality, is psychologically inappropriate for the beginner, because it fails to take account of the realities of the learning process. By concentrating on the technicalities, at the expense of the manner in which the ideas are conceived, it presents only one side of the coin. The practising mathematician does not think purely in a dry and stereotyped symbolism: on the contrary, his thoughts tend to concentrate on those parts of a problem which his experience tells him are the main sources of difficulty. While he is grappling with them, logical rigour takes a secondary place: it is only after a problem has, to all intents and purposes, been solved intuitively that the underlying ideas are filled out into a formal proof. Naturally there are exceptions to this rule: parts of a problem may be fully formalised before others are understood, even intuitively; and some mathematicians seem to *think* symbolically. Nonetheless, the basic force of the statement remains valid.

The aim of this book is to acquaint the student with the way that a practising mathematician tackles his subject. This involves including the standard 'foundations' material; but our aim is to develop the formal approach as a natural outgrowth of the underlying pattern of ideas. A sixth-form student has a broad grasp of many mathematical principles, and our aim is to make use of this, honing his mathematical intuition into a razor-sharp tool which will cut to the heart of a problem. Our point of view is diametrically opposed to that where (all too often) the student is told 'Forget all you've learned up till now, it's wrong, we'll begin again from scratch, only this time we'll get it right'. Not only is such a statement damaging to a student's confidence: it is also untrue. Further, it is grossly misleading: a student who really did forget all he had learned so far would find himself in a very sorry position.

The psychology of the learning process imposes considerable restraints on the possible approaches to a mathematical concept. Often it is simply not appropriate to *start* with a precise definition, because the content of the definition cannot be appreciated without further explanation, and the provision of suitable examples.

The book is divided into four parts to make clear the mental attitude required at each stage. Part I is at an informal level, to set the scene. The first chapter develops the underlying philosophy of the book by examining the learning process itself. It is not a straight, smooth path; it is of necessity a rough and stony one, with side-turnings and blind alleys. The student who realises this is better prepared to face the difficulties. The second chapter analyses the intuitive concept of a real number as a point on the number line, linking this to the idea of an infinite decimal, and explaining the importance of the completeness property of the real numbers.

Part II develops enough set theory and logic for the task in hand, looking in particular at relations (especially equivalence relations and order relations) and functions. After some basic symbolic logic we discuss what 'proof' consists of, giving a formal definition. Following this we analyse an actual proof to show how the customary mathematical style relegates routine steps to a contextual background—and quite rightly so, inasmuch as the overall flow of the proof becomes far clearer. Both the advantages and the dangers of this practice are explored.

Part III is about the formal structure of number systems and related concepts. We begin by discussing induction proofs, leading to the Peano axioms for natural numbers, and show how set-theoretic techniques allow us to construct from them the integers, rational numbers, and real numbers. In the next chapter we show how to reverse this process, by axiomatising the real numbers as a complete ordered field. We prove that the structures obtained in this way are essentially unique, and link the formal structures to their intuitive counterparts of part I. Then we go on to consider complex numbers, quaternions, and general algebraic and mathematical structures, at which point the whole vista of mathematics lies at our feet. A discussion of infinite cardinals, motivated by the idea of counting, leads towards more advanced work. It also hints that we have not yet completed the task of formalising our ideas. Part IV briefly considers this final step: the formalisation of set theory. We give one possible set of axioms, and discuss the axiom of choice, the continuum hypothesis, and Gödel's theorems.

Throughout we are more interested in the ideas behind the formal façade than in the internal details of the formal language used. A treatment suitable for a professional mathematician is often not suitable for a student. (A series of tests carried out by one of us with the aid of first-year undergraduates makes this assertion very clear indeed!) So this is not a rigidly logical development from the elements of logic and set theory, building up a rigorous foundation for mathematics (though by the end the student will be in a position to appreciate how this may be achieved). Mathematicians do not think in the orthodox way that a formal text seems to imply. The mathematical mind is inventive and intricate; it jumps to conclusions: it does not always proceed in a sequence of logical steps. Only when everything is understood does the pristine logical structure emerge. To show a student the finished edifice, without the scaffolding required for its construction, is to deprive him of the very facilities which are essential if he is to construct mathematical ideas of his own.

I.S. and D.T.
Warwick
October 1976

CONTENTS

Part V Strengthening the Foundations

PART I
The Intuitive Background

The first part of the book reflects on the experiences that the reader will have encountered in school mathematics to use it as a basis for a more sophisticated logical approach that precisely captures the structure of mathematical systems.

Chapter 1 considers the learning process itself to encourage the reader to be prepared to think in new ways to make sense of a formal approach. As new concepts are encountered, familiar approaches may no longer be sufficient to deal with them and the pathway may have side-turnings and blind alleys that need to be addressed. It is essential for the reader to reflect on these new situations and to prepare a new overall approach.

Using a 'building' metaphor, we are surveying the territory to see how we can use our experience to build a firm new structure in mathematics that will make it strong enough to support higher levels of development. In a 'plant' metaphor, we are considering the landscape, the quality of the soil, and the climate to consider how we can operate to guarantee that the plants we grow have sound roots and predictable growth.

Chapter 2 focuses on the intuitive visual concept of a real number as a point on a number line and the corresponding symbolic representation as an infinite decimal, leading to the need to formulate a definition for the completeness property of the real numbers. This will lead in the long term to surprising new ways of seeing the number line as part of a wider programme to study the visual and symbolic representations of formal structures that bring together formal, visual, and symbolic mathematics into a coherent framework.

CHAPTER 1

Mathematical Thinking

Mathematics is not an activity performed by a computer in a vacuum. It is a human activity performed in the light of centuries of human experience, using the human brain, with all the strengths and deficiencies that this implies. You may consider this to be a source of inspiration and wonder, or a defect to be corrected as rapidly as possible, as you wish; the fact remains that we must come to terms with it.

It is not that the human mind cannot think logically. It is a question of different kinds of understanding. One kind of understanding is the logical, step-by-step way of understanding a formal mathematical proof. Each individual step can be checked but this may give no idea how they fit together, of the broad sweep of the proof, of the reasons that lead to it being thought of in the first place.

Another kind of understanding arises by developing a global viewpoint, from which we can comprehend the entire argument at a glance. This involves fitting the ideas concerned into the overall pattern of mathematics, and linking them to similar ideas from other areas. Such an overall grasp of ideas allows the individual to make better sense of mathematics as a whole and has a cumulative effect: what is understood well at one stage is more likely to form a sound basis for further development. On the other hand, simply learning how to 'do' mathematics, without having a wider grasp of its relationships, can limit the flexible ways in which mathematical knowledge can be used.

The need for overall understanding is not just aesthetic or educational. The human mind tends to make errors: errors of fact, errors of judgement, errors of interpretation. In the step-by-step method we might not notice that one line is not a logical consequence of preceding ones. Within the overall framework, however, if an error leads to a conclusion that does not fit into the total picture, the conflict will alert us to the possibility of a mistake.

For instance, given a column of a hundred ten-digit numbers to add up, where the correct answer is 137568304452, we might make an arithmetical error and get 137568804452 instead. When copying this answer we might make a second error and write 1337568804452. Both of these errors could escape detection. Spotting the first would almost certainly need a step-by-step check of the calculation. The second error, however, is easily detected because it does not fit into the overall pattern of arithmetic. A sum of 100 ten-digit numbers will be at most a twelve-digit number (since $9999999999 \times 100 = 999999999900$) and the final proposed answer has thirteen.

It is a combination of step-by-step and overall understanding that has the best chance of detecting mistakes; not just in numerical work, but in all areas of human understanding. The student must develop both kinds, in order to appreciate the subject fully and be an effective practitioner. Step-by-step understanding is fairly easy; just take one thing at a time and do lots of 'drill' exercises until the idea sinks in. Overall understanding is much harder; it involves taking a lot of individual pieces of information and making a coherent pattern out of them. What is worse is that having developed a particular pattern which suits the material at one stage, new information may arise which seems to conflict. The new information may be erroneous but it often happens that previous experiences that worked in one situation no longer operate in a new context. The more radical the new information is, the more likely that it does not fit, and that the existing overall viewpoint has to be modified. That is what this first chapter is about.

Concept Formation

When thinking about any area of mathematics, it helps to understand a little about how we learn new ideas. This is especially true of foundational issues, which involve revisiting ideas that we already think we know. When we discover that we do not—more precisely, that there are basic questions that we have not been exposed to—we may feel uncomfortable. If so, it's good to know that we are not alone: it happens to nearly everyone.

All mathematicians were very young when they were born. This platitude has a non-trivial implication: even the most sophisticated mathematician must have passed through the complex process of building up mathematical concepts. When first faced with a problem or a new concept, the mathematician turns it over in the mind, digging into personal experiences to see if it is like something that has been encountered before. This exploratory, creative phase of mathematics is anything but logical. It is only when the pieces begin to fit together and the mathematician gets a 'feel' for the concept, or the problem, that a semblance of order emerges. Definitions are formulated in ways that can be used for deduction, and there is a final polishing phase where the essential facts are marshalled into a neat and economical proof.

As a scientific analogy, consider the concept 'colour'. A dictionary definition of this concept looks something like 'the sensation produced in the eye by rays of decomposed light'. We do not try to teach the concept of colour to a child by presenting them with this definition. ('Now, Angela, tell me what sensation is produced in your eye by the decomposed light radiating from this lollipop . . . ') First you teach the concept 'blue'. To do this you show a blue ball, a blue door, a blue chair, and so on, accompanying each with the word 'blue'. You repeat this with 'red', 'yellow', and so on. After a while the child begins to get the idea; you point to an object they have not seen before and their response is 'blue'. It is relatively easy to refine this to 'dark blue', 'light blue', and so forth. After repeating this procedure many times, to establish the individual colours, you start again. 'The colour of that door is blue. The colour of this box is red. What colour is that buttercup?' If the response is 'yellow' then the concept 'colour' is beginning to develop.

As a child develops and learns scientific concepts they may eventually be shown a spectrum obtained by passing light through a prism. This may lead to learning about the wavelength of light, and, as a fully fledged scientist, being able to say with precision which wavelength corresponds to light of a particular colour. The understanding of the concept 'colour' is now highly refined, but it does not help the scientist to explain to a child what 'blue' is. The existence of a precise and unambiguous definition of 'blue' in terms of wavelength is of no use at the concept-forming stage.

It is the same with mathematical concepts. The reader already has a large number of mathematical concepts established in their mind: how to solve a quadratic equation, how to draw a graph, how to sum a geometric progression. They have great facility in arithmetical calculations. Our aim is to build on this wealth of mathematical understanding and to refine these concepts to a more sophisticated level. To do this we use examples, drawn from the reader's experience, to introduce new concepts. Once these concepts are established, they become part of a richer experience upon which we can again draw to aim even higher.

Although it is certainly possible to build up the whole of mathematics by axiomatic methods starting from the empty set, using no outside information whatsoever, it is also totally unintelligible to anyone who does not already understand the mathematics being built up. An expert can look at a logical construction in a book and say 'I guess that thing there is meant to be "zero", so that thing is "one", that's "two", . . . this load of junk must be the integers, . . . what's that? Oh, I think I see: it must be "addition". . . '.

The non-expert is faced with an indecipherable mass of symbols. It is never sufficient to define a new concept without giving enough examples to explain what it looks like and what can be done with it. Of course, an expert is often in a position to supply their own examples, and may not need much help.

Schemas

A mathematical concept, then, is an organised pattern of ideas that are somehow interrelated, drawing on the experience of concepts already established. Psychologists call such an organised pattern of ideas a 'schema'. For instance, a young child may learn to count ('one, two, three-four-five, once I caught a fish alive') progressing to ideas like 'two sweets', 'three dogs', . . . and eventually discovers that two sweets, two sheep, two cows have something in common, and that something is 'two'. He or she builds a schema for the concept 'two' and this schema involves the experience that everyone has two hands, two feet, last week we saw two sheep in a field, the fish-alive rhyme goes 'one, two, . . . ', and so on. It is really quite amazing how much information the brain has lumped together to form the concept, or the schema.

The child progresses to simple arithmetic ('If you have five apples and you give two away, how many will you have left?') and eventually builds up a schema to handle the problem 'What is five minus two?' Arithmetic has very precise properties. If 3 and 2 make 5, then 5 take away 2 leaves 3. The child discovers these properties by trying to make sense of arithmetic. It then becomes possible to use known facts to derive new facts. If the child knows that 8 plus 2 makes 10, then 8 plus 5 can be thought of as 8 plus 2 plus 3, so the sum is 10 plus 3, which is 13. Over time the child can build up a rich schema of whole number arithmetic.

At this point, if you ask 'What is five minus six?' the response is likely to be 'You can't do it', or perhaps just an embarrassed giggle that an adult should ask such a silly question. This is because the question does not fit the child's schema for subtraction: when thinking about 'five apples, take six away', this simply cannot be done. At a later stage, experiencing negative numbers will give the answer 'minus one'. What has happened? The child's original schema for 'subtraction' has been modified to accommodate new ideas—perhaps by thermometer scales, or the arithmetic of banking, or whatever—and the understanding of the concept changes. During the process of change, confusing problems will arise (what does minus one apple look like?) which may eventually be resolved satisfactorily (apples don't behave like thermometer readings).

A large part of the learning process involves making an existing schema more sophisticated, so that it can take account of new ideas. This process, as we have said, may be accompanied by a state of confusion. If it were possible to learn mathematics without becoming confused, life would be wonderful.

Unfortunately, the human mind does not seem to work that way. More than 2000 years ago, Euclid supposedly told King Ptolemy I that 'There is no royal road to geometry'. The next best thing is to recognise not just the confusion, but also its causes. At various stages in reading this book the reader will be confused. Sometimes, no doubt, the cause will be the authors' sloppiness, but often it will be the process of modifying personal knowledge to make sense of a more general situation. This type of confusion is creative, and it should be welcomed as a sign that progress is being made—unless it persists for too long. By the same token, once the confusion is resolved, a sudden clarity can appear with a feeling of great pleasure that the pieces fit together perfectly like a jigsaw. It is this feeling of perfect harmony that makes mathematics not only a challenge, but also an endeavour that leads to deep aesthetic satisfaction.

An Example

This way to develop new ideas is illustrated by the historical development of mathematical concepts—itself a learning process, but involving many minds instead of one. When negative numbers were first introduced, they met considerable opposition: 'You can't have less than nothing'. Yet nowadays, in this financial world of debits and credits, negative numbers are a part of everyday life.

The development of complex numbers is another example. Like all mathematicians, Gottfried Leibniz knew that the square of a positive number or of a negative number must always be positive. If i is the square root of minus one, then $i^2 = -1$, so i cannot be a positive or a negative number. Leibniz believed that it should therefore be endowed with great mystical significance: a non-zero number neither less than zero nor greater than zero. This led to enormous confusion and distrust concerning complex numbers; it persists to this day in some quarters.

Complex numbers do not fit readily into many people's schema for 'number', and students often reject the concept when it is first presented. Modern mathematicians look at the situation with the aid of an enlarged schema in which the facts make sense.

Imagine the real numbers marked on a line in the usual way:

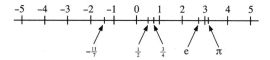

Fig. 1.1 The real numbers

Negative numbers are to the left of zero, positive to the right. Where does i go? It can't go to the left; it can't go to the right. The people whose schema does not allow complex numbers must argue thus: this means that it can't go anywhere. There is no place on the line where we can mark i, so it's not a number.

However, there's an alternative. We can visualise complex numbers as the points of a plane. (In 1758 François Daviet de Foncenex stated that it was pointless to think of imaginary numbers as forming a line at right angles to the real line. Fortunately others disagreed.) The real numbers lie along the 'x-axis', the number i lies one unit above the origin along the 'y-axis', and the number $x + iy$ lies x units along the real line and then y units above it (change directions for negative x or y). The objection to i ('it can't lie anywhere on the line') is countered by the observation that it doesn't. It lies one unit above the line. The enlarged schema can accommodate the disturbing facts without any trouble.

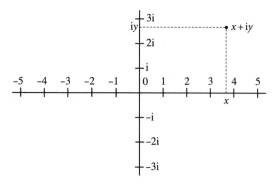

Fig. 1.2 Putting i in its place

This happens quite often in mathematics. When a particular situation is generalised to a new context, some properties operate in the same way as before, such as addition and multiplication both being commutative. But other properties (such as the order properties of real numbers) that work well in the original schema are no longer relevant in the extended schema (in this case the schema of complex numbers).

This is a very general phenomenon; it has happened not only to students, but to mathematicians throughout history, up to the present day. If you work in an established situation where the ideas have been fully sorted out, and the methods used are sufficient to solve all of the usual problems, it is not that difficult to teach an apprentice the trade. All you need is to grasp the current principles and develop fluency in the methods. But when there is a genuine change in the nature of the system, as happened when negative numbers were introduced in a world that only used natural counting numbers, or when complex numbers were encountered solving equations, then there is a genuine period of confusion for everyone. What are these newfangled things? They certainly don't work the way I expected them to!

This can cause deep confusion. Some conquer it by engaging with the ideas in a determined and innovative fashion; others suffer a growing feeling of anxiety, even revulsion and rejection.

One such major occasion began in the final years of the nineteenth century and transformed the mathematics of the twentieth and twenty-first centuries.

Natural and Formal Mathematics

Mathematics began historically with activities such as counting objects and measuring quantities, dealing with situations in the natural world. The Greeks realised that drawing figures and counting pebbles had more profound properties, and they built up the method of Euclidean proof in geometry and the theory of prime numbers in arithmetic. Even though they developed a Platonic form of mathematics that imagined perfect figures and perfect numbers, their ideas were still linked to nature. This attitude continued for millennia. When Isaac Newton studied the force of gravity and the movement of the heavenly bodies, science was known as 'natural philosophy'. He built his ideas about calculus on Greek geometry, and on algebra that generalised the natural operations of arithmetic.

The reliance on 'naturally occurring' mathematics continued until the late nineteenth century, when the focus changed from the properties of objects and operations to the development of formal mathematics based on set-theoretic definition and logical proof. This historical transition from natural to formal mathematics involved a radical change of viewpoint, leading to far more powerful insights into mathematical thinking. It plays an essential role in the shift from school geometry and algebra to formal mathematics at university.

Building Formal Ideas on Human Experience

As mathematics becomes more sophisticated, new concepts often involve some ideas that generalise, but others that operate in new ways. As the transition is made from school mathematics to formal mathematics, it may seem logical to start anew with formal definitions and learn how to make formal deductions from first principles. However, experience over the last half-century has shown that this is not a sensible idea. In the 1960s, schools tried a new approach to mathematics, based on set theory and abstract definitions. This 'new math' failed because, although experts might understand the abstract subtleties, learners need to build up a coherent schema of knowledge to make sense of the definitions and proofs. We now know more about how humans learn to think mathematically. This lets us give examples from practical research to show how students have interpreted ideas in ways that are subtly different from what is intended in the printed text. We mention this to encourage you to think carefully about the precise meanings involved, and to develop strong mathematical links between ideas.

It is helpful to read proofs carefully and to get into the habit of *explaining to yourself* why the definitions are phrased as they are and how each line of a proof follows from previous lines. (See the Appendix on Self-Explanation on page 377.) Recent research [3] has shown that students who make an effort to think through theorems for themselves benefit in the long run. Eye-tracking equipment has been used to study how students read pages from the first edition of this very book. There is a strong correlation between spending longer considering significant steps in a proof and obtaining higher marks on tests administered at a later stage. It's a no-brainer really. A stronger effort at making personal links gives you a more coherent personal schema of knowledge that will be of benefit in the long run.

You need to be sensible about how to proceed. In practice, it is not always possible to give a precise, dictionary definition for every concept encountered. We may talk about a set being 'a well-defined collection of objects', but we will be begging the question, since 'collection' and 'set' mean the same thing.

When studying the foundations of mathematics, we must be prepared to become acquainted with new ideas by degrees, rather than by starting from a watertight definition that can be assimilated at once. As we continue along that path, our understanding of an idea can become more sophisticated. We can sometimes reach a stage where the original vague definition can be reformulated in a rigorous context ('yellow is the colour of light with a wavelength of 5500 Å'). The new definition, seemingly so much better than the vague ideas that led to its formulation, has a seductive charm.

Wouldn't it be so much better to start from this nice, logical definition? The short answer is 'no'.

In this book, we begin in Part I with ideas that you have met in school. We consider the visual number line, and how it is built up by marking various number systems, such as the whole numbers, 1, 2, 3, . . . ; then fractions between adjacent whole numbers; then signed numbers to the right and left of the origin, including signed whole numbers (the integers) and signed fractions (the rationals); then expanding to the real numbers including both rational and irrational numbers. In particular, we focus on natural ways to perform operations such as addition, multiplication, subtraction, and division, using whole numbers, fractions, decimals, and so on, to highlight properties that can be used as a basis for formal axioms for the various number systems.

Part II lays the foundations for set theory and logic, appropriate to the concept of proof used by mathematicians, with a sensible balance of logical precision and mathematical insight. In particular, the reader should note that it is essential to focus not only on what the definitions actually say, but also to be careful not to assume other properties that may arise not from the definition but from mental links set up by previous experience. For instance, students in school meet functions such as $y = x^2$ or $f(x) = \sin 3x$, which are always given by some kind of formula. However, the general notion of a function does not require a formula. All that is needed is that for each value of x (in a specified set) there is a single corresponding value of y. This broader definition applies to sets in general, not just to numbers. The properties that a defined concept must have are deduced from the definition by mathematical proof.

Part III develops the axiomatic structures appropriate for the succession of number systems, starting with axioms for natural numbers and proof by induction. The story continues by demonstrating how successive systems—integers, rationals, and real numbers—can be constructed from first principles using set-theoretic techniques. This process culminates in a list of axioms that defines the system of real numbers, with two operations (addition and multiplication) that satisfy specified properties of arithmetic and order, together with a 'completeness axiom' that states that any increasing sequence bounded above must tend to a limit. These axioms define a 'complete ordered field', and we prove that they specify the real numbers *uniquely*. Real numbers may be pictured as points on a line with the defined operations of addition, multiplication, and order, where the line is filled out to include irrational numbers such as $\sqrt{2}$ or π as infinite decimals that may be computed to any required accuracy as a finite decimal. For instance, $\sqrt{2}$ is 1·414 to 3 decimal places, π is approximately equal to the fraction 22/7, or may be calculated to any desired accuracy as a decimal, say 3·14 to two decimal places or 3·1415926536 to ten places.

Formal Systems and Structure Theorems

This sequence of development, building a formal system from a carefully chosen list of axioms, can be generalised to cover a wide range of new situations. It has a huge advantage compared to dealing with naturally occurring systems that are encountered in everyday life. The theorems that can be deduced from a given list of axioms using formal proof must hold in *any* system that satisfies the axioms—old or new. Formal theorems are *future-proofed*. The theorems apply not only to systems that are already familiar, but also to any new system that satisfies the given axioms. This releases us from the necessity of re-checking our beliefs in every new system we encounter. This is a major step forward in mathematical thinking.

Another more subtle development is that some theorems deduced within a formal system prove that the system has specific properties that allow it to be visualised in a certain way, and other properties that allow its operations to be carried out using symbolic methods. Such theorems are called *structure theorems*. For example, any complete ordered field has a unique structure that may be represented as points on a number line or as decimal expansions.

This shifts formal proof to a new level of power. Not only do we devote lengthy resources to develop a consistent approach to formal proof, ultimately we can develop new ways of thinking that blend together formal, visual, and symbolic ways of operation that combine human ingenuity and formal precision.

Using Formal Mathematics More Flexibly

In Part IV we show how these more flexible methods can be applied in various contexts, first by applying the ideas to group theory and then to two quite different extensions of finite ideas to infinite concepts. One is the extension of counting from finite sets to infinite sets, by saying that two sets have the same *cardinal number* if all their elements can be paired so that each element in one set corresponds to precisely one element in the other. Cardinal numbers have many properties in common with regular counting numbers, but they also have new and unfamiliar properties. For instance, we can take away an infinite subset (such as the even numbers) from an infinite set (such as the natural numbers) to leave an infinite subset (the odd numbers) with the same cardinal number of elements as the original set. By the same token, subtraction cannot be uniquely defined for infinite cardinal numbers, nor can division, so the reciprocal of an infinite cardinal number is not defined as a cardinal number.

The second extension places the real numbers, which form a complete ordered field, inside a larger (but not complete) ordered field. Here, an element k in the larger field may satisfy the order property '$k > r$ for every real number r'. In this sense, k is infinite: in the formally defined order, it is greater than *all* real numbers. Yet this k behaves quite differently from an infinite cardinal number, because it has a reciprocal $1/k$. Moreover, $1/k$ is smaller than any positive real number.

Upon reflection, we should not be surprised by these apparently contradictory possibilities, where an infinite number has a reciprocal in one system but not in another. The system of whole numbers that we use for counting does not provide reciprocals, but the systems of rational and real numbers do. If we select certain properties to generalise different systems, we should not be surprised if the generalisations are also different.

This brings us to an important conclusion. Mathematics is a living subject, in which seemingly impossible ideas may become possible in a new formal context, determined by stating appropriate axioms.

Writing over a century ago, when the new formal approach to mathematics was becoming widespread, Felix Klein [4] wrote:

> Our standpoint today with regard to the foundations is different from that of the investigators of a few decades ago; and what we today would state as ultimate principles, will certainly be outstripped after a time.

On the same page he noted:

> Many have thought that one could, or that one indeed must, teach all mathematics *deductively* throughout, by starting with a definite number of axioms and deducing everything from these by means of logic. This method, which some seek to maintain on the authority of Euclid, certainly does not correspond to the historical development of mathematics. In fact, mathematics has grown like a tree, which does not start from its tiniest roots and grow merely upward, but rather sends its roots deeper and deeper at the same time and rate that its branches and leaves are spreading upwards. Just so—if we may drop the figure of speech—mathematics began its development from a certain standpoint corresponding to normal human understanding and has progressed, from that point, according to the demands of science itself and of the then prevailing interests, now in one direction toward new knowledge, now in the other through the study of fundamental principles.

We follow this development throughout the book by starting from the experiences of students in school, digging deeper in Part II to find fundamental ideas that we use in Part III to build into formal structures for number systems, and expanding the techniques to wider formal structures in Part IV. In Part V, we close this introduction to the foundations of mathematics by reflecting on the deeper development of fundamental logical principles that become necessary to support more powerful mathematical growth in the future.

Exercises

The following examples are intended to stimulate you into considering your own thought processes and your present mathematical viewpoint. Many of them do not have a 'correct' answer, however it will be most illuminating for you to write out solutions and keep them in a safe place to see how your opinions may change as you read the text. Later in the book (at the end of chapters 6 and 12) you will be invited to reconsider your responses to these questions to see how your thinking has changed. Don't be afraid at this time to say that some of the ideas do not make sense to you at the moment. On the contrary, it is to your advantage to acknowledge any difficulties you may have. The intention of this book is that the ideas will become much clearer as you develop in sophistication.

1. Think how you think about mathematics. If you meet a new problem which fits into a pattern that you recognise, your solution may follow a time-honoured logical course, but if not, then your initial attack may be anything but logical. Try these three problems and do your best to keep track of the steps you take as you move towards a solution.
   (a) John's father is three times as old as John; in ten years he will only be twice John's age. How old is John now?
   (b) A flat disc and a sphere of the same diameter are viewed from the same distance, with the plane of the disc at right angles to the line of vision. Which looks larger?
   (c) Two hundred soldiers stand in a rectangular array, in ten rows of twenty columns. The tallest man in each row is selected and of these ten, S is the shortest. Likewise the shortest in each column is singled out and T is the tallest of these twenty. Are S and T one and the same? If not, what can be deduced about the relative size of S and T?

   Make a note of the way that you attempted these problems, as well as your final solution, if you find one.

2. Consider the two following problems:
   (a) Nine square metres of cloth are to be divided equally between five dressmakers; how much cloth does each one get?
   (b) Nine children are available for adoption and are to be divided equally between five couples; how many children are given to each couple?

   Both of these problems translate mathematically into:

   'Find x such that $5x = 9$'.

   Do they have the same solution? How can the mathematical formulation be qualified to distinguish between the two cases?

3. Suppose that you are trying to explain negative numbers to someone who has not met the concept and you are faced with the comment:

   'Negative numbers can't exist because you can't have less than nothing.'

   How would you reply?

4. What does it mean to say that a decimal expansion 'recurs'? What fraction is represented by the decimal $0 \cdot 333 \ldots$? What about $0 \cdot 999 \ldots$?

5. Mathematical use of language sometimes differs from colloquial usage. In each of the following statements, record whether you think that they are true or false. Keep them for comparison when you read chapter 6.
   (a) All of the numbers 2, 5, 17, 53, 97 are prime.
   (b) Each of the numbers 2, 5, 17, 53, 97 is prime.
   (c) Some of the numbers 2, 5, 17, 53, 97 are prime.
   (d) Some of the numbers 2, 5, 17, 53, 97 are even.
   (e) All of the numbers 2, 5, 17, 53, 97 are even.
   (f) Some of the numbers 2, 5, 17, 53, 97 are odd.

6. 'If pigs had wings, they'd fly.'
   Is this a logical deduction?

7. 'The set of natural numbers 1, 2, 3, 4, 5, ... is infinite.' Give an explanation of what you think the word 'infinite' means in this context.

8. A formal definition of the number 4 might be given in the following terms.
   First note that a set is specified by writing its elements between curly brackets { } and that the set with no elements is denoted by $\varnothing$. Then we define

   $$4 = \{\varnothing, \{\varnothing\}, \{\varnothing, \{\varnothing\}\}, \{\varnothing, \{\varnothing\}, \{\varnothing, \{\varnothing\}\}\}\}.$$

Can you understand this definition? Do you think that it is suitable for a beginner?

9. Which, in your opinion, is the most likely explanation for the equality

$$(-1) \times (-1) = +1?$$

(a) A scientific truth discovered by experience.

(b) A definition formulated by mathematicians as being the only sensible way to make arithmetic work.

(c) A logical deduction from suitable axioms.

(d) Some other explanation.

Give reasons for your choice and retain your comments for later consideration.

10. In multiplying two numbers together, the order does not matter, $xy = yx$. Can you justify this result

(a) when x, y are both whole numbers?

(b) when x, y are any real numbers?

(c) for any numbers whatever?

CHAPTER 2

Number Systems

The reader will have built up a coherent understanding of the arithmetic of the various number systems: counting numbers, negative numbers, and so on. But he or she may not have subjected the processes of arithmetic to close logical scrutiny. Later, we place these number systems in a precise axiomatic setting. In this chapter we give a brief review of how the reader may have developed their ideas about these systems. Although constant use of the ideas will have smoothed out many of the difficulties that were encountered when the concepts were being formed, these difficulties tend to reappear in the formal treatment and have to be dealt with again. It is therefore worth spending a little time to recall the development, before we plunge into the formalities.

The experienced reader may feel tempted to skip this chapter because of the very simple level of the discussion. Please don't. Every adult's ideas have been built up from simple beginnings as a child. When trying to understand the foundations of mathematics, it is important to be aware of the genesis of your own mathematical thought processes.

Natural Numbers

The natural numbers are the familiar counting numbers $1, 2, 3, 4, 5, \ldots$. Young children learn the names of these, and the order in which they come, by rote. Contact with adults leads the children to an awareness of the meaning that adults attach to phrases such as 'two sweets', 'four marbles'. Use of the word 'zero' and the concept 'no sweets' is more subtle and follows later.

To count a collection of objects, we point to them in turn while reciting 'one, two, three, ...' until we have pointed to all of the objects, once each.

Next we learn the arithmetic of natural numbers, starting with addition. At this stage the basic 'laws' of addition (which we can express algebraically as the commutative law $a + b = b + a$, and the associative law

$a + (b + c) = (a + b) + c$) may or may not be 'obvious', depending on the approach used. If addition is introduced in terms of combining collections of real-world objects and then counting the result, then these two laws depend only on the tacit assumption that rearranging the collection does not alter the number of things in it. Similarly, one modern approach using coloured rods whose lengths represent the numbers (which are added by placing them end to end) makes commutativity and associativity so obvious that it is almost confusing to have them pointed out. However, if a child is taught addition by 'counting on', the story is quite different. To calculate 3 + 4, he or she starts at 3 and counts on four more places: 4, 5, 6, 7. The calculation 4 + 3 starts at 4 and counts on three places: 5, 6, 7. That the two processes yield the same answer is now much more mysterious. In fact children taught this way often have difficulty doing a calculation such as 1 + 17, but find 17 + 1 trivial!

Next we come to the concept of place-value. The number 33 involves two threes, but they don't mean the same thing. It must be emphasised that this is purely a matter of notation, and has nothing to do with the numbers themselves. But it is a highly useful and important notation. It can represent (in principle) arbitrarily large numbers, and is very well adapted to calculation. However, a precise mathematical description of the general processes of arithmetic in Hindu-Arabic place notation is quite complicated (which is why children take so long to learn them all) and not well adapted to, say, a proof of the commutative law. (This can be done, but it's harder than we might expect.) Sometimes a more primitive system has some advantages. For instance, the ancient Egyptians used the symbol | to represent 1, a hoop ∩ to represent 10, the end of a scroll ◎ for 100, with other symbols for 1000, etc. A number was written by repeating these symbols: thus 247 would have been written

$$◎◎∩∩∩∩||||||$$

Adding in Egyptian is easy: all we do is to put the symbols together. Now the commutative and associative laws are obvious again. But the notation is less suited to computation. To recover place-notation from Egyptian we must supply some 'carrying rules', such as $||||||||||= ∩$ and insist that we never use any particular symbol more than nine times.

Before proceeding, we introduce a small amount of notation. We write **N** for the set of all natural numbers. The symbol $\in$ will mean 'is an element of' or 'belongs to'. So the symbols

$$2 \in \mathbf{N}$$

are read as '2 belongs to the set of natural numbers', or in more usual language, '2 is a natural number'.

Fractions

Fractions are introduced into arithmetic to make division possible. It is easy to divide 12 into 3 parts: $12 = 4 + 4 + 4$. It is not possible to divide, say, 11 into 3 equal parts if we insist that these parts are natural numbers. Hence we are led to define fractions as m/n where $m, n \in \mathbf{N}$ and $n \neq 0$. This introduces a new idea, that different fractions such as 2/4 and 3/6 can involve two different processes, where the first divides an object into 4 equal pieces and takes 2 of them to get 2 fourths while the second would divide the object into 6 equal pieces and take 3 to get 3 sixths. The processes are different, but the quantity produced is the same (a half). These fractions are said to be *equivalent*. Equivalent fractions, when marked on a number line, are marked at the same point.

This observation proves to be seminal throughout this book: equivalent concepts at one stage are often reconsidered as single entities later on. In this case equivalent fractions are considered as a single rational number.

Operations of addition and multiplication on the set $\mathbf{F}$ of fractions can be defined algebraically by the rules

$$\frac{m}{n} + \frac{p}{q} = \frac{mq + np}{nq},$$

$$\frac{m}{n} \times \frac{p}{q} = \frac{mp}{nq}.$$

It is straightforward (but somewhat tedious) to prove that if the fractions are replaced by equivalent fractions, these formulas for the operations yield equivalent results.

Integers

What fractions do for division, integers do for subtraction. A subtraction sum like $2 - 7 = ?$ cannot be answered in $\mathbf{N}$. To do so, we introduce negative numbers. Children are often introduced to negative numbers in terms of a 'number line': a straight line with equally spaced points marked on it. One of them is called 0; then natural numbers $1, 2, 3, \ldots$ are marked successively to the right, and negative numbers $-1, -2, -3, \ldots$ to the left.

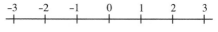

Fig. 2.1 The integers

This gives an extended number system called the 'integers'. An integer is either a natural number n, or a symbol $-n$ where n is a natural number, or 0. We use **Z** to denote the set of integers. (Z is the initial letter of 'Zahlen', the German for integers.)

In your own learning, you met counting numbers **N** before the integers **Z** were introduced. This step is usually motivated by thinking of a negative number as a 'debt'. Then we can see why we have the rule that 'minus times minus makes plus', because taking away a debt has the same result as giving the corresponding credit.

Sometimes in school mathematics, a distinction may initially be made between counting numbers, $1, 2, 3, \ldots$, and positive integers $+1, +2, +3, \ldots$ with their negative counterparts $-1, -2, -3, \ldots$. There are times when this distinction is useful or necessary. Indeed, later we start with counting numbers and show how to construct integers formally. In this process there *is* a difference between the two. However, if we carry on maintaining such distinctions, we will only be making unnecessary work for ourselves. For example, the symbolic statement $4 - (+2)$ (taking away $+2$ from 4) involves a different operation from $4 + (-2)$ (adding -2 to 4). However, it is clearly sensible to say that both equal $4 - 2$.

In the same way, later we start with counting numbers and use set theory to construct integers. This process leads to a different symbolism for counting numbers and positive integers; however, they clearly have the same properties, so it is sensible to think of them as being the same.

In set-theoretic notation, the symbol $\subseteq$ means 'is a subset of'. We then have

$$\mathbf{N} \subseteq \mathbf{Z},$$

where every natural number is also a (positive) integer. Similarly

$$\mathbf{N} \subseteq \mathbf{F}.$$

Rational Numbers

The system **Z** is designed to allow subtraction in all cases; the system **F** allows division (except by zero). However, in neither system are both operations always possible. To get both working at once we move into the system of rational numbers **Q** (for 'quotients'). This is obtained from **F** by introducing 'negative fractions' in much the same way that we obtained **Z** from **N**.

We can still represent **Q** by points on a number line, by marking fractions at suitably spaced intervals between the integers, with negative ones to the left of 0 and positive ones to the right. For example, 4/3 is marked one third of the way between 1 and 2, like this:

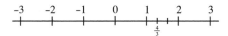

Fig. 2.2 Marking a rational number

The rules for adding and multiplying rational numbers are the same as for fractions, but now m, n, p, q are allowed to be integers rather than natural numbers.

Both **Z** and **F** are subsets of **Q**. We can summarise the relations between the four number systems so far encountered by the diagram:

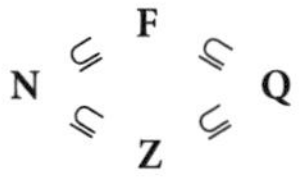

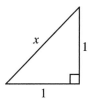

Fig. 2.3 Four number systems

Real Numbers

Numbers can be used to measure lengths or other physical quantities. However, the Greeks discovered that there exist lines whose lengths, in theory, cannot be measured exactly by a *rational* number. They were magnificent geometers, and one of their simple but profound results was Pythagoras' theorem. Applied to a right-angled triangle whose two shorter sides have lengths 1, this implies that the hypotenuse has length x, where $x^2 = 1^2 + 1^2 = 2$.

Fig. 2.4 Pythagoras and $\sqrt{2}$

However, x cannot be rational, because there is no rational number m/n such that $(m/n)^2 = 2$. To see why, we use the result that any natural number can be factorised uniquely into primes. For instance, we can write

$$360 = 2 \times 2 \times 2 \times 3 \times 3 \times 5$$

or

$$360 = 5 \times 2 \times 3 \times 2 \times 3 \times 2,$$

but however we write the factors we will always have one 5, two 3s, and three 2s. Using index notation we write

$$360 = 2^3 \times 3^2 \times 5.$$

We shall prove this unique factorisation theorem formally in chapter 8 but for the moment we assume it without further proof.

If we factorise any natural number into primes and then square, each prime will occur an even number of times. For instance,

$$360^2 = (2^3 \times 3^2 \times 5)^2 = 2^6 \times 3^4 \times 5^2,$$

and the indices 6, 4, 2 are all even. A general proof is not hard to find.

Now take any rational number m/n and square it. (Since m/n has the same square as $-m/n$, we may assume m and n positive.) Factorise m^2 and n^2 and cancel factors top and bottom if possible. Whenever a prime p cancels, then since all primes occur to even powers it follows that p^2 cancels. Hence, after cancellation, all primes still occur to even powers. But $(m/n)^2$ is supposed to equal 2, which has one prime (namely 2) which only occurs once (which is an odd power).

It follows that no rational number can have square 2, so the hypotenuse of the given triangle does not have rational length.

With a little more algebraic symbolism we can tidy up this proof and present it as a formal argument, but the above is all that we really need. The same argument shows that numbers like 3, 3/4, or 5/7 do not have rational square roots.

The implication is clear. If we want to talk of lengths like $\sqrt{2}$, we must enlarge our number system further. Not only do we need rational numbers, we need 'irrational' ones as well.

Using Hindu-Arabic notation this can be done by introducing decimal expansions. We construct a right-angled triangle with sides of unit length, and using drawing instruments transfer the length of its hypotenuse to the number line. We then obtain a specific point on the number line that we call $\sqrt{2}$. It lies between 1 and 2 and, on subdividing the unit length from 1 to 2 into ten equal parts, we find that $\sqrt{2}$ lies between 1·4 and 1·5.

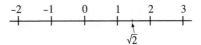

Fig. 2.5 Marking $\sqrt{2}$

By further subdividing the distance between 1·4 and 1·5 into ten equal parts we might hope to obtain a better approximation to $\sqrt{2}$.

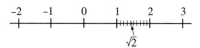

Fig. 2.6 Marking more accurately

Already in a practical situation we are reaching the limit of accuracy in drawing. We might imagine that in an accurate diagram we can look sufficiently close, or magnify the picture, to give the next decimal place. If we were to look at an actual picture under a magnifying glass, not only would the lengths be magnified, but so would the thickness of the lines in the drawing. This would not be a very satisfactory way to obtain a better estimate for $\sqrt{2}$.

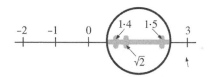

Fig. 2.7 Using a magnifying glass

Practical drawing is in fact extremely limited in accuracy. A fine drawing pen marks a line $0 \cdot 1$ millimetres thick. Even if we use a line 1 metre long as a unit length, since $0 \cdot 1$ mm $= 0 \cdot 0001$ metres, we could not hope to be accurate to more than four decimal places. Using much larger paper and more refined instruments gives surprisingly little increase in accuracy in terms of the number of decimal places we can find. A light year is approximately $9 \cdot 5 \times 10^{15}$ metres. As an extreme case, suppose we consider a unit length 10^{18} metres long. If a light ray started out at one end at the same time that a baby was born at the other, the baby would have to live to be over 100 years old before seeing the light ray. At the lower extreme of vision, the wavelength of red light is approximately 7×10^{-7} metres, so a length of 10^{-7} metres is smaller than the wavelength of visible light. Hence an ordinary optical microscope cannot distinguish points which are 10^{-7} metres apart. On a line where the unit length is 10^{18} metres we cannot distinguish numbers which are less than $10^{-7}/10^{18} = 10^{-25}$ apart. This means that we cannot achieve an accuracy of 25 decimal places by a drawing. Even this is a gross exaggeration in practice, where three or four decimal places is often the best we can really hope for.

Inaccurate Arithmetic in Practical Drawing

The inherent inaccuracy in practice leads to problems in arithmetic. If we add two inaccurate numbers, the errors also add. If we cannot distinguish errors less than some amount e, then we cannot tell the difference, in practice, between a and $a + \frac{3}{4}e$ and between b and $b + \frac{3}{4}e$. But adding, we can distinguish between $a + b$ and $a + b + \frac{3}{2}e$. When we come to multiplication, errors can increase even more dramatically. We cannot hope to get answers to the same degree of accuracy as the numbers used in the calculation.

If we use arithmetic to calculate all answers correct to a certain number of decimal places, the errors involved lead to some disturbing results. Suppose, for example, that we work to two decimal places ('rounding up' if the third place is 5 or more and down if it is less). Given two real numbers a and b, we denote their product correct to two decimal places by $a \otimes b$. For example, $3{\cdot}05 \otimes 4{\cdot}26 = 12{\cdot}99$ because $3{\cdot}05 \times 4{\cdot}26 = 12{\cdot}993$. Using this law of multiplication we find that

$$(1{\cdot}01 \otimes 0{\cdot}5) \otimes 10 \neq 1{\cdot}01 \otimes (0{\cdot}5 \otimes 10).$$

The left-hand side reduces to $0{\cdot}51 \otimes 10 = 5{\cdot}1$, whilst the right-hand side becomes $1{\cdot}01 \otimes 5 = 5{\cdot}05$. This is by no means an isolated example, and it shows that the associative law does not hold for $\otimes$.

If we further define $a \oplus b$ to be the sum correct to two decimal places, we will find other laws that do not hold, including the distributive law

$$a \otimes (b \oplus c) \overset{?}{=} (a \otimes b) \oplus (a \otimes c).$$

A Theoretical Model of the Real Line

We have just seen that if our measurement of numbers is not precise, then some of the laws of arithmetic break down. To avoid this we must make our notion of real number exact.

Suppose we are given a real number x on a theoretical real line, and we try to express it as a decimal expansion. As a starting point, we see that x lies between two integers.

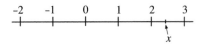

Fig. 2.8 Marking a real number

In the above example x is between 2 and 3, so x is 'two point something'. Next we divide the interval between 2 and 3 into ten equal parts.

Again, x lies in some sub-interval. In the picture, x lies between $2{\cdot}4$ and $2{\cdot}5$, so x is '$2{\cdot}4$ something'. To obtain a still better idea, we divide the interval between $2{\cdot}4$ and $2{\cdot}5$ into ten equal parts and repeat the process to find the next figure in the decimal expansion. Already, in a practical situation, we are reaching the limit of accuracy in drawing.

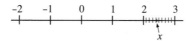

Fig. 2.9 Marking more accurately

For our theoretical picture we must imagine that we can look sufficiently closely, or magnify the picture, to read off the next decimal place. If we looked at an actual picture under a magnifying glass, not only would the lengths be magnified, but so would the thickness of the lines.

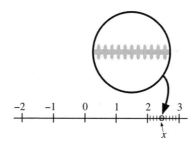

Fig. 2.10 Magnifying

This is not very satisfactory for getting a better estimate. We must, in the theoretical case, assume that the lines have no thickness, so that they are not made wider when the picture is magnified. We can represent this as a practical picture by drawing the magnified lines with the same drawing implements as before, and making them as fine as possible. In this case x lies between 2·43 and 2·44, so x is '2·44 something'.

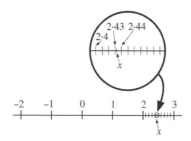

Fig. 2.11 Magnifying more accurately

Using this method we can, in theory, represent any real number as a decimal expansion to as many figures as we require. If we are careful to define this expansion to avoid ambiguity, two numbers will be different if, by calculating sufficiently many terms, we eventually obtain different answers for some decimal place.

We can express this theoretical method as follows in more mathematical terms.

(i) Given a real number x, find an integer a_0 such that

$$a_0 \leq x < a_0 + 1.$$

(ii) Find a whole number a_1 between 0 and 9 inclusive such that

$$a_0 + \frac{a_1}{10} \leq x < a_0 + \frac{a_1 + 1}{10}.$$

(iii) After finding $a_0, a_1, \ldots, a_{n-1}$, where $a_1, \ldots, a_{n-1}$ are integers between 0 and 9 inclusive, find the integer a_n between 0 and 9 inclusive for which

$$a_0 + \frac{a_1}{10} + \cdots + \frac{a_n}{10^n} \leq x < a_0 + \frac{a_1}{10} + \cdots + \frac{a_n + 1}{10^n}.$$

This gives an inductive process which at the nth stage determines x to n decimal places:

$$a_0 \cdot a_1 a_2 \ldots a_n \leq x < a_0 \cdot a_1 a_2 \ldots a_n + 1/10^n.$$

The theoretically exact representation of the number x requires a decimal expansion

$$a_0 \cdot a_1 a_2 a_3 a_4 a_5 a_6 \ldots$$

that goes on forever. (Of course, if all a_n from some point on are zero, we omit them in normal notation; instead of $1066 \cdot 31700000000 \ldots$ we write $1066 \cdot 317$.) An infinite decimal is called a *real number*. The set of all real numbers is denoted by **R**.

In most practical situations we will need only a few decimal places. Earlier we saw that 25 decimal places are sufficient for all ratios of lengths within human visual capacity, and that two or three places are usually sufficient for many practical purposes.

Different Decimal Expansions for Different Numbers

If we expand a number x as above in an endless decimal, we say that $a_0 \cdot a_1 a_2 \ldots a_n$ is the expansion of x to the first n decimal places (without 'rounding up'). If two real numbers x and y have the same decimal expansion to n places then

$$a_0 \cdot a_1 \ldots a_n \leq x < a_0 \cdot a_1 \ldots a_n + 1/10^n,$$
$$a_0 \cdot a_1 \ldots a_n \leq y < a_0 \cdot a_1 \ldots a_n + 1/10^n.$$

The second line of inequalities can be rewritten as

$$-a_0 \cdot a_1 \ldots a_n - 1/10^n < -y \leq -a_0 \cdot a_1 \ldots a_n.$$

Adding this to the first line we obtain

$$-1/10^n < x - y < 1/10^n.$$

In other words, if two real numbers have the same decimal expansion to n places, then they differ by at most $1/10^n$.

If x and y are different numbers on the line and we wish to distinguish between them, all we need do is find n such that $1/10^n$ is less than their difference: then their expansion to n places will differ. This again exposes the deficiencies of practical drawing, where x and y might be too close to distinguish. In our theoretical concept of the real line, this distinction must always be possible. It is so important that it is worth giving it a name. The great Greek mathematician Archimedes stated a property that is equivalent to what we want, so we shall name our condition after him:

Archimedes' Condition: Given a positive real number ε, there exists a positive integer n such that $1/10^n < \varepsilon$.

Rationals and Irrationals

As we have seen, the real number $\sqrt{2}$ is irrational: so are many others. It is not always easy to prove a given number irrational. (It's moderately easy for e, less so for π, and there are many interesting numbers which mathematicians have been convinced for centuries are irrational, but have never proved them to be.) But just the fact that $\sqrt{2}$ is irrational implies that between any two rational numbers there exist irrational numbers. First we need:

Lemma 2.1: If m/n and r/s are rational, with $r/s \neq 0$, then $m/n + (r/s)\sqrt{2}$ is irrational.

Proof: Suppose that $m/n + (r/s)\sqrt{2}$ is rational, equal to p/q where p, q are integers. Solve for $\sqrt{2}$ to obtain

$$\sqrt{2} = (pn - mq)s/qnr$$

which is rational, contrary to the irrationality of $\sqrt{2}$. $\qquad\square$

Proposition 2.2: Between any two distinct rational numbers there exists an irrational number.

Proof: Let the rational numbers be m/n and r/s, where $m/n < r/s$. Then

$$m/n < m/n + \frac{\sqrt{2}}{2}(r/s - m/n) < r/s$$

(because $\sqrt{2}/2 < 1$), and the number in the middle is irrational by the lemma. □

There is a corresponding result with 'rational' and 'irrational' interchanged:

Proposition 2.3: Between any two distinct irrational numbers there exists a rational number.

Proof: Let the irrational numbers be a, b with $a < b$. Consider their decimal expansions, and let the nth decimal place be the first in which they differ. Then

$$a = a_0 \cdot a_1 \ldots a_{n-1}a_n \ldots,$$
$$b = a_0 \cdot a_1 \ldots a_{n-1}b_n \ldots,$$

where $a_n \neq b_n$. Let $x = a_0 \cdot a_1 \ldots a_{n-1}b_n$. Then x is rational and $a < x \leq b$. But since b is irrational, $x \neq b$, so we must have $a < x < b$. □

In fact, the exercises at the end of this chapter show that the rational and irrational numbers are mixed up in a very complicated way. One should not make the mistake of thinking that they 'alternate' along the real line.

The rational numbers may be characterised as those whose decimal expansions repeat at regular intervals (though we shall omit the proof). To be precise, say that a decimal is repeating if, from some point on, a fixed sequence of digits repeats indefinitely. For example, $1 \cdot 5432174174174174\ldots$ is a repeating decimal. We shall write it as $1 \cdot 5432\dot{1}7\dot{4}$, with dots over the end digits of the block that repeats.

The Need for Real Numbers

The Greeks' belief that all numbers are rational (enshrined in the mystic philosophy of the cult of Pythagoreans) led them to a logical impasse. Viewing the real numbers as infinite decimals helps to overcome this mental block, because it makes it clear that rational numbers, whose expansions repeat, do not exhaust the possibilities.

However, we have also seen that for practical purposes we do not need infinite decimals, nor even very long finite ones. Why go to all the trouble? One reason we have already noted: the arithmetic of decimals of limited length fails to obey the familiar laws which integers and rational numbers obey. A perhaps more serious reason arises in analysis.

Consider the function f given by

$$f(x) = x^2 - 2 \quad (x \in \mathbf{R}).$$

This is negative at $x = 1$, positive at $x = 2$. In between, it is zero at $x = \sqrt{2}$. However, if we restrict x to take only rational values, the function

$$f(x) = x^2 - 2 \quad (x \in \mathbf{Q})$$

is also negative at $x = 1$, positive at $x = 2$, but is not zero at any rational x in between, because $x^2 = 2$ has no rational solution.

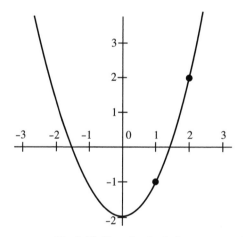

Fig. 2.12 No rational solution

This is a nuisance. A fundamental theorem in analysis asserts that if a continuous function is negative at one point and positive at another, then it must be zero in between. This is true for functions over the real numbers, but not for functions over the rationals. A civilisation such as that of the ancient Greeks, with no satisfactory method for handling irrational numbers, cannot build a theory of limits, or invent calculus.

Arithmetic of Decimals

The idea of infinite decimals representing real numbers is a useful one, but it is not well suited to numerical manipulations, nor to theoretical investigations beyond an elementary level. We add two finite decimals by starting at the right-hand end, but infinite decimals do not have right-hand ends, so there is nowhere to start.

We can instead start at the left-hand end, adding the first decimal places, then the first two, then the first three, and so on. To see what happens, try adding $2/3 = 0 \cdot \dot{6}$ and $2/7 = 0 \cdot \dot{2}8571\dot{4}$ in this way.

·6	+ ·2	= ·8
·66	+ ·28	= ·94
·666	+ ·285	= ·951
·6666	+ ·2857	= ·9523
·66666	+ ·28571	= ·95237
·666666	+ ·285714	= ·952380.

The actual answer is $2/3 + 2/7 = 20/21 = \cdot\dot{9}5238\dot{0}$. Notice that adding the first decimal places does not give the answer to one decimal place, nor does adding the first two places give the first two places of the answer. This is precisely because of the possibility of 'carried' digits from later places affecting earlier ones.

However, in this example, successive terms increase and get closer and closer to the actual answer. The sequence of numbers ·8, ·94, ·951, ·9523, . . . is an increasing sequence of real numbers, and it 'tends to' 20/21 in the sense that the error can be made as small as we please by calculating enough decimal places.

In the next few sections we shall examine in detail the ideas required to make this concept precise. For theoretical purposes it is often easier to use increasing sequences (of approximations to a real number) rather than decimal expansions.

Sequences

A *sequence* of real numbers can be thought of as an endless list

$$a_1, a_2, a_3, a_4, \ldots, a_n, \ldots$$

where each term a_n is a real number. (Using set theory we shall give a more formal definition in chapter 5.)

Examples 2.4:

(1) The sequence of squares: $1, 4, 9, 16, \ldots$ where $a_n = n^2$.

(2) The sequence of decimal approximations to $\sqrt{2}$ is $1 \cdot 4$, $1 \cdot 41$, $1 \cdot 414, \ldots$ where $a_n = \sqrt{2}$ to n places.

(3) The sequence $1, 1\frac{1}{2}, 1\frac{5}{6}, \ldots$ where $a_n = 1 + \frac{1}{2} + \frac{1}{3} + \ldots + \frac{1}{n}$.

(4) The sequence $3, 1, 4, 1, 5, 9, \ldots$ where $a_n =$ the nth digit in the decimal expansion of π.

We often use the shorthand notation

$$(a_n)$$

for the sequence $a_1, a_2, \ldots$, where the nth term is placed in round brackets. Thus example (1) could be written (n^2).

Notice how general the concept of a sequence is. We can consider *any* endless list of numbers. It is not necessary that the nth term be defined by a 'nice formula', as long as we know what each a_n is supposed to be.

Sequences can be added, subtracted, or multiplied. It is necessary to define what we mean by this: the simplest way is to perform the operations on each pair of terms in corresponding positions. In other words, to add the sequences

$$a_1, a_2, \ldots$$

and

$$b_1, b_2, \ldots$$

means to form the sequence

$$a_1 + b_1, a_2 + b_2, \ldots.$$

For example, if $a_n = n^2$ and $b_n = 1 + \frac{1}{2} + \cdots + \frac{1}{n}$, then the nth term of $(a_n) + (b_n)$ is

$$n^2 + 1 + \frac{1}{2} + \cdots + \frac{1}{n}.$$

Since the nth term of the sequence $(a_n) + (b_n)$ is $a_n + b_n$, we can express the rule for addition as

$$(a_n) + (b_n) = (a_n + b_n).$$

Similarly the rules for subtraction and multiplication are

$$(a_n) - (b_n) = (a_n - b_n),$$
$$(a_n)(b_n) = (a_n b_n).$$

In the case of division we put

$$(a_n)/(b_n) = (a_n/b_n),$$

noting that this division can be carried out only when *all* terms b_n are non-zero.

Example 2.5: If $a_n = \sqrt{2}$ to n decimal places, and $b_n =$ the nth decimal place in π, then the first few terms of $(a_n)(b_n)$ are

$$1\cdot4 \times 3 = 4\cdot2$$
$$1\cdot41 \times 1 = 1\cdot41$$
$$1\cdot414 \times 4 = 5\cdot656$$
$$1\cdot4142 \times 1 = 1\cdot4142.$$

If you were given the sequence $4\cdot2$, $1\cdot41$, $5\cdot656$, $1\cdot4142$, could you have guessed the rule for the nth term? This drives home the point that in order to specify a sequence we must know in principle how to calculate *all* of its terms. In general, it is not enough to write down the first few terms and a few dots. The sequence $3, 1, 4, 1, 5, 9, \ldots$ certainly looks as if it consists of the digits of π. However, it might just as well be the sequence of digits of the number $355/113$, which starts off the same way. This is why, in example (4), we specify the general rule for finding the nth term.

Nevertheless, you will often find mathematicians writing things like $2, 4, 8, 16, 32, \ldots$ and expecting you to infer that the nth term is 2^n. One aspect of learning mathematics is to understand how mathematicians actually work, and what their idiosyncrasies are: you should be prepared to accept slight differences in notation provided that the idea is clear from the context.

Order Properties and the Modulus

We digress to introduce an important concept. If x is a real number we define the modulus or absolute value of x to be

$$|x| = \begin{cases} x & \text{if } x \geq 0, \\ -x & \text{if } x < 0. \end{cases}$$

The graph of $|x|$ against x looks like:

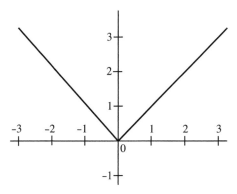

Fig. 2.13 The modulus function

The value of $|x|$ tells us how large or small x is, ignoring whether it is positive or negative. Perhaps the most useful fact about the modulus is the triangle inequality, so called because its generalisation to complex numbers expresses the fact that each side of a triangle is shorter than the other two put together. It is:

Proposition 2.6 (Triangle Inequality): If x and y are real numbers, then

$$|x + y| \leq |x| + |y|.$$

Proof: The visual idea is that $|x + y|$ says how far from the origin $x + y$ is, and this is at most the sum of the distances $|x|$ and $|y|$ of x and y from the origin, being less if x and y have opposite sign. (Draw some pictures to check this.) The easiest way to prove it logically is to divide into cases, according to the signs and relative sizes of x and y.

(i) $x \geq 0, y \geq 0$. Then $x + y \geq 0$, so

$$|x + y| = x + y = |x| + |y|.$$

(ii) $x \geq 0, y < 0$. If $x + y \geq 0$ then

$$|x + y| = x + y < x - y = |x| + |y|.$$

On the other hand, if $x + y < 0$ then

$$|x + y| = -(x + y) = -x - y < |x| + |y|.$$

(iii) $x < 0, y \geq 0$ follows as in case (ii) with x and y interchanged.

(iv) $x < 0, y < 0$. Then $x + y < 0$, so

$$|x + y| = -x - y = |x| + |y|. \qquad \Box$$

Be on the lookout for variations on this theme, such as

$$|x - y| + |y - z| \geq |x - z|,$$

which follows since $x - z = (x - y) + (y - z)$, so that $|x - y| + |y - z| \geq |x - z|$.

The modulus is most useful for expressing certain inequalities succinctly. For example,

$$a - \varepsilon < x < a + \varepsilon$$

can be written

$$-\varepsilon < x - a < \varepsilon,$$

which translates into

$$|x - a| < \varepsilon.$$

Convergence

Now we are ready to consider the general notion of representing a real number as a 'limit' of a sequence, rather than just being a particular decimal expansion. As an exercise, the reader should mark, to as large a scale and as accurately as possible, the numbers 1·4, 1·41, 1·414, 1·4142, $\sqrt{2}$, on the interval between 1 and 2.

The numbers 1·4, 1·41, 1·414, 1·4142, get closer and closer together, until they become indistinguishable from each other and from $\sqrt{2}$, up to the accuracy of the drawing. By drawing a more accurate picture we must go further along the sequence of decimal approximations to $\sqrt{2}$ before this happens. If we work to an accuracy of 10^{-8}, then from the eighth term onwards all points of the sequence are indistinguishable from $\sqrt{2}$.

This observation motivates the theoretical concept of convergence. Let ε be any positive real number (ε is the Greek letter epsilon, for 'e', and may be thought of as the initial letter of 'error'). For practical convergence of a sequence (a_n) to a limit l, if we are working to an accuracy ε, we require there to be some natural number N such that the difference between a_n and l has size less than ε when $n > N$. In other words, $|a_n - l| < \varepsilon$. In the following diagram we cannot distinguish points less than ε apart; in this case $N = 7$ and a_n is indistinguishable from l when $n > 7$.

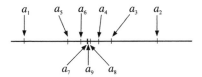

Fig. 2.14 Practical convergence

For theoretical convergence we ask that a similar phenomenon should occur for *all* positive ε. This is on the explicit understanding that smaller values of ε may require larger values of N. In this sense, N is allowed to depend on ε. Thus we reach:

Definition 2.7: A sequence (a_n) of real numbers *tends to a limit* l if, given any $\varepsilon > 0$, there is a natural number N such that

$$|a_n - l| < \varepsilon \text{ for all } n > N.$$

Mathematicians use various pieces of shorthand notation to express this concept. To say 'the sequence (a_n) tends to the limit l' we write

$$\lim_{n \to \infty} a_n = l,$$

or

$$a_n \to l \text{ as } n \to \infty.$$

The symbol '$n \to \infty$' is read as 'n tends to infinity' and is meant to remind us that we are interested in the behaviour of a_n as n becomes large (namely $n > N$ for an appropriately large number N).

The symbol ∞ has historical connotations that can have a variety of different meanings. We will return to these in chapters 14 and 15 to see that ideas that occurred in history and in the minds of growing students can be interpreted formally in very interesting ways. Until then, we will usually refrain from using the symbol and just write

$$\lim a_n = l.$$

Example 2.8: The sequence $1 \cdot 1$, $1 \cdot 01$, $1 \cdot 001$, $1 \cdot 0001$, $\dots$, for which $a_n = 1 + 10^{-n}$, tends to the limit 1. For, given $\varepsilon > 0$, we have to make

$$|1 + 10^{-n} - 1| < \varepsilon \text{ for } n > N$$

by finding a suitable N. But this follows from Archimedes' condition: if we find N to make $10^{-N} < \varepsilon$, then for all $n > N$ we have $10^{-n} < 10^{-N} < \varepsilon$. (If the theory of logarithms is available, we take $N > \log_{10}(l/\varepsilon)$.)

Definition 2.9: A sequence (a_n) which tends to a limit l is called *convergent*. If no limit exists, it is said to be *divergent*.

A convergent sequence can tend to only one limit. For suppose $a_n \to l$ and $a_n \to m$, where $l \neq m$. Take $\varepsilon = \frac{1}{2}|l - m|$. For large enough n,

$$|a_n - l| < \varepsilon, \quad |a_n - m| < \varepsilon.$$

From the triangle inequality, $|l - m| < 2\varepsilon = |l - m|$, which is not the case.

In other words, if all the terms a_n must eventually be very close to l, they cannot also be very close to m, because this requires them to be in two different places at the same time.

Completeness

Definition 2.10: A sequence (a_n) is *increasing* if each $a_n \leq a_{n+1}$, so that

$$a_1 \leq a_2 \leq a_3 \leq \ldots.$$

Suppose that (a_n) is an increasing sequence. Either the terms a_n increase without limit, eventually becoming as large as we please, or else there must be some real number k such that $a_n \leq k$ for all n. An example of a sequence of the first type is $1, 4, 9, 16, 25, \ldots$; one of the latter type is the sequence of decimal approximations to e: $2{\cdot}7, 2{\cdot}71, 2{\cdot}718, 2{\cdot}7182, \ldots$, every term of which is less than 3.

Definition 2.11: If there exists a real number k such that $a_n < k$ for all n we say that (a_n) is *bounded*.

If we draw the points of a bounded increasing sequence on a part of the real line we need only draw the interval between a_1 and k, since all the other points lie inside this. So a typical picture is:

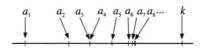

Fig. 2.15 A bounded increasing sequence

It seems visually evident that the terms become increasingly squashed together, and tend to some limit $l \leq k$. This intuition is correct if we consider sequences of real numbers and real limits, but it is wrong for sequences of rational numbers and *rational* limits. In fact the sequence of decimal approximations to $\sqrt{2}$ is an increasing sequence of rational numbers with no rational number as limit.

The fact that every bounded sequence of real numbers tends to a real number as limit as known as the *completeness* property of the real numbers. The origin of the name is that the rational numbers are 'incomplete' because numbers like $\sqrt{2}$ are 'missing'. As we consider a formal approach to the real numbers, we will see this idea in a new light.

We can make the completeness property of the reals very plausible in terms of our ideas about decimals. Let (a_n) be an increasing sequence of real numbers, with $a_n \leq k$ for all k.

The set of integers between $a_1 - 1$ and k is finite, so there is an integer b_0 that is the largest integer for which some term a_{n_0} of the sequence is $\geq b_0$. Now all terms a_n are less than $b_0 + 1$.

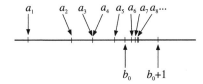

Fig. 2.16 Later terms between successive integers

We subdivide the interval from b_0 to $b_0 + 1$ into ten parts, and find b_1 so that some term $a_{n_1} \geq b_0 + b_1/10$, but no term $a_n \geq b_0 + (b_1 + 1)/10$. Continuing in this way we get a sequence of decimals

$$b_0, b_0 \cdot b_1, b_0 \cdot b_1 b_2, \ldots$$

such that for $n > n_r$ the term a_n lies between $b_0 \cdot b_1 b_2 \ldots b_r$ and $b_0 \cdot b_1 b_2 \ldots b_r + 1/10^r$. Then the real number

$$l = b_0 \cdot b_1 b_2 \ldots$$

has the property that $|a_n - l| < 1/10^r$ for all $n > n_r$. Hence $a_n \to l$ as $n \to \infty$.

It is easy to check that this l is less than or equal to k.

Decreasing Sequences

There is no need to be obsessed with increasing sequences.

Definition 2.12: A sequence (a_n) is *decreasing* if $a_n \geq a_{n+1}$ for all n. If it satisfies $a_n \geq k$ for all n then k is a *lower bound* and the sequence is *bounded below*. (To avoid ambiguity with increasing sequences we can now say 'bounded above' instead of 'bounded'.) There is a similar theorem concerning decreasing sequences, but instead of copying out the proof again and changing the inequalities we use a trick. If (a_n) is decreasing, then $(-a_n)$ is increasing. If $a_n \geq k$ for all n then $-a_n \leq -k$ for all n, so $(-a_n)$ is bounded above, hence tends to a limit l. It follows easily that $a_n \to -l$. Hence any decreasing sequence of real numbers bounded below by k tends to a limit $-l \geq k$.

Different Decimal Expansions for the Same Real Number

Previously we expanded a real number x as an infinite decimal, $x = a_0 \cdot a_1 a_2 \ldots$, by using the inequalities

$$a_0 + \frac{a_1}{10} + \cdots + \frac{a_n}{10^n} \leq x < a_0 + \frac{a_1}{10} + \cdots + \frac{a_n + 1}{10^n},$$

where a_0 is an integer and a_n is an integer from 0 to 9 for $n \geq 1$. This condition can be written

$$a_0 \cdot a_1 a_2 \ldots a_n \leq x < a_0 \cdot a_1 a_2 \ldots a_n + 1/10^n. \tag{2.1}$$

This, used successively for $n = 1, 2, 3, \ldots$, gives a unique decimal expansion for any real number, and different real numbers have different decimal expansions. However, this is not quite the whole story since certain decimal expansions do not occur when we use condition (2.1). For example the expansion $0 \cdot 999999 \ldots$, where $a_0 = 0$ and $a_n = 9$ for all $n \geq 1$, does not occur.

Why does this happen? Suppose there were a real number x with decimal expansion (according to (2.1)) $0 \cdot 999999 \ldots$. Then

$$0 \cdot 999 \ldots 9 \leq x < 0 \cdot 999 \ldots 9 + 1/10^n,$$

where there are n 9s each time. Therefore

$$1 - (1/10^n) \leq x < 1,$$

or

$$0 < 1 - x \leq 1/10^n$$

for all $n \in \mathbf{N}$. But this is impossible by Archimedes' condition: since $1 - x > 0$ there must exist n with $1/10^n < 1 - x$.

The reason why this sequence of 9s cannot occur is our choice of inequalities in (2.1). If instead we use

$$a_0 \cdot a_1 a_2 \ldots a_n < x \le a_0 \cdot a_1 a_2 \ldots a_n + 1/10^n \tag{2.2}$$

then we get an equally useful definition of the decimal expansion, and it is easy to see that the expansion of the number $x = 1$ now takes the form $0 \cdot 999999 \ldots$.

However, the second rule (2.2) will now never give us the expansion $1 \cdot 000000 \ldots$.

These are the only possibilities. For example, if a number x has two different decimal expansions, then, without loss in generality, we can take

$$x = a_0 \cdot a_1 \ldots a_{n-1} a_n \ldots = a_0 \cdot a_1 \ldots a_{n-1} b_n \ldots \text{ where } a_n < b_n.$$

Multiply through by 10^n to get

$$a_0 a_1 \ldots a_{n-1} a_n \cdot a_{n+1} \ldots = a_0 a_1 \ldots a_{n-1} b_n \cdot a_{n+1} \ldots \text{ where } a_n < b_n.$$

Subtracting the whole number $a_0 a_1 \ldots a_{n-1} a_n$ gives

$$0 \cdot a_{n+1} \ldots = k \cdot b_{n+1} \ldots \text{ where } k = b_{n+1} - a_{n+1} > 0 \text{ is a positive integer.}$$

But the first decimal is $0 \cdot a_{n+1} \ldots < 0 \cdot 999 \ldots \le 1$ and the second exceeds the positive integer k. So they can be equal only if $k = 1$ and both decimals represent the same limiting value 1. In this case, $a_{n+1} = a_{n+2} = \cdots = 9$, $b_{n+1} = b_{n+2} = \cdots = 0$ and $b_n = a_n + 1$.

For example, $3 \cdot 14999 \ldots$ equals $3 \cdot 15000 \ldots$.

This proves that an infinite decimal expansion is unique, *except* when one representation is finite, given by (2.1), and the other ends in an infinite number of 9s, given by (2.2).

It is important not to think that $0 \cdot 99 \ldots 9 \ldots$ is a number 'infinitely smaller' than 1. They are just two different ways of writing the same real number.

It is convenient to allow both notations because under certain circumstances a calculation may give rise to the infinite sequence of 9s. This will happen using the method given earlier to find the decimal expansion of the limit of a bounded increasing sequence.

Example 2.13: Suppose $a_1 = 1$ and in general $a_{n+1} = a_n + \left(\frac{1}{2}\right)^{n-1}$, then trivially (a_n) is increasing and a calculation gives $a_n = 2 - \left(\frac{1}{2}\right)^{n-1}$, so the sequence is bounded above by 2. Using the same method to calculate the decimal expansion using definition (2.2) instead of (2.1), the limit of the sequence (a_n) is then found to be

$$b_0 \cdot b_1 b_2 \ldots b_n \ldots = 1 \cdot 99 \ldots 9 \ldots.$$

To cover all cases, we introduce the following:

Definition 2.14: The value of an infinite decimal $a_0 \cdot a_1 a_2 \ldots a_n \ldots$ is the limit l of the sequence (d_n) of decimals to n decimal places, where $d_n = a_0 \cdot a_1 a_2 \ldots a_n$.

Using this definition, $0\cdot333\ldots$ *is* 1/3, and $0\cdot999\ldots$ *is* 1.

COMMENT. Research has shown that most people initially believe that 0.999... is 'just less than 1'. The psychological reason seems to be that we think of a sequence (a_n) not as a list of numbers but as a 'variable quantity' that varies as n varies. For example, if $a_n = 1/n$, then we tend to think of the nth term as varying with n and becoming dynamically smaller and smaller. The variable term in this case gets closer and closer to zero, but never equals zero. This dynamic intuition makes us believe that $0\cdot999\ldots$ is 'just less than one' rather than equal to one. It can lead to resistance to accepting the definition of an infinite decimal being *defined* as the limiting value.

One of us taught an introductory course [5] on convergence using computers for students to investigate the numerical convergence of sequences to get the sense that if a sequence converged, then, to a given number of places, the sequence stabilised onto a fixed value. They were introduced to the idea that the limit was the *precise* value that the sequence stabilised on, leading to the formal definition of the limit l of a sequence (a_n), including the specific example that if $a_n = 1 - 1/10^n$ then the limit l equals 1. Before the course, as expected, 21 out of 23 stated that $0\cdot\dot{9}$ was just less than one and only two said that it was equal to 1. After the course, the students remained of the same opinion. In a class discussion, the general opinion of the students was that they *knew* that the repeating decimal never reached 1, so trying to *define* it equal to one was not possible.

In order to make sense of formal mathematics, it is essential to get to know the definitions and to be aware of precisely what they say. Only then will it become possible to build up a coherent formal theory. In this case, the limit of a sequence (a_n) is defined to be *the fixed number l* that it approaches, as formulated in the definition.

Bounded Sets

By drawing the picture of a bounded increasing sequence we can actually see the limit process in action, as later terms in the sequence pack together inside a rapidly decreasing space. We now consider not just a sequence, but an arbitrary subset $S \subseteq \mathbf{R}$ which is bounded above by some k. This means that $s \leq k$ for all $s \in S$. Is there some concept analogous to the limit?

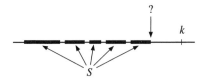

Fig. 2.17 A set S bounded above

The naive thing to expect is that S has a greatest element, a number $s_0 \in S$ such that $s_0 \geq s$ for every $s \in S$. Unfortunately, this is not quite right. For example, if S is the set of all elements of $\mathbf{R}$ that are strictly less than 1, then S is bounded above—for example by $k = 1$. However, there is no element in S that is greater than all the others. For suppose that y were a greatest element. Then $y \in S$ so $y < 1$, and then

$$y < \tfrac{1}{2}(y + 1) < 1.$$

So $\tfrac{1}{2}(y + 1) \in S$, but is greater than the supposedly greatest element y.

However, all is not lost: we just have to be more subtle.

We need some terminology:

Definition 2.15: A non-empty subset $S \subseteq \mathbf{R}$ is *bounded above* by $k \in \mathbf{R}$ if $s \leq k$ for all $s \in S$. The number k is called an *upper bound* for S.

In the previous example the set S has many upper bounds: in fact any $k \geq 1$ is an upper bound for S. Now the set of all upper bounds does have a least element. In fact in this example it is 1. In other words, not only is 1 an upper bound, but every other upper bound is bigger.

Definition 2.16: A subset $S \subseteq \mathbf{R}$ has a *least upper bound* $\lambda \in \mathbf{R}$ if:

   (i) λ is an upper bound for S,
  (ii) if $k \in \mathbf{R}$ is any other upper bound for S, then $\lambda \leq k$.

Although upper bounds are ten a penny, a least upper bound must be unique. For if λ and μ are least upper bounds for S, then (ii), applied to each of them, tells us that $\lambda \leq \mu$ and $\mu \leq \lambda$, so that $\lambda = \mu$.

Examples 2.17:

(1) If S is the set of all integers, then S has no upper bounds, so certainly no least upper bound.
(2) If S is the set of all real numbers less than or equal to 49, then 49 is the least upper bound of S.
(3) If S is the set of all decimal approximations $1 \cdot 4$, $1 \cdot 41$, $1 \cdot 414, \ldots$ to $\sqrt{2}$, then the least upper bound of S is $\sqrt{2}$.
(4) If S is the set of all rational numbers r such that $r^2 < 2$, then $\sqrt{2}$ is the least upper bound of S.

In example (2) the least upper bound is an element of S, but in examples (3) and (4) it is not. So even when least upper bounds exist, they may not be members of the original set.

There is once more a parallel set of concepts.

Definition 2.18: A subset S is *bounded below* if there exists $k \in \mathbf{R}$ with $k \le s$ for all $s \in S$, and k is then called a lower bound. The number $\mu \in \mathbf{R}$ is a greatest lower bound for S if:

(i) μ is a lower bound for S,
(ii) if k is another lower bound for S, then $\mu \ge k$.

A similar trick to that used on decreasing sequences allows us to refer all problems on greatest lower bounds back to least upper bounds. In fact all the basic properties of upper bounds hold for lower bounds, provided that we interchange $\ge$ and $\le$.

COMMENT. One student, who continued to think that 'zero point nine repeating is just less than one' despite all efforts to convince him otherwise, also believed that the least upper bound of a set was always a member of the set. He was invited to consider the set S of real numbers less than one. He declared that this proved his point because the least upper bound of S was, in his view, equal to zero point nine repeating, which is just less than one [1]. This belief may prove to be very difficult to overcome. As another student commented at the end of his first course: 'I understand it should be 1 . . . and that the limit of the sequence is actually 1. It's down to notation. It's just a bit hard to let go of $0 \cdot 9999$ recurring. . . ' [6].

Strong beliefs based on human intuition can impede the appreciation of a more formal approach using definitions. However, you will not make progress in formal mathematics unless you build carefully on the definitions as given. The limit is the *fixed value* to which the terms of the sequence approximate. For example, the notation $1 \cdot 414 \ldots$ in the context of the sequence of decimal expansions of '$\sqrt{2}$ to n decimal places' denotes the limit of the sequence, which is *the fixed real number $\sqrt{2}$*.

To make progress in building mathematics from definition and proof, it is important to *know* the definition and how to make deductions from it. Only then will formal mathematics build into a coherent structure. For example, using formal definition and deduction, the completeness property of the real numbers can be used as a foundation to prove more general properties of the real numbers, such as:

Proposition 2.19: Every non-empty subset of **R** that is bounded above has a least upper bound.

Note the careful use of the adjective 'non-empty'. This is necessary because any number is an upper bound for a set with no elements. The proof of the above proposition can be made plausible by using decimal expansions in the same sort of way that we dealt with increasing sequences. It is more straightforward to deal with lower bounds and then use the trick to convert to upper bounds. This means we look at:

Proposition 2.20: Every non-empty subset of **R** which is bounded below has a greatest lower bound.

Proof: Let $S \subseteq \mathbf{R}$ and let a_0 be the largest integer that is a lower bound for S. Let a_1 be the largest integer between 0 and 9 for which $a_0 \cdot a_1$ is a lower bound for S. Then, generally, let a_n be the greatest integer between 0 and 9 for which $a_0 \cdot a_1 a_2 \ldots a_n$ is a lower bound for S. We claim that

$$\lambda = a_0 \cdot a_1 a_2 \ldots$$

is the greatest lower bound. The proof is mainly a matter of unravelling decimal notation, and is complicated by the occurrence of 'carry digits' in arithmetic.

First, we show that λ is a lower bound. If not, there exists $s \in S$ such that $s < \lambda$. By Archimedes' condition there exists $n \in \mathbf{N}$ such that $10^{-n} < \lambda - s$. Therefore a_n can be reduced by 1 in the definition of λ; or, if $a_n = 0$, some earlier $a_m > 0$ can be reduced by 1. But this contradicts the definition of λ.

Then we show that every lower bound μ is less than or equal to λ. If not, $\mu > \lambda$, so by Archimedes' condition there exists $n \in \mathbf{N}$ such that $10^{-n} < \mu - \lambda$. Therefore a_n can be increased by 1 in the definition of λ; or, if $a_n = 9$, some earlier $a_m < 9$ can be increased by 1. But this contradicts the definition of λ. $\square$

We have remarked that it is not possible to make a drawing sufficiently accurate to distinguish rational numbers from irrational ones. But questions of upper and lower bounds expose a vital theoretical difference between real and rational numbers. Examples (3) and (4) above are bounded sets of rationals with no rational least upper bound. In this sense, **R** is complete but **Q** is not. It is this property that will play a vital role when we come to a formal definition of the real numbers later in this book.

Exercises

1. Assuming any results you need about prime factorisation of natural numbers, show that every positive rational number can be written in exactly one way as a product

$$r = p_1^{\alpha_1} p_2^{\alpha_2} \cdots p_s^{\alpha_s}$$

where $p_1 = 2, p_2 = 3, \ldots$ are the primes in increasing order and each α_k is an integer (positive, negative, or zero).

Write the following rationals in this manner: 14/45, 3/8, 2, 20/45.

Show that $\sqrt{r}$ is rational precisely when all of the powers $\alpha_1, \alpha_2, \ldots$ are even. Deduce that for a positive integer n, $\sqrt{n}$ is irrational if and only if n is not the square of an integer.

2. Extend the result of exercise 1 to find those rational numbers r such that $\sqrt[3]{r}$ (cube root of r) is irrational. Show that $\sqrt[n]{\frac{3}{8}}$ is irrational for all natural numbers $n \geq 2$.

3. Which of the following statements are true?
   (a) If x is rational and y is irrational, then $x + y$ is irrational.
   (b) If x is rational and y is rational, then $x + y$ is rational.
   (c) If x is irrational and y is rational, then $x + y$ is rational.
   (d) If x is irrational and y is irrational, then $x + y$ is irrational.
   Prove the true ones and give examples to disprove the false ones.

4. Prove that between any two distinct real numbers there exist infinitely many rational numbers and infinitely many distinct irrational numbers. (Here, 'infinitely many' means that given any natural number n, there exist at least n numbers with the required property.)

5. For real numbers a, r and natural number n, let $s_n = a + ar + \cdots + ar^n$. Show that $rs_n - s_n = a(r^{n+1} - 1)$ and deduce that

$$\left| s_n - \frac{a}{1-r} \right| = \left| \frac{r^{n+1}}{1-r} \right| \quad \text{for } r \neq 1.$$

For $|r| < 1$ deduce that $s_n \to a/(1 - r)$ as $n \to \infty$.

6. Prove that an infinite decimal $x = a_0 \cdot a_1 a_2 a_3 \ldots$ is a rational number if and only if it is 'eventually recurring', that is, after some n onwards it repeats the same block of digits indefinitely.

$$x = a_0 \cdot a_1 \ldots a_n \underbrace{a_{n+1} \ldots a_{n+k}}\ \underbrace{a_{n+1} \ldots a_{n+k}}\ \underbrace{a_{n+1} \ldots a_{n+k}} \ldots$$

(*Hint*: One way round, use question 5 with $a = a_{n+1} \ldots a_{n+k}/10^{n+k}$ with $r = 1/10^k$.)

7. Let

$$y = 0 \cdot 12345678910111213141516171819 20 \ldots,$$

whose digits are the natural numbers in decimal form, strung end to end. Prove that y is irrational.

Is

$$0 \cdot 101001000100001 \ldots,$$

where each successive string of 0s has one more digit, rational or irrational?

8. Say whether each of the following sequences (a_n) tends to a limit, and if so, what the limit is. Use the $\varepsilon - N$ definition to prove your answers are correct.

(a) $a_n = n^2$

(b) $a_n = 1/(n^2 + 1)$

(c) $a_n = 1 + \frac{1}{3} + \frac{1}{9} + \cdots + \left(\frac{1}{3}\right)^n$

(d) $a_n = (-1)^n$

(e) $a_n = \left(-\frac{1}{2}\right)^n$

PART II
The Beginnings of Formalisation

The next five chapters develop the techniques we need to place mathematical reasoning on a firmer logical basis. We still permit the use of our intuitive ideas, but now only as *motivation* for the concepts introduced, and no longer as an integral part of the reasoning.

In the 'building' metaphor, we are getting together the bricks, cement, timber, tiles, pipes, and other materials, and assembling a workforce of brick-layers, plasterers, joiners, and plumbers to put them together in the right way. In the 'plant' metaphor, it is a question of flowerpots, stakes, forks, and trowels, and a good stock of insecticide to keep the bugs off.

We concentrate on two main ideas: the use of set theory as a source of raw material, and the use of mathematical logic to ensure that the proofs of theorems are rigorous and sound. There are three chapters on sets and related topics, followed by two on logic. We approach both from the point of view of a practical mathematician who is more interested in using them to do mathematics than in their own internal workings.

CHAPTER 3

Sets

In accordance with the point of view stated in chapter 1, we make no attempt to give a precise *definition* of the concept 'set'. This will not prevent us from explaining what a set is. A set is any collection of objects whatsoever. The word 'collection' is not intended to imply anything about the number of objects in the set: it may be finite or infinite; there may be just one object, or even none. Nor is there any intention to imply any uniformity in the type of object used to make up the set: a perfectly good set might consist of three numbers, two triangles, and a function. Obviously such a broad concept allows vast scope for whimsical examples. However, the sets of interest in mathematics consist only of mathematical objects. At an elementary level we encounter sets of numbers, sets of points in the plane, sets of geometrical curves, sets of equations. In more advanced mathematics, there is an enormous variety of sets; in fact almost all the concepts of interest are built up from a set-theoretical standpoint.

Nowadays the concept 'set' is considered to be fundamental to the whole of mathematics—even more fundamental than the concept 'number', which earlier ages plumped for. There are many reasons for this. One is that the solution of equations usually yields a set of solutions, rather than just one; quadratic equations, for example, usually have two solutions. Again, modern mathematics places emphasis on generality. Interesting theorems tend to apply to a variety of cases. Pythagoras' theorem is important, not because it applies to one particular right-angled triangle, but because it applies to *all* of them. It thereby expresses a property of the set of all right-angled triangles. The concept of a 'group' (which we describe later, particularly in chapter 13) appears in many guises throughout the whole of mathematics. The language of sets helps us formulate general properties of a group, which therefore apply to *all* of its realisations. It is this power of expression of general concepts using set-theoretic language that gives modern mathematics its distinctive flavour.

To deal with all of the sets that arise in mathematics, it is easiest to develop first those general properties common to all sets, and then to apply them in more special situations. For the rest of this chapter we concentrate on various natural ways to combine and modify sets to form other sets. The systematic study of these methods leads to a kind of 'algebra' of sets, in the same way that a systematic study of the general properties of numbers and operations such as addition, subtraction, multiplication, and division leads to an algebra of numbers.

Members

The objects that together make up a given set are called the *members* or *elements* of the set. The members themselves are said to *belong* to the set. To express symbolically that an element x belongs to a set S, we write

$$x \in S.$$

If x does not belong to S, we write

$$x \notin S.$$

In order to know which set is under consideration, we must know exactly which objects are members. Conversely, if we know the exact membership, we know which set the members form. Being pedantic about this is not as silly as it might seem, because we often describe the same set in different ways; we can be sure that we are dealing with the same set by looking at its members. For example, if A is the set of solutions of the equation

$$x^2 - 6x + 8 = 0,$$

and B is the set of even integers between 1 and 5, then A and B both have precisely two members, 2 and 4. This means that A and B are the same set. It is sensible, therefore, to say that two sets are *equal* if they have the same members. Equality of two sets S and T is expressed in the usual way by

$$S = T,$$

and if S and T are not equal, we write

$$S \neq T.$$

This apparently trite criterion for equality of sets has some interesting consequences, as we see in a moment.

The simplest way to specify a set is to list its members (if that is possible). The standard notation is to enclose the list in braces { }. So

$$S = \{1, 2, 3, 4, 5, 6\}$$

means that S is the set whose members are the numbers 1, 2, 3, 4, 5, 6, *and only these*. As another example, if

$$T = \{79, \pi^2, \sqrt{(5 + \sqrt{7})}, \tfrac{4}{5}\}$$

then the members of T are the numbers 79, π^2, $\sqrt{(5 + \sqrt{7})}$, and $\tfrac{4}{5}$.

Two features of this notation should be emphasised; both are consequences of our notion of equality of sets. First, it is immaterial in which order we write the list of members. The set $\{5, 4, 3, 2, 6, 1\}$ is the *same* as the set S above, and so is the set $\{3, 5, 2, 1, 6, 4\}$. Why? Because in all three cases, we have the same members, namely 1, 2, 3, 4, 5, and 6. The order within the braces arises not from any mathematical cause, but from our conventions about writing from left to right. Second, if elements are repeated in the list, this does not alter the set either. For instance, $\{1, 2, 3, 4, 6, 1, 3, 5\}$ is just our old friend S again. Once more, there is a reason for this seemingly peculiar convention. We might combine two lists to give the set consisting of, say, all the proper divisors of 12, namely 1, 2, 3, 4, 6, together with the odd numbers less than 6, which are 1, 3, 5. Just writing one list after the other gives precisely what we have written. In this case it would have been quite easy to go through and cross out repeats, but in general, it is better to retain flexibility of notation and allow repeats. Our convention about a set being specified by its members implies that all of the various ways of specifying S have precisely the members 1, 2, 3, 4, 5, 6 and no others.

These peculiarities of notation have no great conceptual significance. We are used to the fact that when writing fractions, we can get different symbols for the same number: $\tfrac{1}{2} = \tfrac{2}{4} = \tfrac{3}{6}$ and so on. In fact this is one of the most common usages of the equality sign: when we write $x = y$ we mean that the two symbols on either side of the sign are two different names for the same thing. For instance, $2 + 2 = 12 \div 3 = 5 - 1 = +\sqrt{16} = 4$. We use the same convention when we write $S = T$ for equality of sets. Having understood this, there is no essential difficulty here; we have just raised these questions in order to dispose of them.

When specifying a set, it may not be convenient, or even possible, to write down a complete list of all the members. The set of prime numbers is better described precisely by that phrase, rather than by the list

$$\{2, 3, 5, 7, 11, 13, 17, 19, \ldots\}.$$

Indicating a few terms of an infinite set in this manner is open to the same sort of misinterpretation as writing the first few terms of a sequence, only slightly worse. A sequence is thought of *in order*, but according to our conventions about sets, the elements inside braces are *not* in any specific order. So the list above might also be written

$$\{7, 17, 37, 47, 2, 11, 3, 5, \ldots\}.$$

Who could sort out this jumble and say, with hand on heart, that this is the set of all primes? We admit that there are occasions when mathematicians do use the bracket notation for infinite sets. We do so ourselves sometimes. In such cases, it is always clear what is intended.

In the given case we could be more precise by writing

$$P = \{\text{all prime numbers}\},$$

which is self-explanatory. A slight variation on this, which is very useful, is

$$P = \{p \mid p \text{ is a prime number}\}.$$

Here the braces are read as 'the set of all . . .', the vertical line as 'such that', and the whole symbol reads 'the set of all p such that p is a prime number', which obviously means 'the set of all prime numbers'. In general a definition of the type

$$Q = \{x \mid \text{something or other involving } x\}$$

means that Q is the set of all x for which the something or other involving x is true.

To see how useful this notation is, suppose we want to define S to be the set of solutions of the quadratic equation

$$x^2 - 5x + 6 = 0.$$

We could, of course, solve, and define $S = \{2, 3\}$. Much easier, since it avoids solving the equation, is to write

$$S = \{x \mid x^2 - 5x + 6 = 0\}.$$

This gives a precise and unequivocal definition of S. Of course it is no help in solving the equation! But that is the point of the whole exercise: we can specify the set S without actually doing any calculations.

There is room for ambiguity in this notation. If we are thinking about integers only, then the set

$$\{x \mid 1 \leq x \leq 5\}$$

consists of the numbers 1, 2, 3, 4, 5. But if we are thinking about real numbers, all the other real numbers between 1 and 5 are included as well. The best way out of this impasse is to specify a set Y from which the elements are to be chosen. The notation

$$X = \{x \in Y \mid \text{something or other involving } x\}$$

means that X is the set of those members x of the given set Y such that something or other involving x is true. This is the same as

$$X = \{x \mid x \in Y \text{ and something or other involving } x\},$$

but the first notation is preferable because it emphasises the role of Y.

If $\mathbf{Z}$ is the set of integers and $\mathbf{R}$ the set of real numbers, then

$$\{x \in \mathbf{Z} \mid 1 \leq x \leq 5\}$$

has members 1, 2, 3, 4, 5, while every $a \in \mathbf{R}$ satisfying $1 \leq a \leq 5$ is a member of

$$\{x \in \mathbf{R} \mid 1 \leq x \leq 5\}.$$

There is an even more serious reason for specifying a set Y from which the members of the set X are chosen: to make sure that the 'something or other involving x' makes sense for all $x \in Y$. The 'something or other' needs to be a property that is clearly true or false for every $x \in Y$. Then the set X selected by this property comprises those members of Y for which the property is true.

In English grammar, a sentence is divided into two parts: the *subject* of the sentence, and the rest, called the *predicate*, which tells us about the subject.

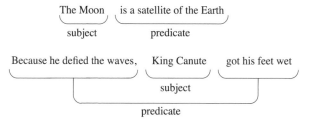

Fig. 3.1 Subject and predicate

A mathematician who customarily uses a symbol like x to denote an unknown might say that the predicate in the first sentence is

$$x \text{ is a satellite of the earth}$$

and the predicate in the second is

> Because he defied the waves, x got his feet wet.

The beauty of this description is that it specifies the position of the subject in the sentence. To get the original sentence back again, we simply substitute the appropriate subject in place of x.

This motivates the mathematical definition:

Definition 3.1: A *predicate* is a sentence involving a symbol x so that when we substitute an element $a \in Y$ for x, the resultant statement is clearly either true or false. We say that the predicate is 'valid for the set Y' if this is so.

For instance, the sentence

$$1 \leq x \leq 5$$

is a predicate which is valid for the set **Z**. It is also valid for the set **R**. Substitute any integer or real number and we get a statement that is either true or false.

$$1 \leq 3 \leq 5 \text{ is true,}$$
$$1 \leq 57 \leq 5 \text{ is false,}$$

and so on.

The set $\{x \in \mathbf{Z} \mid 1 \leq x \leq 5\}$ is just the set of $x \in \mathbf{Z}$ such that the predicate $1 \leq x \leq 5$ is true.

A predicate need not be restricted just to sets of numbers. For instance, if T is the set of triangles in the plane, then the sentence

> x is right-angled

is a predicate valid for the set T and

$$\{x \in T \mid x \text{ is right-angled}\}$$

is just the set of right-angled triangles in the plane.

We could go on giving examples galore of predicates, but plenty will turn up in the text anyway. The reader should make it clear in their own mind that whenever the symbolism

$$\{x \in Y \mid P(x)\}$$

is used, then $P(x)$ is a predicate in x which is valid for all $x \in Y$.

Subsets

Within any given set A there exist other sets, obtained by omitting some of the elements of A. These are called subsets of A. More formally:

Definition 3.2: B is a *subset* of A if every element of B is an element of A. We write

$$B \subseteq A$$

or

$$A \supseteq B.$$

We also say that B is *contained in*, or *included in*, A and that A *contains* or *includes* B. With this definition we have $A \subseteq A$, for trivial reasons. If $B \subseteq A$ and $B \neq A$ then we say that B is a *proper subset* of A and write

$$B \subsetneq A.$$

Many mathematicians use $\subset$ where we have used $\subseteq$ and others write $\subset$ where we have chosen $\subsetneq$. We use $\subseteq$ because it is unambiguous.

The criterion for equality of sets leads to a trivial but useful result:

Proposition 3.3: Let A and B be sets. Then $A = B$ if and only if $A \subseteq B$ and $B \subseteq A$.

Proof: If $A = B$ then, since $A \subseteq A$, it follows that $A \subseteq B$ and $B \subseteq A$. Conversely, suppose that $A \subseteq B$ and $B \subseteq A$. Then each element of A is an element of B, and each element of B is an element of A. Hence A and B have the same elements, so $A = B$. $\qquad\square$

In practice this proposition is used to prove equality of two sets when each is defined by a predicate. We start with a typical element in A (given in terms of the appropriate predicate) and show that this element is also a member of B. This verifies that $A \subseteq B$. Then we carry out a similar argument to show that $B \subseteq A$. We will see plenty of examples of this procedure soon (in propositions 3.8, 3.9, and 3.10, for instance).

A basic property of subsets is that a subset of a subset is itself a subset:

Proposition 3.4: If A, B, C are sets with $A \subseteq B$ and $B \subseteq C$, then $A \subseteq C$.

Proof: Every element of A is an element of B and every element of B is an element of C. Therefore every element of A is an element of C, so $A \subseteq C$. $\square$

WARNING: It is important not to confuse subsets and members: the two concepts are quite different. The *members* of $\{1, 2\}$ are 1 and 2. The subsets of $\{1, 2\}$ are $\{1, 2\}$, $\{1\}$, $\{2\}$, and a fourth subset which for the moment is best written as $\{\ \}$.

Further, proposition 3.4 becomes false if we change '$\subset$' to '$\in$'. Members of members need not be members. For example, let $A = 1$, $B = \{1, 2\}$, $C = \{\{1, 2\}, \{3, 4\}\}$. Then $A \in B$ and $B \in C$. But the members of C are $\{1, 2\}$ and $\{3, 4\}$, so $A = 1$ is not a member.

Now let us return to that set $\{\ \}$.

Definition 3.5: A set is *empty* if it has no members.

For instance, the set

$$\{x \in \mathbf{Z} \mid x = x + 1\}$$

is empty, because the equation $x = x + 1$ has no solutions in $\mathbf{Z}$.

An empty set has remarkable properties (remarkable at first sight, that is) by default. For instance, if E is an empty set and X is any set whatsoever, then $E \subseteq X$. Why? We have to show that every element of E is an element of X. The only way that this can fail is if E has some element e which does not belong to X. But E, being empty, has no elements at all, so cannot contain any such element.

This (curious but logical) argument is an example of 'vacuous reasoning', because it discusses properties of something that does not exist. Vacuous reasoning is rarely encountered in everyday argument. However, for mathematicians, it has a unifying feature that allows logical arguments to be used in cases where everyday intuition may not apply. Really, we are discussing properties that something *would have* if it did exist, with the aim of obtaining a contradiction. Then we conclude that it does not exist. So it is useful to allow statements about non-existent objects.

For instance, suppose we have two empty sets E and E'. The above tells us that $E \subseteq E'$ and $E' \subseteq E$. Then proposition 3.3 tells us that $E = E'$. *All empty sets are equal.* Hence there is a *unique* empty set. We therefore give it a special symbol: we write

$$\varnothing$$

to denote *the* empty set.

This is hardly surprising. In the absence of any elements whatsoever, we have no way to distinguish two empty sets. In the words of [31]: 'the contents of two empty paper bags are equal'.

Is There a Universe?

Just as there is an empty set $\varnothing$ that contains no elements, we might ask whether there is a very large set Ω that includes absolutely everything. This turns out to be far too fanciful. Such a set would have to be an incredibly vast rag-bag; if it contained everything it would include all numbers, all elements of every set, all sets, all places in the universe, the Declaration of Independence, Winston Churchill, the year 1066, the wit of Oscar Wilde, . . . If we dare contemplate such an Ω, then Ω itself must be an acceptable concept and we would have to include it in the collection of absolutely everything. So $\Omega \in \Omega$. Most sensible sets do not belong to themselves; in fact you could while away an interesting half an hour trying to find such a set.

However, there is a nastier problem. If we select from the putative set Ω the subset comprising everything that is a set but does not belong to itself, we get:

$$S = \{A \in \Omega \mid A \notin A\}.$$

Now ask the key question: is $S \in S$?

If $S \in S$, then, according to the defining predicate, $S \notin S$.
If $S \notin S$, then S satisfies the defining predicate, so $S \in S$.

Our flight of fancy in assuming the existence of a universe Ω has led to a paradox. Therefore there cannot be a universal set.

Can we salvage the situation by removing all the whimsical things and concentrating on a universal set for the realm of mathematics? No: this too has its pitfalls. If we try to contemplate a set Ω_M of all mathematical objects (whatever that means), then we reach the same contradiction when we consider the subset of Ω_M consisting of all mathematical objects that do not belong to themselves.

To avoid such paradoxes, it is essential for the sets we consider to be clearly defined, in the sense that in principle we know precisely which objects are members and which are not.

The non-existence of a universal set is another reason why the notation

$$\{x \in Y \mid P(x)\},$$

where Y is a known set and $P(x)$ is a predicate, is preferable to

$$\{x \mid P(x)\}.$$

Having specified the set Y, we can investigate the predicate $P(x)$ and make sure it is valid for all elements of Y before selecting those elements in Y for which the predicate is true. Used indiscriminately, the notation $\{x \mid P(x)\}$, which allows us to try absolutely *any* object x to see if it is a member, is like considering $\{x \in \Omega \mid P(x)\}$. But there is no universal set Ω. If we don't specify a set Y, then we have unlimited choice of objects x to try in $P(x)$. We might consider an element that had not been intended at the outset, and end up with a paradox again.

Here is an example. If $\mathbf{Z}$ is the set of integers, $\mathbf{R}$ the set of real numbers, and T the set of all triangles in the plane, then $\mathbf{Z} \notin \mathbf{Z}, \mathbf{R} \notin \mathbf{R}, T \notin T$. If Y is the set whose members are $\mathbf{Z}, \mathbf{R}, T$, then

$$\{x \in Y \mid x \notin x\} = \{\mathbf{Z},\ \mathbf{R},\ T\}.$$

On this set Y, the property $x \notin x$ is a perfectly acceptable predicate. However, if we consider

$$\{x \mid x \notin x\}$$

with no restrictions on x, imagination can run riot. Considering $S = \{x \mid x \notin x\}$ itself, we end up with the same contradiction as before: $S \in S$ if and only if $S \notin S$.

The moral is that set theory is a system of notation, not a magical prescription. As such, it is as good as the manner in which it is used. When used sensibly, it behaves well. But if you use it badly, things can go wrong.

Union and Intersection

Two important methods for combining sets are known as the union and intersection.

Definition 3.6: The *union* of two sets A and B is the set whose elements are those of A together with those of B. We write $A \cup B$ to denote the union of A and B. Now

$$A \cup B = \{x \mid x \in A \text{ or } x \in B \text{ (or both)}\}.$$

For example, if

$$A = \{1, 2, 3\}$$
$$B = \{3, 4, 5\}$$

then the union is $\{1,\ 2,\ 3,\ 4,\ 5\}$.

Definition 3.7: The *intersection* of A and B is the set whose elements belong both to A and to B. The symbol for the intersection is $A \cap B$. In this case,

$$A \cap B = \{x \mid x \in A \text{ and } x \in B\}.$$

For example, with A and B as above, their intersection is $\{3\}$, because only 3 belongs to both of them.

The intersection can also be written as

$$A \cap B = \{x \in A \mid x \in B\},$$

so we can think of it being the subset of A selected using the predicate $x \in B$. (Equivalently we could think of it as being the subset of B which satisfies the predicate $x \in A$.) The union, on the other hand, involves constructing a new set which is (usually) bigger than both A and B, so here we have an example of a set construction that does not select elements from a previously prescribed set Y.

The operations of union and intersection obey certain standard laws. Most of them are obvious, but for convenience we list them in the next three propositions.

Proposition 3.8: Let A, B, C be sets. Then

(a) $A \cup \varnothing = A$
(b) $A \cup A = A$
(c) $A \cup B = B \cup A$
(d) $(A \cup B) \cup C = A \cup (B \cup C)$.

Proof: Only (d) is remotely difficult, so we leave the first three as an exercise. Before trying them, however, read the proof of (d).

Suppose that $x \in (A \cup B) \cup C$. Then either $x \in A \cup B$ or $x \in C$. If $x \in C$, then $x \in B \cup C$, so $x \in A \cup (B \cup C)$. If not, then $x \in A \cup B$, so either $x \in A$ or $x \in B$. In either case $x \in A \cup (B \cup C)$. So we have proved that if $x \in (A \cup B) \cup C$ then $x \in A \cup (B \cup C)$, that is:

$$(A \cup B) \cup C \subseteq A \cup (B \cup C).$$

A similar argument shows that

$$A \cup (B \cup C) \subseteq (A \cup B) \cup C.$$

Using proposition 3.3 we obtain equality. □

This proof is more complicated than the situation really warrants, because it is obvious that $(A \cup B) \cup C$ is the set whose members are those of A, those of B, and those of C together. Clearly this is the same set as $A \cup (B \cup C)$. Once we know this, it is possible to omit the brackets altogether and write just

$$A \cup B \cup C.$$

Similar results hold for intersections:

Proposition 3.9:

(a) $A \cap \varnothing = \varnothing$
(b) $A \cap A = A$
(c) $A \cap B = B \cap A$
(d) $(A \cap B) \cap C = A \cap (B \cap C)$.

The proofs are analogous to those in proposition 3.3. □

Finally, there are two equations that mix up unions and intersections:

Proposition 3.10:

(a) $A \cup (B \cap C) = (A \cup B) \cap (A \cup C)$
(b) $A \cap (B \cup C) = (A \cap B) \cup (A \cap C)$.

Proof: Let $x \in A \cup (B \cap C)$. Then either $x \in A$ or $x \in B \cap C$. If $x \in A$ then certainly $x \in A \cup B$ and $x \in A \cup C$, hence $x \in (A \cup B) \cap (A \cup C)$. Alternatively, $x \in B \cap C$ gives $x \in B$ and $x \in C$. Hence $x \in A \cup B$ and $x \in A \cup C$, so $x \in (A \cup B) \cap (A \cup C)$. This proves that

$$A \cup (B \cap C) \subseteq (A \cup B) \cap (A \cup C). \tag{3.1}$$

Conversely, suppose $y \in (A \cup B) \cap (A \cup C)$. Then $y \in A \cup B$ and $y \in A \cup C$. There are two cases to consider: when $y \in A$ and when $y \notin A$. If $y \in A$, then certainly $y \in A \cup (B \cap C)$. On the other hand, if $y \notin A$ then, since $y \in A \cup B$, we must have $y \in B$; similarly $y \in C$. Thus $y \in B \cap C$, which again implies $y \in A \cup (B \cap C)$. Therefore

$$(A \cup B) \cap (A \cup C) \subseteq A \cup (B \cap C).$$

Together with (3.1), this yields the desired result.
The proof of (b) is analogous. □

Proposition 3.10 is a pair of 'distributive laws', which should be compared with the way that multiplication of numbers is distributive over addition:

$$a \times (b + c) = (a \times b) + (a \times c).$$

With numbers, however, the interchange of the two operations does not give a new rule:

$$a + (b \times c) = (a + b) \times (a + c)$$

is *not* true in general.

The operations ∪ and ∩ on sets behave in a much more symmetrical way: each is distributive over the other.

One way to visualise these various set theoretic identities is to draw *Venn diagrams*. The identity

$$A \cup (B \cap C) = (A \cup B) \cap (A \cup C)$$

can be represented by drawing three overlapping discs, supposed to represent the sets A, B, C, and proceeding as follows:

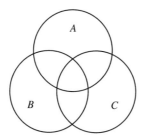

Fig. 3.2 Three overlapping sets

$B \cap C$ is the shaded region common to B and C:

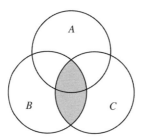

Fig. 3.3 $B \cap C$

and the union of this with A is:

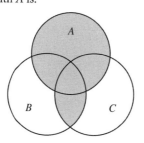

Fig. 3.4 $A \cup (B \cap C)$

On the other hand, $A \cup B$ is:

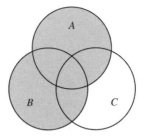

Fig. 3.5 $A \cup B$

and $A \cup C$ is:

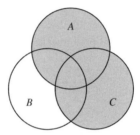

Fig. 3.6 $A \cup C$

so $(A \cup B) \cap (A \cup C)$ is the region common to both, which is:

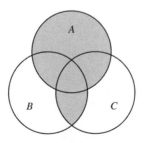

Fig. 3.7 $(A \cup B) \cap (A \cup C)$

This is the same as before.

You may wish to try your hand at illustrating the other identities in propositions 3.8, 3.9, and 3.10 by drawing Venn diagrams. Such pictorial devices, if well chosen, help most people to get a coherent idea of what is going on. To obtain the most general picture, the diagram must be drawn with care. With one set A, there are two distinct regions involved, inside A and outside A:

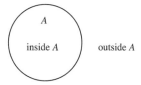

Fig. 3.8 Inside and outside a set

With two sets A and B there are four regions, (1) outside both, (2) inside A but not B, (3) inside B but not A, and (4) inside both of them:

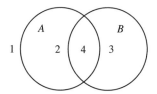

Fig. 3.9 Two sets, four regions

With three sets, A, B, and C, there are eight regions:

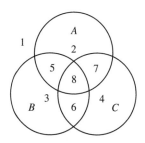

Fig. 3.10 Three sets, eight regions

If we add a fourth set D so that D meets each of these eight regions and the area outside D meets each of the regions, then we get sixteen regions in all.

There is no way that this can be achieved if A, B, C, and D are all drawn as circles. Try to draw a fourth circle in the last diagram above to meet this prescription, and you will see what we mean. It can be done, but not with a circle:

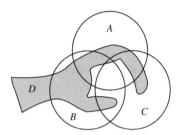

Fig. 3.11   Four sets, . . .

Many more elegant diagrams, which in principle can handle any finite number of sets, have been devised (see [2]). Venn himself was aware of such limitations when he first drew the diagrams. They can be ironed out by using more complicated shapes to represent the sets. This problem illustrates the need to shift from the use of pictures as an aid to mental processes, to a general proof that works in all cases in a manner analogous to those in propositions 3.8, 3.9, and 3.10. This is an important aspect of the journey in this book, from intuitive beginnings that may be imagined as pictures, to formal definitions and proofs that work in general. Initially, proofs should be verbalised in terms of definitions, to establish relationships that can be used with confidence in new settings.

There is a general connection between unions, intersections, and subsets:

Proposition 3.11: If A and B are sets, the following are equivalent:

(a) $A \subseteq B$
(b) $A \cap B = A$
(c) $A \cup B = B$.

Proof: Equation (b) says that the elements common to A and B are all the elements of A, so every element of A belongs to B, which implies $A \subseteq B$. The converse is obvious, so (a) and (b) are equivalent.

Equation (c) says that if we add to B the elements of A, we still get B. Therefore no element of A can fail to belong to B, and again $A \subseteq B$. The converse is once more obvious, so (a) and (c) are equivalent. $\qquad\square$

Complements

Let A and B be sets.

Definition 3.12: The set-theoretic *difference* $A\backslash B$ is defined to be the set of all those elements of A that do not belong to B. In symbols,

$$A\backslash B = \{x \in A \mid x \notin B\}.$$

In a Venn diagram, $A\backslash B$ is the shaded region in:

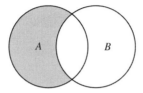

Fig. 3.12 The set-theoretic difference

If B is a subset of A, then we call $A\backslash B$ the *complement of B relative to A*.

Fig. 3.13 The complement of B relative to A

It would be nice to forget about A entirely, thus defining the complement of B to consist of everything not belonging to B. However, this is too much to ask, because it would mean that B and its supposed complement would make up a set Ω which contains absolutely everything, and we have already shown that such a set cannot exist.

In a particular piece of mathematics, however, there may be a set U that includes all of the elements that we wish to consider. We call this set the *universe of discourse* or *universal set* (universal, that is, for current purposes). When dealing with the set of integers, for example, we might take the universal set to be $U = \mathbf{Z}$. Of course, $U = \mathbf{R}$ would do just as well. The important thing is that the universal set should be sufficiently all-embracing to include all of the elements under discussion. As one of us says elsewhere, 'In a discussion about dogs, when thinking about all non-sheepdogs, it is pointless to worry about camels'.

Definition 3.13: Having agreed upon U, we define the *complement* B^c of every subset B of U by

$$B^c = U \backslash B.$$

Thus B^c is the complement of B relative to U. But because U is agreed upon, we can omit it from the notation, which is the object of the exercise.

Of course, the operation ^{c} obeys some simple laws. They include:

Proposition 3.14: If A and B are subsets of the universal set U, then

(a) $\varnothing^c = U$
(b) $U^c = \varnothing$
(c) $(A^c)^c = A$
(d) If $A \subseteq B$ then $A^c \supseteq B^c$. □

In view of (c) we can write $A^{cc} = (A^c)^c = A$. Less elementary, but highly interesting, are:

Proposition 3.15 (De Morgan's Laws): If A and B are subsets of the universal set U, then

(a) $(A \cup B)^c = A^c \cap B^c,$
(b) $(A \cap B)^c = A^c \cup B^c.$

Proof: Let $x \in (A \cup B)^c$. Then $x \notin A \cup B$. This implies that $x \notin A$ and $x \notin B$, so $x \in A^c$ and $x \in B^c$, so $x \in A^c \cap B^c$. Therefore $(A \cup B)^c \subseteq A^c \cap B^c$. To obtain the reverse inclusion, reverse the steps in the argument. This proves (a).

Equation (b) can be proved similarly. Alternatively, we can replace A by A^c and B by B^c in (a), which gives

$$\left(A^c \cup B^c\right)^c = A^{cc} \cap B^{cc} = A \cap B.$$

Taking complements,

$$A^c \cup B^c = \left(A^c \cup B^c\right)^{cc} = (A \cap B)^c,$$

and this is (b). □

These laws explain a phenomenon that the alert reader may have observed already: set-theoretic laws come in pairs, so that if we start with one and change all unions to intersections and all intersections to unions, we obtain another. We could formulate this as follows:

Theorem 3.16 (De Morgan Duality Principle): If in any valid set-theoretic identity involving only the operations $\cup$ and $\cap$ the operations $\cup$ and $\cap$ are interchanged throughout, the result is another valid identity.

Proof: To prove this in general is not hard, but it needs an induction argument. The following example is a typical case. Start with the identity

$$A \cup (B \cap C) = (A \cup B) \cap (A \cup C).$$

Take complements of both sides and use De Morgan's laws to get

$$A^c \cap (B \cap C)^c = (A \cup B)^c \cup (A \cup C)^c,$$

then use De Morgan again to get

$$A^c \cap (B^c \cup C^c) = (A^c \cap B^c) \cup (A^c \cap C^c).$$

Already we have interchanged $\cup$ and $\cap$. Now systematically replace A by A^c, B by B^c, C by C^c. Since the equation is true for *any* sets A, B, C, this is legitimate. We get

$$A \cap (B \cup C) = (A \cap B) \cup (A \cap C).$$

This is the original law, with $\cup$s and $\cap$s interchanged. $\qquad\square$

QUESTION: How does the presence of the operation ^{c} affect the argument? (Try the identity

$$B \cup (A \cap A^c) = B$$

and use the same approach. What happens?)

Sets of Sets

It may happen that all the elements of a given set S are themselves sets. Indeed, it is often a useful device to consider a set of sets. For instance, we may have $S = \{A, B\}$ where $A = \{1, 2\}$, $B = \{2, 3, 4\}$. A more sophisticated example is to take any set X and let $\mathbb{P}(X)$ be the set of all subsets of X. This is called the *power set of X* and satisfies the property:

$$Y \in \mathbb{P}(X) \text{ if and only if } Y \subseteq X.$$

For example, if $X = \{0, 1\}$, then $\mathbb{P}(X) = \{\varnothing, \{0\}, \{1\}, \{0, 1\}\}$. In cases like these, where every member of S is itself a set, we can go a level further and consider the elements belonging to these members. This gives us generalisations of the notions of union and intersection:

$$\bigcup S = \{x \mid x \in A \text{ for some } A \in S\}$$
$$\bigcap S = \{x \mid x \in A \text{ for every } A \in S\}.$$

The taller versions of the symbols remind us that these are related to the operations $\cup$ and $\cap$ but now apply to sets of sets. We call $\bigcup S$ the 'union of S' and $\bigcap S$ the 'intersection of S'. Put into words, the union of S consists of all the elements in the members of S and the intersection of S consists of those elements common to all members of S. For instance,

$$\bigcup\{\{1,2\}, \{2,3,4\}\} = \{1,2,3,4\}$$
$$\bigcap\{\{1,2\}, \{2,3,4\}\} = \{2\}.$$

In general, for any set X,

$$\bigcup \mathbb{P}(X) = X$$
$$\bigcap \mathbb{P}(X) = \varnothing.$$

Although this notation may seem a little strange at first, it is extremely economical and it does act as a genuine extension of the usual concepts. For instance, given two sets A_1, A_2, let $S = \{A_1, A_2\}$. Then

$$\bigcup S = A_1 \cup A_2$$
$$\bigcap S = A_1 \cap A_2.$$

More generally,

$$\bigcup\{A_1, A_2, \ldots, A_n\} = A_1 \cup A_2 \cup \ldots \cup A_n$$
$$\bigcap\{A_1, A_2, \ldots, A_n\} = A_1 \cap A_2 \cap \ldots \cap A_n.$$

Alternative (and much more used) notations for these last two concepts are

$$A_1 \cup A_2 \cup \ldots \cup A_n = \bigcup_{r=1}^{n} A_r$$
$$A_1 \cap A_2 \cap \ldots \cap A_n = \bigcap_{r=1}^{n} A_r.$$

We return to generalised unions and intersections at the end of chapter 5.

Exercises

1. Which of the following sets are the same?
   (a) $\{-1, 1, 2\}$
   (b) $\{-1, 2, 1, 2\}$

(c) $\{n \in \mathbf{Z} \mid |n| \leq 2 \text{ and } n \neq 0\}$

(d) $\{2, 1, 2, -2, -1, 2\}$

(e) $\{2, -2\} \cup \{1, -1\}$

(f) $\{-2, -1, 1, 2\} \cap \{-1, 0, 1, 2, 3\}$.

2. Prove that for all sets A, B,

$$(A \backslash B) \cup (B \backslash A) = (A \cup B) \setminus (A \cap B).$$

If A is the set of even integers, and B is the set of integers that are multiples of 3, describe $(A \backslash B) \cup (B \backslash A)$.

3. Write out the proofs of propositions 3.8(a), 3.8(b), 3.8(c), and all of proposition 3.9. Draw Venn diagrams to illustrate these results.

4. Draw a Venn diagram suitable for all formulas involving five different sets.

5. If $S = \{$all subsets $X \subseteq \mathbf{Z}$ such that $0 \in X\}$, find $\bigcap S$, $\bigcup S$.

6. If $S = S_1 \cup S_2$, prove that $\bigcup S = \left(\bigcup S_1 \right) \cup \left(\bigcup S_2 \right)$.

7. If A has n elements ($n \in \mathbf{N}$), calculate the number of subsets of A. If you are acquainted with proof by induction, prove your result by this technique.

8. If A, B, C are finite sets and $|A|$ denotes the number of elements in A, show that

$$|A \cup B \cup C| = |A| + |B| + |C| - |A \cap B| - |B \cap C| - |C \cap A| + |A \cap B \cap C|.$$

Draw a Venn diagram.

9. In each of the following statements, if we replace S by one of $\mathbf{N}$, $\mathbf{Z}$, $\mathbf{Q}$, $\mathbf{R}$, then we get a true statement. Find the appropriate set in each case:

(a) $\{x \in S \mid x^3 = 5\} \neq \varnothing$

(b) $\{x \in S \mid -1 \leq x \leq 1\} = \{1\}$

(c) $\{x \in S \mid 2 < x^2 < 5\} \backslash \{x \in S \mid x > 0\} = \{-2\}$

(d) $\{x \in S \mid 1 < x \leq 4\} = \{x \in S \mid x^2 = 4\} \cup \{3, 4\}$

(e) $\{x \in S \mid 4x^2 = 1\} \backslash \{x \in S \mid x < 0\} = \{x \in S \mid 5x^2 = 3\} \cup \{x \in S \mid 2x = 1\} \neq \varnothing$.

10. The equation $x + y = z$ has many solutions $x, y, z \in \mathbf{N}$; the equation $x^2 + y^2 = z^2$ has solutions including $x = 3, y = 4, z = 5$.

Let $F = \{n \in \mathbf{N} \mid x^n + y^n = z^n \text{ has a solution where } x, y, z \in \mathbf{N}\}$.

What must be done to show $F = \{1, 2\}$? What does this tell us about verifying equality between sets in general?

CHAPTER 4

Relations

The aim of this chapter is to introduce one of the most important concepts in set theory. The notion of a relation is found throughout mathematics and applies in many situations outside the subject as well. Examples of relations involving numbers include 'greater than', 'less than', 'divides', 'is not equal to'; examples from the realms of set theory include 'is a subset of', 'belongs to'; examples from other areas include 'is the brother of', 'is the son of'. What all these have in common is that they refer to two things, and the first is either related to the second in the manner described, or not. Thus the statement $a > b$, where a and b are integers, is either true or not true ($2 > 1$ is true, $1 > 2$ is false).

The two things that are related must be taken in a specific order, for instance the statement $a > b$ is quite different from $b > a$. So the first thing we do in this chapter is to set up some machinery about ordered pairs.

Relations can occur between elements in different sets; that is, we can have a relation between elements in a set A and those in a set B. Most of the examples we mentioned concern objects from the same set, but we slipped one into the set-theoretic list which was 'belongs to'. If A is a set of elements and B is a set whose members are themselves sets, then we can determine whether $x \in Y$ for each member $x \in A$ and each member $Y \in B$. Since $x \in Y$ is either true, or not true, for every $x \in A$ and $Y \in B$, this defines a relation between A and B in the sense to be described in this chapter. The beauty of the description given is that it can be formulated entirely in set-theoretic terms.

In the latter part of the chapter we develop a detailed theory of two particularly important types of relation: equivalence relations and order relations.

Ordered Pairs

We have said that for sets, the order in which we write the elements in a list makes no difference, so that for a set with two elements, $\{a, b\} = \{b, a\}$. This is all very well, but there are occasions on which it is essential to distinguish the order. For instance, in coordinate geometry we think of all the points of the plane as being represented by pairs (x, y) of real numbers. The order is crucial; for example the points $(1, 2)$ and $(2, 1)$ are different:

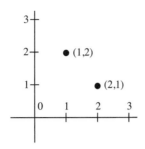

Fig. 4.1 Order matters

We are thus led to the concept of an *ordered pair* (x, y), round brackets being used to make a distinction from $\{x, y\}$. The important property that we require of this new concept is:

The Ordered Pair Property:

$$(x, y) = (u, v) \text{ if and only if } x = u \text{ and } y = v. \qquad \text{(OPP)}$$

This notion is used throughout set theory.

This is all very well; the only problem is that we haven't actually said precisely what we mean by an ordered pair. What is (x, y)? If $A = B = \mathbf{R}$, then we can think of an ordered pair (x, y) as a point in the coordinate plane, using Cartesian coordinates. This, indeed, is where the notion of ordered pair arose. In this sense we can refer to the plane as $\mathbf{R} \times \mathbf{R}$ (or, in more usual mathematical shorthand, as $\mathbf{R}^2$). But what happens if A is a set like {apple, orange, grapefruit} and B is {knife, fork}, what then is $A \times B$? It certainly consists of the ordered pairs:

(apple, knife), (apple, fork), (orange, knife), (orange, fork), (grapefruit, knife), (grapefruit, fork).

However, that doesn't answer the question: what is the ordered pair (apple, knife)?

The solution lies not in 'what is it?', but in 'how do we get it?'. The answer is that to obtain (x, y) in general, we first select x from A, then we select y from B. That's all.

The mathematician Kasimierz Kuratowski saw in this process a possible abstract definition of (x, y) using only set-theoretic notions that we have already described. Having selected x from A, we have the singleton set $\{x\}$; then selecting y from B we arrive at the set $\{x, y\}$. Kuratowski defined the ordered pair (x, y) to consist of these sets:

Definition 4.1 (Kuratowski): The *ordered pair* (x, y) of two elements x, y is defined to be the set

$$(x, y) = \{\{x\}, \{x, y\}\}.$$

Notice that we get a set here. This peculiar-looking definition has the advantage that it satisfies the ordered pair property (OPP):

Proposition 4.2: With Kuratowski's definition,

$$(x, y) = (u, v) \text{ if and only if } x = u \text{ and } y = v.$$

Proof: If $x = u, y = v$, then the definition gives $(x, y) = (u, v)$. In the other direction, suppose that $(x, y) = (u, v)$. If $x \neq y$, then $(x, y) = \{\{x\}, \{x, y\}\}$ has two distinct members, $\{x\}$ and $\{x, y\}$, which must each belong to $(u, v) = \{\{u, \}, \{u, v\}\}$. This means that the members $\{u\}$ and $\{u, v\}$ must be different also, implying $u \neq v$. Now we must have $\{x\} = \{u\}$ or $\{x\} = \{u, v\}$, and the latter is clearly impossible (because it would mean that u, v both belonged to $\{x\}$, implying $u = x = v$, contradicting $u \neq v$). So $\{x\} = \{u\}$ and $x = u$. In a similar fashion, $\{x, y\} = \{u, v\}$, and since $x = u, x \neq y$ and $y \in \{u, v\}$, we deduce that $y = v$. Thus $x = u$ and $y = v$, as required.

If $x = y$, the set-theoretic construction collapses somewhat to give

$$(x, y) = \{\{x\}, \{x, y\}\} = \{\{x\}, \{x, x\}\} = \{\{x\}, \{x\}\} = \{\{x\}\},$$

so (x, y) has only *one* member, namely $\{x\}$. If $(x, y) = (u, v)$, then (u, v) has only one member also, implying $\{u\} = \{u, v\}$, so $u = v$ and $(u, v) = \{\{u\}\}$. The equality $(x, y) = (u, v)$ then becomes $\{\{x\}\} = \{\{u\}\}$, which reduces successively to $\{x\} = \{u\}$ and then $x = u$. Thus this case reduces to $x = y = u = v$ and the proof is complete. □

By being a little more sophisticated, we can prove this result much more quickly. In the notation of the last section of chapter 3,

$$\bigcap \{\{x\}, \{x, y\}\} = \{x\} \text{ and } \bigcup \{\{x\}, \{x, y\}\} = \{x, y\}.$$

So $\bigcap(x, y) = \{x\}$, $\bigcup(x, y) = \{x, y\}$. If $(x, y) = (u, v)$, then comparing intersections and unions, $\{x\} = \{u\}$, $\{x, y\} = \{u, v\}$. The first gives $x = u$ and from this (whether $x = y$ or not), the second gives $y = v$.

Where does this get us? First the good news: we have a definition of the ordered pair (x, y) involving only established set-theoretic concepts. Then the bad news: the definition does not correspond to the intuitive notion of ordered pairs in coordinate geometry. Indeed, if any mathematician were asked to visualise (2, 1), he or she would, like as not, think of it as a point in the plane; it is most unlikely that their thoughts would revolve around the idea $\{\{2\}, \{2, 1\}\}$.

The pragmatic solution is to let Kuratowski's definition fade into the background, safe in the knowledge that it is there should we ever be asked to give a rigorous foundation. The important notion is the ordered pair property (OPP).

Here we meet a fundamental idea that underpins the whole of formal mathematics. What is important is not what a mathematical object *is*, but what its *properties* are. Formal mathematical concepts are specified by definitions that state their required properties in terms of set theory. Other properties of a given concept are then deduced by mathematical proof as theorems. This principle has the powerful consequence that the theorems proved must be valid in *any* context that satisfies the specified definitions. This is true not only of situations that are familiar, but also in any situation we meet in the future where the definitions are satisfied.

Mathematical Precision and Human Insight

The situation occurring for ordered pairs happens throughout formal mathematics. Essentially the same underlying mathematics can be expressed in a variety of different ways. For example we now have (at least) three different ways of thinking about an ordered pair (x, y) for elements $x, y \in S$. We can represent it *symbolically* as (x, y), *visually* as a point in the plane (when S is the set of real numbers), and *formally* as the Kuratowski definition. All of these have the same property that $(x, y) = (x', y')$ if and only if $x = x'$ and $y = y'$. When we *think* about ordered pairs in our everyday working, we almost always use a visual or symbolic representation rather than the Kuratowski definition.

More generally, we often write the same thing in different ways. For instance, we can write the fraction 1/2 as 2/4 or 3/6 and say that all these fractions are 'equivalent'. As processes of calculation, these fractions are different, but they all produce the same result. There are many other instances where 'equivalent' things are essentially the same. For instance, the algebraic expressions $3(x + 2)$ and $3x + 6$ are different processes ('treble the result of adding x and 2' and 'three times x plus 6'), but they have the same result. If we use the concept of a function, the two functions $f(x) = 3(x + 2)$ and $g(x) = 3x + 6$ are the *same* function, according to the set-theoretic definition.

As mathematics gets more sophisticated, we realise that various ways of thinking about a particular concept can be conceived as a single idea. The Greeks sought 'the essence' of mathematical concepts in arithmetic and geometry. For example, when considering a circle, they started from physical circles with different locations and different sizes. From these examples they extracted a single Platonic object: the locus (now spoken of as the 'set') of all points in a plane that are equidistant from a given point, the centre.

Likewise, equivalent fractions represent an underlying rational number, and equivalent algebraic expressions represent the same algebraic object written in different ways.

Something that we can hold in our minds in various different ways, all of which essentially represent the same underlying idea, is called a *crystalline concept* (see [35]). The term 'crystalline' does not mean that the concept looks like a crystal with regular faces; it means that it has strong links that relate its properties in a coherent and inevitable way. For instance, the sum of two numbers is a crystalline concept in the sense that $2 + 3$ is 5 in the context of our usual number notation, *and* that if we take 3 from 5 then the result can only be 2. In the same way, if we have a triangle drawn in Euclidean geometry that has two equal sides, then it must have two equal angles. Now we have *one* concept, isosceles triangle, defined in two ways. Not two *differently* defined concepts that happen to be equivalent.

A 'crystalline concept' formulates how we *think* about sophisticated mathematical ideas, rather than offering a formal definition of a mathematical concept. In the natural world, different concepts can have the same essential structure. For example, there is a huge difference between 3 ducks each with 2 legs and 2 ducks with 3 legs. But in calculating the number of legs, the products 2×3 and 3×2 both give the same result. Similarly, taking away two \$10 bills is different, as an operation, from adding two debts of \$10, but the effect on your finances is the same.

As we become more sophisticated, we do not say '-2 times -3 is *equivalent* to 2 times 3'; we say '$(-2) \times (-3)$ *equals* 2×3'. Formal mathematics takes these ideas to a higher level, defining the properties that a particular formal structure must have and deducing all its other properties by mathematical proof. Definitions that at one level are considered 'equivalent' concepts may be imagined at a higher level as a single crystalline concept. For instance, equivalent algebraic expressions become conceptualised as the same function.

The notion of 'crystalline concept' is a psychological term rather than a mathematical one, so you will not find it in current mathematics textbooks. However, it represents a breakthrough in cognitive psychology that enables us to think about mathematics in a more powerful way. We could plough on, always talking about 'equivalence' at every stage. However, it soon becomes apparent that it is more useful for our human minds to build ideas on the underlying crystalline concept. Mathematicians speak of this as 'identifying' equivalent concepts to create a single idea. This procedure will become more apparent in later chapters as the mathematics becomes more sophisticated.

Alternative Ways to Conceptualise Ordered Pairs

Definition 4.3: The *Cartesian product* $A \times B$ is the set of all ordered pairs:

$$A \times B = \{(x, y) \mid x \in A, y \in B\}.$$

In the case of $\mathbf{R} \times \mathbf{R}$, visualising ordered pairs as points in the plane remains a most useful one; it certainly satisfies the ordered pair property (OPP). This interpretation of $A \times B$ is also useful when A and B are subsets of $\mathbf{R}$. For instance, if $A = \{1, 2, 3\}$, and $B = \{5, 7\}$, then $A \times B$ is the set

$$\{(1, 5), (1, 7), (2, 5), (2, 7), (3, 5), (3, 7)\}.$$

Thinking in terms of Cartesian coordinates, we can draw a picture:

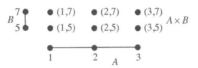

Fig. 4.2 The Cartesian product

When A and B are not subsets of $\mathbf{R}$ this sort of picture is less appropriate, but it can still be useful. For example, if $A = \{a, b, c\}$ and $B = \{u, v\}$, then

$$A \times B = \{(a, v), (a, v), (b, u), (b, v), (c, u), (c, v)\}$$

and the structure is represented by

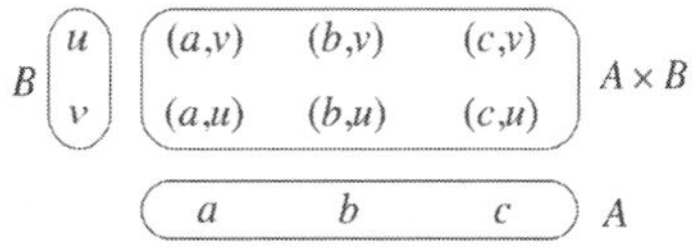

$$B \begin{pmatrix} u \\ v \end{pmatrix} \begin{pmatrix} (a,v) & (b,v) & (c,v) \\ (a,u) & (b,u) & (c,u) \end{pmatrix} \quad A \times B$$
$$\begin{pmatrix} a & b & c \end{pmatrix} \quad A$$

Fig. 4.3 A more general Cartesian product

In general, $A \times B \neq B \times A$. For example, with A and B as above,

$$B \times A = \{(u, a), (u, b), (u, c), (v, a), (v, b), (v, c)\}$$

which is not the same as $A \times B$. However, the Cartesian product does obey some general laws:

Proposition 4.4: For any sets A, B, C,

   (a) $(A \cup B) \times C = (A \times C) \cup (B \times C)$
   (b) $(A \cap B) \times C = (A \times C) \cap (B \times C)$
   (c) $A \times (B \cup C) = (A \times B) \cup (A \times C)$
   (d) $A \times (B \cap C) = (A \times B) \cap (A \times C)$.

Proof: All are easy, and the argument is similar in each case, so we prove only (a), leaving the remainder as an exercise. Let $(u, v) \in (A \cup B) \times C$. Then $u \in A \cup B, v \in C$. So $u \in A$ or $u \in B$. If $u \in A$ then $(u, v) \in A \times C$; if $u \in B$ then $(u, v) \in B \times C$. Either way, $(u, v) \in (A \times C) \cup (B \times C)$. Therefore

$$(A \cup B) \times C \subseteq (A \times C) \cup (B \times C).$$

Now let $x = (y, z) \in (A \times C) \cup (B \times C)$. Either $x \in A \times C$ or $x \in B \times C$. In the first case, $y \in A$ and $z \in C$. In the second, $y \in B$ and $z \in C$, so $x = (y, z) \in (A \cup B) \times C$. This shows

$$(A \times C) \cup (B \times C) \subseteq (A \cup B) \times C.$$

Putting the two parts together finishes the proof.   □

This can be illustrated by the following diagram:

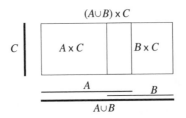

Fig. 4.4 $A \times B \cap C = (A \times B) \cap (A \times C)$.

Proposition 4.5: For all sets A, B, C, D,

$$(A \times B) \cap (C \times D) = (A \cap C) \times (B \cap D).$$

Proof: Let $x = (y, z) \in (A \times B) \cap (C \times D)$. Then $y \in A, z \in B$, and $y \in C, z \in D$. So $y \in A \cap C, z \in B \cap D$, so $x \in (A \cap C) \times (B \cap D)$. Hence

$$(A \times B) \cap (C \times D) \subseteq (A \cap C) \times (B \cap D).$$

Conversely let $x = (y, z) \in (A \cap C) \times (B \cap D)$. Then $y \in A$ and $y \in C, z \in B$, and $z \in D$, so $x \in (A \times B) \cap (C \times D)$. Therefore

$$(A \cap C) \times (B \cap D) \subseteq (A \times B) \cap (C \times D). \qquad \square$$

Pictorially:

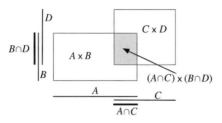

Fig. 4.5 The intersection of Cartesian products

The same picture should make it clear why a theorem like this does *not* hold for unions in place of intersections.

Having got ordered pairs, it is easy to go on to define ordered triples, quadruples, etc. by setting

$$(a, b, c) = ((a, b), c)$$
$$(a, b, c, d) = (((a, b), c), d)$$

and so on. These are elements of repeated Cartesian products, defined by

$$A \times B \times C = (A \times B) \times C$$
$$A \times B \times C \times D = ((A \times B) \times C) \times D.$$

Later we find a better way to formulate the general concept of an ordered n-tuple

$$(a_1, a_2, \ldots, a_n)$$

for any natural number n. At our present stage we can do this for any *particular* n by repeating the process used for triples or quadruples. These generalisations have similar properties to the main property (OPP) of pairs. For example, $(a, b, c) = (u, v, w)$ if and only if $a = u, b = v, c = w$. The proof of this follows from repeated use of (OPP).

Relations

Intuitively a relation between two mathematical objects a and b is some condition involving a and b that is either true or false for particular values of a and b. For example 'greater than' is a relation between natural numbers. Using the usual symbol $>$ we have

$$2 > 1 \quad \text{is true}$$
$$1 > 2 \quad \text{is false}$$
$$3 > 17 \quad \text{is false}$$

and so on. The relation is some sort of property of the pairs of elements a, b. In fact we must use the *ordered* pair (a, b), since, for instance, $2 > 1$ but not $1 > 2$.

If we know for which ordered pairs (a, b) that $a > b$ is true, then, to all intents and purposes, we have specified exactly what we mean by the relation 'greater than'. In other words, a relation may be defined by using a set of ordered pairs:

Definition 4.6: Let A and B be sets. A *relation between A and B* is a subset R of $A \times B$.

If $A = B$ we talk of a relation *on A*, which is a subset of $A \times A$.

This definition requires elucidation. For example, the relation 'greater than' on **N** is the set of all ordered pairs (a, b) where $a, b \in$ **N** and (in the usual sense) $a > b$. We might illustrate this set as follows:

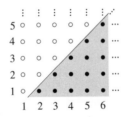

Fig. 4.6 $a > b$ for $a, b \in$ **N**

If R is a relation between sets A and B, then we say that $a \in A$ and $b \in B$ are *related* by R if $(a, b) \in R$. More commonly we use the notation

$$a \, R \, b$$

to mean $(a, b) \in R$. Then $(a, b) \notin R$ will be written $a\not R b$. This allows some sleight of hand. If we denote the relation 'greater than' by the usual symbol $>$, then, letting R be $>$, we find that $a > b$ (in the above sense) means the same as $(a,\ b) \in >$, and this by definition means that $a > b$ in the usual sense. On the other hand, if $(a, b) \notin >$ we write $a \not> b$, which again corresponds to normal usage. Thus we 'recover' the standard symbolism by an unscrupulous trick of notation. This is an excellent idea—at least, mathematicians seem pleased by it—and in future we use the $a R b$ notation.

We consider more examples. The relation $\geq$ on $\mathbf{N}$:

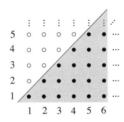

Fig. 4.7 $a \geq b$ for $a, b \in \mathbf{N}$

The relation $=$ on $\mathbf{N}$:

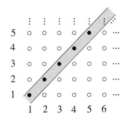

Fig. 4.8 $a = b$ for $a, b \in \mathbf{N}$

In fact, the relation $=$ on $\mathbf{N}$ is the set $\{(x, x) \mid x \in \mathbf{N}\}$.

For a final example, let $X = \{1, 2, 3, 4, 5, 6\}$ and let '$|$' be the relation 'is a divisor of', so that $a|b$ means 'a is a divisor of b'. As a set of ordered pairs,

$$| = \{(1, 1), (1, 2), (1, 3), (1, 4), (1, 5), (1, 6), (2, 2), (2, 4),$$
$$(2, 6), (3, 3), (3, 6), (4, 4), (5, 5), (6, 6)\}.$$

In pictures:

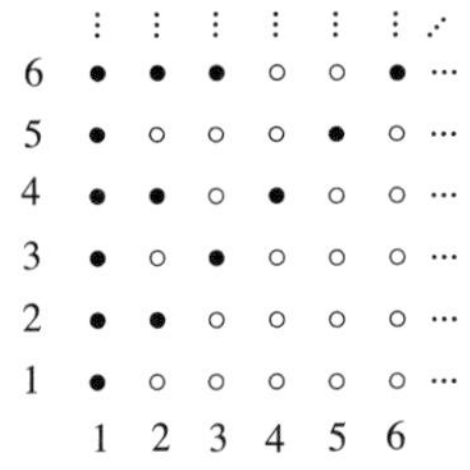

```
⋮  ⋮  ⋮  ⋮  ⋮  ⋮  ⋰
6  •  •  •  ○  ○  •  ⋯
5  •  ○  ○  ○  •  ○  ⋯
4  •  •  ○  •  ○  ○  ⋯
3  •  ○  •  ○  ○  ○  ⋯
2  •  •  ○  ○  ○  ○  ⋯
1  •  ○  ○  ○  ○  ○  ⋯
   1  2  3  4  5  6
```

Fig. 4.9 $a|b$ for numbers up to 6

Given a relation R between sets A and B, and subsets A', B' of A and B respectively, we can define a relation R' between A' and B' by

$$R' = \{(a, b) \in R \mid a \in A' \text{ and } b \in B'\}.$$

In fact, set-theoretically,

$$R' = R \cap (A' \times B').$$

We call R' the *restriction* of R to A' and B'. As far as the elements of A' and B' go the relations R and R' say the same thing. The only difference is that R' says nothing about elements not in A' and B'.

Equivalence Relations

The *odd* integers are those of the form $2n + 1$ for an integer n, namely . . . , $-5, -3, -1, 1, 3, 5, \ldots$ and the *even* integers are those of the form $2n$, namely . . . , $-4, -2, 0, 2, 4, \ldots$. In both elementary and advanced mathematics, the distinction between odd and even integers is often important. The set $\mathbf{Z}$ of all integers splits into two disjoint subsets

$$\mathbf{Z}_{\text{odd}} = \{\text{all odd integers}\}$$
$$\mathbf{Z}_{\text{even}} = \{\text{all even integers}\}.$$

We can summarise this statement as

$$\mathbf{Z}_{\text{odd}} \cap \mathbf{Z}_{\text{even}} = \varnothing, \ \mathbf{Z}_{\text{odd}} \cup \mathbf{Z}_{\text{even}} = \mathbf{Z}.$$

There is another way to split $\mathbf{Z}$ into these two pieces, using a *relation*, which for the moment we call by the noncommittal name '$\sim$'. Define, for $m, n \in \mathbf{Z}$,

$$m \sim n \text{ if and only if } m - n \text{ is a multiple of } 2.$$

Then

all even integers are related by $\sim$,
all odd integers are related by $\sim$,
no even integer is related to an odd integer,
no odd integer is related to an even integer.

These statements are a consequence of some general properties of $\sim$, and we shall analyse the situation in general to see what is required.

Imagine a set X broken up into a number of disjoint pieces.

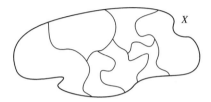

Fig. 4.10 A set divided into disjoint pieces

We can define a relation $\sim$ by

$x \sim y$ if and only if x and y are both in the same piece.

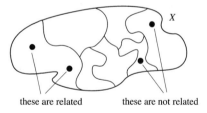

these are related   these are not related

Fig. 4.11 Defining a relation

Conversely, we can reconstruct the pieces from the relation $\sim$: the piece to which $x \in X$ belongs is

$$E_x = \{y \in X \mid x \sim y\}.$$

If we try this with a different relation $\sim$, all sorts of things can go wrong. In particular we may not get *disjoint* pieces. Consider the relation $\mid$ on integers for which $a \mid b$ means 'a divides b without remainder'. If we take $\sim$ to be the relation $\mid$ on $\{1, 2, 3, 4, 5, 6\}$, then

$$E_1 = \{1, 2, 3, 4, 5, 6\}$$
$$E_2 = \{2, 4, 6\}$$
$$E_3 = \{3, 6\}$$
$$E_4 = \{4\}$$
$$E_5 = \{5\}$$
$$E_6 = \{6\}.$$

So the set splits up according to:

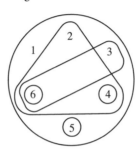

Fig. 4.12 What happens if the pieces overlap

If instead we use the relation $>$ on $\mathbf{N}$, we do not even get $x \in E_x$, so E_x is in no sense 'the piece to which x belongs'.

What is it that makes the original relation $\sim$ work, whereas the others go wrong? We must take account of three very simple statements:

(i) x belongs to the same piece as x;

(ii) if x belongs to the same piece as y then y belongs to the same piece as x;

(iii) if x belongs to the same piece as y and y belongs to the same piece as z, then x belongs to the same piece as z.

Clearly any relation $\sim$ with the property that $x \sim y$ if and only if x and y belong to the same piece must have the three corresponding properties, which we formalise as (E1), (E2), (E3) of the next definition.

Definition 4.7: A relation $\sim$ on a set X is an *equivalence relation* if it has the following properties for $x, y, z \in X$:

(E1) $x \sim x$ for all $x \in X$ ($\sim$ is *reflexive*),

(E2) If $x \sim y$ then $y \sim x$ ($\sim$ is *symmetric*),

(E3) If $x \sim y$ and $y \sim z$ then $x \sim z$ ($\sim$ is *transitive*).

If we break X into disjoint pieces, the relation 'is in the same piece as' is an equivalence relation. We now show that *every* equivalence relation arises in this way from a suitable choice of pieces. In fact, there is an intimate connection between the two concepts. First we need a formal definition of 'breaking into disjoint pieces'.

Definition 4.8: A *partition* of a set X is a set P whose members are non-empty subsets of X, subject to the conditions

(P1) Each $x \in X$ belongs to some $Y \in P$,
(P2) If $X, Y \in P$ and $X \neq Y$, then $X \cap Y = \emptyset$.

The elements of P are our 'pieces'. Condition (P1) says that X is the union of all the pieces, so that each element of X lies in some piece; (P2) says that distinct pieces don't overlap. It follows that no element of X can belong to two distinct pieces.

Given an equivalence relation $\sim$ on X we define the *equivalence class* (with respect to $\sim$) of $x \in X$ to be the set

$$E_x = \{y \in X \,|\, x \sim y\}.$$

Theorem 4.9: Let $\sim$ be an equivalence relation on a set X. Then $\{E_x \,|\, x \in X\}$ is a partition of X. The relation 'belongs to the same piece as' is the same as $\sim$.

Conversely, if P is a partition of X, let $\sim$ be defined by $x \sim y$ if and only if x and y lie in the same piece. Then $\sim$ is an equivalence relation, and the corresponding partition into equivalence classes is the same as P.

PRE-PROOF REMARK: This theorem lets us pass at will from an equivalence relation to a partition or back again, by a procedure which, when done twice, leads back to where we started.

Proof: Since $x \in E_x$, condition (P1) is satisfied. To verify (P2), suppose that $E_x \cap E_y \neq \emptyset$. Then we can find $z \in E_x \cap E_y$. Then $x \sim z$ and $y \sim z$. By symmetry $z \sim y$, and then transitivity implies $x \sim y$. We show that this implies $E_x = E_y$. For if $u \in E_x$ then $x \sim u$ and $y \sim x$, so $y \sim u$; hence $E_x \subseteq E_y$. Similarly $E_y \subseteq E_x$. This shows that $E_x = E_y$. Thus we have proved that $E_x \cap E_y = \emptyset$ or $E_x = E_y$. But this statement is logically equivalent to (P2).

Now define $x \approx y$ to mean 'x and y are in the same equivalence class'. Then

$$x \approx y \quad \text{if and only if } x, y \in E_z \text{ for some } z$$
$$\text{if and only if } z \sim x \text{ and } z \sim y \text{ for some } z$$
$$\text{if and only if } x \sim y.$$

Hence $\approx$ and $\sim$ are the same.

The second part of the theorem is proved in a similar manner, but is easier. We leave that to you. $\square$

Example: Arithmetic Modulo n

We use the equivalence relation concept to generalise the distinction between odd and even integers, and to set up what is often called (in schools) 'modular arithmetic' or (in universities) 'the integers mod n'.

To begin with, we specialise to $n = 3$. Define the relation $\equiv_3$ of *congruence modulo* 3 on **Z** by

$$m \equiv_3 n \text{ if and only if } m - n \text{ is a multiple of } 3.$$

Proposition 4.10: $\equiv_3$ is an equivalence relation on **Z**.

Proof:

(E1): $m - m = 0 = 3 \cdot 0$.

(E2): If $m - n = 3k$ then $n - m = 3(-k)$.

(E3): If $m - n = 3k$, $n - p = 3l$, then $m - p = 3(k + l)$. $\square$

We know that the equivalence classes (known as *congruence classes* mod 3) partition **Z**. What are they? It is easiest to see this with the help of examples.

$$E_0 = \{y \mid 0 \equiv_3 y\}$$
$$= \{y \mid y - 0 \text{ is a multiple of } 3\}$$
$$= \{y \mid y = 3k \text{ for some } k \in \mathbf{Z}\}.$$
$$E_1 = \{y \mid 1 \equiv_3 y\}$$
$$= \{y \mid y - 1 = 3k\}$$
$$= \{y \mid y = 3k + 1 \text{ for some } k \in \mathbf{Z}\}.$$
$$E_2 = \{y \mid 2 \equiv_3 y\}$$
$$= \{y \mid y = 3k + 2 \text{ for some } k \in \mathbf{Z}\}.$$
$$E_3 = \{y \mid y = 3k + 3 \text{ for some } k \in \mathbf{Z}\}.$$

However, $3k + 3 = 3(k + 1)$, so $E_3 = E_0$. Similarly, $E_4 = E_1$, $E_5 = E_2$, $E_{-1} = E_2$, $E_{-2} = E_1$, and so on. Every integer is either of the form $3k$, $3k + 1$, or $3k + 2$

(according as it leaves remainder 0, 1, or 2 on division by 3) so we get exactly three equivalence classes:

$$E_0 = \{\ldots, -9, -6, -3, 0, 3, 6, 9, \ldots\}$$
$$E_1 = \{\ldots, -8, -5, -2, 1, 4, 7, 10, \ldots\}$$
$$E_2 = \{\ldots, -7, -4, -1, 2, 5, 8, 11, \ldots\}.$$

So much for the equivalence relation. More intriguing is the possibility of doing *arithmetic* with these equivalence classes.

To make the notation more transparent in general, we denote the equivalence class of n by n_3 instead of E_n. In this notation, the three classes above become 0_3, 1_3, and 2_3. Let $\mathbf{Z}_3 = \{0_3, 1_3, 2_3\}$ and define operations of addition and multiplication on $\mathbf{Z}_3$ by

$$m_3 + n_3 = (m + n)_3, \tag{4.1}$$
$$m_3 n_3 = (mn)_3. \tag{4.2}$$

For example, $1_3 + 2_3 = 3_3 = 0_3$; $2_3 2_3 = 4_3 = 1_3$.

This may look pointless: such an impression is erroneous, as will soon be seen. It may also look harmless: certain subtleties must be noticed before worrying that something may go wrong, and a little hard thinking put in to see that, after all, it doesn't.

Here is the subtle problem: the *same* class has several different names; thus $1_3 = 4_3 = 7_3 = \ldots, 2_3 = 5_3 = 8_3 = \ldots$. For all we know at the moment, the definitions (4.1), (4.2) might give different answers to the same question, depending on which names we use. Thus we have seen that $1_3 + 2_3 = 0_3$. But since $1_3 = 7_3$, $2_3 = 8_3$, we also have $1_3 + 2_3 = 7_3 + 8_3 = 15_3$. By a stroke of good fortune, $15_3 = 0_3$, and we can breathe again.

What happens in general? If $i_3 = i'_3$ then $i - i' = 3k$ for some k, and if $j_3 = j'_3$ then $j - j' = 3l$ for some l. Now rule (4.1) gives two possible answers:

$$i_3 + j_3 = (i + j)_3, \quad i'_3 + j'_3 = (i' + j')_3.$$

However,

$$i + j = i' + 3k + j' + 3l = (i' + j') + 3(k + l),$$

so

$$(i + j)_3 = (i' + j')_3.$$

Hence we get the same answer both ways, and (4.1) makes sense as a definition of addition.

Similarly we must check that the multiplication rule is unambiguous. With i, j, i', j' as above, we have

$$i_3 j_3 = (ij)_3, \quad i'_3 j'_3 = (i'j')_3.$$

But

$$ij = (i' + 3k)(j' + 3l) = i'j' + 3(i'l + j'k + 3kl),$$

so

$$(ij)_3 = (i'j')_3,$$

which is what we want.

This problem always arises when we try to define operations on sets by a rule of the type 'select elements from the sets, operate on these, then find the set to which the result belongs'. When, as here, the notation conceals such a process, we must be careful to think what the notation *means* rather than just manipulating symbols blindly. We must check that different choices give the same answer.

It might appear that such checks can be dispensed with, on the grounds that everything nice will work. But consider defining powers in $\mathbf{Z}_3$. The natural way to do this is to mimic (4.1) and (4.2) to define

$$m_3^{n_3} = (m^n)_3 .$$

For example, $2_3^{2_3} = (2^2)_3 = 4_3 = 1_3$. Using this 'definition' we can even prove theorems about the laws of exponentiation, for example

$$m_3^{n_3 + p_3} = (m^{n+p})_3 = (m^n m^p)_3 = (m^n)_3 (m^p)_3 = m_3^{m_3} m_3^{p_3}. \tag{4.3}$$

However, we would be living in a fool's paradise. For, since $2_3 = 5_3$, rule (4.3) also tells us that

$$2_3^{2_3} = 2_3^{5_3} = (2^5)_3 = (32)_3 = 2_3.$$

Since $1_3 \neq 2_3$ this shows that (4.3) is nonsense—but clever and plausible nonsense, the most dangerous kind.

In common parlance, we must check that the operations are 'well defined'. Really this is over-polite: what we are checking is that they are 'defined' at all!

Having digressed at length, let us return to the arithmetic of $\mathbf{Z}_3$. We can write out addition and multiplication tables:

+	0_3	1_3	2_3		×	0_3	1_3	2_3
0_3	0_3	1_3	2_3		0_3	0_3	0_3	0_3
1_3	1_3	2_3	0_3		1_3	0_3	1_3	2_3
2_3	2_3	0_3	1_3		2_3	0_3	2_3	1_3

It can be verified that many of the usual laws of arithmetic hold (such as $x + y = y + x$, $x(y + z) = xy + xz$), although there are some surprises such as

$$((1_3 + 1_3) + 1_3 + 1_3) = 1_3.$$

Instead of 3, we can use any integer n and do arithmetic modulo n. We define a relation $\equiv_n$ on $\mathbf{Z}_n$ by

$$x \equiv_n y \text{ if and only if } x - y \text{ is a multiple of } n.^1$$

We get n distinct equivalence classes $0_n, 1_n, 2_n, \ldots, (n-1)_n$; while $n_n = 0_n$, $(n + 1)_n = 1_n$, and so on; now x_n consists of those integers that leave remainder x on division by n. The set $\mathbf{Z}_n$ of equivalence classes admits operations of arithmetic defined in the same way as (4.1) and (4.2).

We discuss these ideas further in chapter 10.

Subtle Aspects of Equivalence Relations

Although the definition of equivalence relation seems simple, and in our experience virtually all students can write down the three properties, there are subtle aspects that are not apparent without deeper consideration.

For instance, (E1) requires that $x \sim x$ *for all $x \in X$*. Some examples, such as lines being parallel in Euclidean geometry, seem to be equivalences, but they do not satisfy (E1) because technically a line cannot be parallel to itself. (Parallel lines have no point in common.) Parallel lines satisfy (E2), and if x, y, z are all different (E3) is satisfied. We could, if we wished, define the relation $x \sim y$ for lines in the plane to mean 'x is parallel to y or $x = y$'. In this case we *would* have an equivalence relation.

In general, we must check the full meaning of a definition *very* carefully. The precise meaning of a definition is a recurring theme in the rest of this book. A definition means what it says, no more and no less.

Non-Example 4.11: The following question was set in a first-year university examination:

Given a set S with three distinct elements a, b, c, is the relation where only the following hold

$$a \sim a, \ b \sim b, \ a \sim b, \ b \sim a$$

an equivalence relation?

---

[1] The standard notation is $x = y \pmod{n}$. The symbol $=_3$ is used here for consistency with our notation for a relation.

Think about it and write down your response before you read the next paragraph.

You may very likely get the correct answer 'no'. The reason is that we are not told that $c \sim c$, so (E1) is not satisfied. However, when this question was given to well-qualified students, many did not notice this. Instead, they focused on (E3), which says

$$\text{If } x \sim y \text{ and } y \sim z \text{ then } x \sim z,$$

and many declared it to be false claiming that it requires *three* elements, x, y, z, while the set S has only two: a, b. However, in set theory different letters may represent the same element. So we could take $x = a$, $y = b$, $z = b$, or any other combination to show that, in every case, (E3) is true.

It is also easy to come to the belief that the equivalence relations met in mathematics, such as the integers modulo n, have equivalence classes that are similar in some way. For instance the equivalence classes for the relation of integers modulo 3 are all infinite sets with a natural mapping between them. Examples like this may lead unconsciously to the expectation that equivalence classes are all like this in some way. In general, however, the partition theorem shows that equivalence relations break a set up into (non-empty) subsets of *any* size. The subsets chosen in a set do not need to have any special properties, other than every element in the set must lie in precisely one of the subsets in the partition.

An interesting example is equivalence of infinite decimals. Two infinite decimals lie in the same equivalence class if they have the same value. As we have shown in chapter 1, each decimal expansion is either unique, or it is a finite decimal that can be written in exactly two ways: as an infinite number of nines or zeros (such as $3 \cdot 47 = 3 \cdot 46999\ldots$). Here some equivalence classes contain one element, and all others contain two.

Order Relations

A second kind of relation, whose properties are quite different from those of an equivalence relation, arises when dealing with the order between numbers, as exemplified by the statements $4 < 5, 7 > 2\pi, x^2 \geq 0, 1 - x^2 \leq 1$ for any real number x.

Fortunately, the various relations $<, >, \leq, \geq$ are all connected with each other:

$x < y$ means the same as $y > x$,
$x \leq y$ means the same as $y \geq x$,
$x \leq y$ means the same as $x < y$ or $x = y$,
$x < y$ means the same as $x \leq y$ and $x \neq y$.

Therefore we need study only one of them and translate the results to the others.

When handling numbers, it might be preferable to consider one of the strict relations $<$ or $>$. In general we would not write $2 + 2 \geq 4$, simply because we know something more precise: $2 + 2$ *equals* 4. Likewise, we normally write $2 + 2 > 3$, because that contains more exact information than $2 + 2 \geq 3$. But when passing to general statements, the situation changes. For instance, it is true that if $a_n \to a$, $b_n \to b$ and $a_n \geq b_n$, then $a \geq b$, but $a_n > b_n$ does not imply $a > b$. (A counterexample is given by $a_n = 1/n$, $b_n = 0$.) Here there is a slight preference towards the weak inequalities $\leq$ and $\geq$.

We begin with the latter.

Definition 4.12: A relation R on a set A is a *weak order* if

(WO1) $a \, R \, b$ and $b \, R \, c$ imply $a \, R \, c$
(WO2) either $a \, R \, b$ or $b \, R \, a$ (or both)
(WO3) $a \, R \, b$ and $b \, R \, a$ imply $a = b$.

These properties evidently hold for both relations $\leq$ and $\geq$ on the set of real numbers, which may seem rather strange since one means 'bigger' and the other 'smaller'. But looking at the real numbers as points on a line, we see that, by a reflection, we can turn the order round, interchanging left and right, and this interchanges $\leq$ and $\geq$. It is only when we start doing arithmetic, and require $a \geq 0$, $b \geq 0$ to imply $ab \geq 0$, that we find a property of $\geq$ that does not hold for $\leq$. We will postpone this consideration until we study arithmetic in chapter 9.

Weak order relations naturally come in pairs. Given a weak order R, we can define R', the *reverse* of R, by

$$a \, R' b \text{ means } b \, R \, a.$$

The reverse R' is also a weak order relation, and reversing again we get $R'' = R$.

Example 4.13: If $A = \{a, b, c\}$, where a, b, c are all distinct, then a weak order on A can be defined by $a \, R \, b$, $a \, R \, c$, $b \, R \, c$, $a \, R \, a$, $b \, R \, b$, $c \, R \, c$. We can visualise this by considering a, b, c in a row, with $x \, R \, y$ meaning x is to the left of y or $x = y$.

$$a \ b \ c$$
$$\circ \ \circ \ \circ$$

The reverse of R simply puts the elements in order c, b, a.

Example 4.14: Define an order relation R on the plane by: $(x_1, y_1) R (x_2, y_2)$ means

$$\text{either } y_1 = y_2 \text{ and } x_1 \leq x_2 \text{ (both together)},$$

$$\text{or } y_1 < y_2.$$

This at first looks bizarre, but in a picture we see that $(x_1, y_1) R (x_2, y_2)$ means that either (x_1, y_1) and (x_2, y_2) are on the same horizontal line with (x_1, y_1) to the left of (or equal to) (x_2, y_2), or (x_1, y_1) is on a horizontal line strictly below the one through (x_2, y_2).

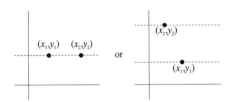

Fig. 4.13 An order on the plane

Example 4.15: $A = \{\{0\}, \{0, 1\}, \{0, 1, 2\}, \{0, 1, 2, 3\}\}$,

$$x R y \text{ means } x \subseteq y \text{ for } x, y \in A.$$

Here $\{0\} \subseteq \{0, 1\} \subseteq \{0, 1, 2\} \subseteq \{0, 1, 2, 3\}$.

Inclusion of sets satisfies

$$X \subseteq Y \quad \text{and} \quad Y \subseteq Z \quad \text{implies } X \subseteq Z,$$

$$X \subseteq Y \quad \text{and} \quad Y \subseteq X \quad \text{implies } X = Y,$$

but for arbitrary sets X, Y we may have $X \not\subseteq Y$ and $Y \not\subseteq X$. This means that set inclusion in general satisfies (WO1), (WO3), but not (WO2). A relation R on a set A satisfying (WO1) and (WO3) is said to be a *partial order*, and A a *partially ordered set*, or, with some loss in dignity, a *poset*. Given any set A of sets, then inclusion always yields a partial order.

Let R be a weak order on a set A; then the corresponding strict order S is given by:

$$x S y \text{ means precisely } x R y \text{ and } x \neq y.$$

For example if R is $\leq$, then S is $<$.

Proposition 4.16: A *strict order* S on a set A satisfies:

(SO1) $a\,S\,b$ and $b\,S\,c$ imply $a\,S\,c$
(SO2) Given $a, b \in A$, then precisely one of the following hold (and not the other two):

$$a\,S\,b,\ b\,S\,a,\ a = b.$$

Proof: Suppose that $a\,S\,b$ and $b\,S\,c$. Then $a\,R\,b$ and $b\,R\,c$, and (WO1) implies that $a\,R\,c$. We cannot have $a = c$, for substituting in $b\,R\,c$ we get $b\,R\,a$, and by (WO3) this and $a\,R\,b$ gives $a = b$, contradicting $a\,S\,b$. This verifies (SO1). By (WO2) $a, b \in A$ implies $a\,R\,b$ or $b\,R\,a$, so $a\,S\,b$ or $a = b$ or $b\,S\,a$. But no two of these can hold simultaneously because $a = b$ contradicts the definitions of both $a\,S\,b$ and $b\,S\,a$, and were $a\,S\,b$, $b\,S\,a$ to hold simultaneously, then $a\,R\,b$, $b\,R\,a$ would hold, so (WO3) gives $a = b$, contradicting $a\,S\,b$ once more. This verifies (SO2). $\qquad\square$

Condition (SO2) in proposition 4.16 is usually referred to as the *trichotomy law*. (Just as a dichotomy is two mutually exclusive possibilities, a trichotomy is three, in this case $a\,S\,b$, $b\,S\,a$, or $a = b$.) For the strict order $<$ on the real numbers the three mutually exclusive possibilities are $a < b$, $b < a$, $a = b$.

We remarked earlier that we could return to the weak order $\leq$ from $<$ through the connection

$$a \leq b \text{ means precisely } a < b \text{ or } a = b.$$

The same happens for any strict order. Given a relation S on a set A satisfying (SO1) and (SO2), define

$$a\,R\,b \text{ to mean } a\,S\,b \text{ or } a = b.$$

It is easy to verify that R satisfies (WO1)–(WO3), and that we can pass freely from a weak order to the corresponding strict order and back again. In this manner the notions of weak and strict order are interchangeable. Although we have taken (WO1)–(WO3) as the basic axioms and proved the properties (SO1), (SO2), we could just as easily reverse their status by taking (SO1), (SO2) as basic axioms and deducing (WO1)–(WO3).

Exercises

1. Write out the proofs of propositions 4.4(b), 4.4(c), and 4.4(d).

2. Prove that

$$(\cup S) \times (\cup T) \subseteq \cup\{X \times Y \mid X \in S, \; Y \in T\}$$
$$(\cap S) \times (\cap T) = \cap\{X \times Y \mid X \in S, \; Y \in T\}$$

for all sets S, T of sets. Show that in the first formula we cannot replace '$\subseteq$' by '$=$'.

3. If $A = \varnothing$, show that $A \times B = \varnothing = B \times A$ for every set B. When $A \neq \varnothing$, show that $A \times B = A \times C$ implies $B = C$. Given $A \times B = B \times A$, what can be deduced about A and B?

4. Let $A = \mathbf{N} \times \mathbf{N}$. Define the relation R on A by

$$(m, n) \; R \, (r, s) \text{ means } m + s = r + n.$$

Show that R is an equivalence relation.

If $B = \{(x, y) \in \mathbf{Z} \times \mathbf{Z} \mid y \neq 0\}$, and S is the relation on B given by

$$(a, \, b) S(c, \, d) \text{ if and only if } ad = bc,$$

is S an equivalence relation? Prove your assertion.

5. How many distinct equivalence relations exist on $\{1, \, 2, \, 3, \, 4\}$?

6. Recall the properties (E1), (E2), (E3) of an equivalence relation. Which of these properties is satisfied by the relation between $x, y \in \mathbf{R}$ given by:
   (a) $x < y$
   (b) $x \geq y$
   (c) $|x - y| < 1$
   (d) $|x - y| \leq 0$
   (e) $x - y$ is rational
   (f) $x - y$ is irrational
   (g) $(x - y)^2 < 0$.

7. Is there a mistake in the following proof that (E2) and (E3) imply (E1)? If so, what is it?

   Let $a \sim b$.
   By (E2), $b \sim a$. By (E3), if $a \sim b$ and $b \sim a$, then $a \sim a$.
   This proves (E1).

8. Give examples of relations (the more elegant, the better) satisfying
   (a) none of the properties (E1), (E2), (E3)
   (b) (E1) but not (E2) or (E3)
   (c) (E2) but not (E1) or (E3)
   (d) (E3) but not (E1) or (E2)

(e) (E2) and (E3) but not (E1)

(f) (E1) and (E3) but not (E2)

(g) (E1) and (E2) but not (E3).

9. Write out addition and multiplication tables for the integers mod 4, mod 5, and mod 6.

Find all a, $b \in \mathbf{Z}_{12}$ such that $ab = 0_{12}$.

10. Define a relation R on $\mathbf{N}$ by

$$a \, R \, b \text{ means } a \text{ divides } b,$$

that is, $b = ac$ for some $c \in \mathbf{N}$. Is R an order relation? If so, is it a weak order or a strict order?

11. Let $X = \{1, 2, 6, 30, 210\}$ and define a relation S on X by

$$a \, S \, b \text{ means } a \text{ divides } b.$$

Is S an order relation? If so, is it a weak order or a strict order?

12. Let A be a set with a (strict) order relation S and B a set with (strict) order relation T. Define the lexicographic relation L on $A \times B$ by

$$(a, b) \, L \, (c, d) \text{ means: either } a \, S \, c, \text{ or } a = c \text{ and } b \, T \, d.$$

Is this an order relation? What is the connection between this and a dictionary?

CHAPTER 5

Functions

Functions are of enormous importance throughout the whole of modern mathematics, at all levels. The concept first became prominent in calculus, which is *about* functions: how to differentiate them or integrate them. Early attempts to develop a *general* concept of functions were somewhat confused and unsatisfactory, largely because they tried to do too much at once. The function concept as it is now understood evolved gradually from these attempts: it has great generality and great simplicity. In fact, the current concept is so general that when doing calculus, extra conditions must be imposed to restrict the class of functions under consideration to those that *can* be differentiated or integrated. Thus the desired object is achieved by taking a very general definition of 'function' and then selecting more special types of function by imposing extra conditions.

In this chapter we develop the general concept of a function gradually, starting from familiar examples and extracting general principles. We discuss some general properties that functions can have. We introduce the graph of a function, and relate it both to the formal definition and the traditional picture.

Some Traditional Functions

Traditionally, a function is defined by introducing a 'variable' x, usually supposed to be a real number, and talking of a 'function $f(x)$ of x'. The most significant feature of such a definition is that in principle we should be able to work out the value of $f(x)$ for any given x (possibly under restrictions such as $x \neq 0$, $x > 0$, depending on the function involved). Here are some familiar examples:

- The exponential function takes value e^x for any real number x (where $e = 2{\cdot}71828\dots$).

- The sine, cosine, and tangent functions take values $\sin(x)$, $\cos(x)$, $\tan(x)$ for all real x, except that for the tangent we have to assume x is not an odd integer multiple of $\pi/2$ in order for the definition $\tan(x) = \sin(x)/\cos(x)$ to make sense. (Often the parentheses are omitted and we write $\sin x$, $\cos x$, and so on.)
- The logarithmic function takes the value $\log(x)$ when x is real and $x > 0$. (The notation $\log x$ is also common.)
- The reciprocal function takes value $1/x$ for real $x \neq 0$.
- The square function takes value x^2 for any real x.
- The factorial function $x!$ is defined only for x a positive integer.

What do all of these examples have in common? It is our ability, in principle, to calculate the value of the function corresponding to the relevant values of x. In other words, a function associates to each relevant real number x a value $f(x)$ which is also a real number. In the above examples, respectively, $f(x) = e^x$, $\sin(x)$, $\cos(x)$, $\tan(x)$, $\log(x)$, $1/x$, x^2, $x!$.

We should not confuse the *values* of the function with the function itself. It is not $\log(x)$ that is the function: it is the rule 'take the logarithm of', which allows us to *work out* the value. In a sense, the function itself is the symbol 'log'. So we think of a function f as some 'rule' which, for any real number x (perhaps subject to restrictions), defines another real number $f(x)$. The definition of $f(x)$ should be *unique*; a rule that gives two different answers to the same question is not especially useful. This means we must be careful with such functions as 'square root', specifying whether we mean the positive square root or the negative one. Don't worry about this now; we'll return to it later when the basic ideas are established.

The General Function Concept

The most general definition of a function comes from the traditional one by relaxing the requirement that x and $f(x)$ should be real numbers. In fact, even in traditional mathematics, complex numbers are permitted; indeed, a wide variety of non-numerical things as well. For example, the area of a triangle is a function defined on triangles. The easiest and most satisfactory assumption is not to impose restrictions of any kind on the nature of x or $f(x)$. However, we must then be more precise about what we mean by a rule, because traditional formulas are too limited in scope.

In our examples of functions, x ranges over some set of possible choices, and so does $f(x)$. The natural choices for these sets are often different; for example the logarithmic function requires $x > 0$, whereas $\log(x)$ can be any real number.

We therefore begin with two arbitrary sets A and B. As a preliminary definition we formulate:

Preliminary Definition 5.1: A *function f from A to B* is a rule that assigns to each $a \in A$ a *unique* element $f(a) \in B$.

This definition is very broad. It includes all of the previous examples: take A to be a suitable subset of $\mathbf{R}$ and B to be $\mathbf{R}$. Thus:

For the exponential: $A = \mathbf{R}$, $B = \mathbf{R}$, and $f(x) = e^x$.
For the logarithm: $A = \{x \in \mathbf{R} \mid x > 0\}$, $B = \mathbf{R}$, and $f(x) = \log(x)$.
For the reciprocal: $A = \{x \in \mathbf{R} \mid x \neq 0\}$, $B = \mathbf{R}$, and $f(x) = 1/x$.
For the factorial: $A = \{x \in \mathbf{R} \mid x > 0\}$, $B = \mathbf{N}$, and $f(x) = x!$.

Examples of rather different types of function that this definition allows include:

$A = \{$all circles in the plane$\}$, $B = \mathbf{R}$, $f(x) =$ the radius of x.

$A = \{$all circles in the plane$\}$, $B = \mathbf{R}$, $f(x) =$ the area of x.

$A = \{$all subsets of $\{0, 2, 4\}\}$, $B = \mathbf{N}$, $f(x) =$ the smallest element of x.

$A = \{$all subsets of $\{0, 1, 2, 3, 4, 5, 6, 7\}\}$, $B = \{0, 1, 2, 3, 4, 5, 6, 7, 8\}$,

$\qquad f(x) =$ the number of elements of x.

$A = \mathbf{N}$, $B = \{0, 1, 2\}$, $f(x) =$ the remainder on dividing x by 3.

$A = \{$camel, lion, elephant$\}$, $B = \{$January, March$\}$,

$\qquad f($camel$) =$ March, $f($lion$) =$ January, $f($elephant$) =$ March.[1]

Definition 5.2: We call A the *domain* of f and B the *codomain*. We write

$$f : A \to B$$

to mean 'f is a function with domain A and codomain B'.

The main item still on the agenda is that troublesome word 'rule'. We obtain a formal definition in exactly the same way that we obtained one for 'relation' in chapter 4, by judicious use of ordered pairs. We want to associate to each $x \in A$ an element $f(x) \in B$. One way to do this is to stick them together in an ordered pair $(x, f(x))$. The 'rule' is then the entire

---

[1] This is of course a pretty silly function, with no mathematical importance. It illustrates that quite arbitrary definitions of $f(x)$ may be made. Actually, this one isn't quite so arbitrary as it may seem. A certain zoo has three animal-houses: the camel-house, the lion-house, and the elephant-house. Once a year the houses are redecorated: the lion-house in January, the others in March. Now $f(x) =$ the month in which the x-house is redecorated.

set of ordered pairs $(x, f(x))$ as x runs through A, and this is of course a subset of the Cartesian product $A \times B$.

In the previous chapter, we defined a subset of $A \times B$ to be a relation from A to B. This means that a function can be viewed a special kind of relation: it relates x to $f(x)$.

The requirement that $f(x)$ is defined for every $x \in A$ translates as the requirement that for any $x \in A$ there is *some* element (x, y) in the set. Uniqueness of $f(x)$ translates as the requirement that for each x, the corresponding element y should be *unique*. Now we see how a set of pairs can capture a rule: to find $f(x)$, look in the set for a pair (x, y). This exists and is unique, so we put $f(x) = y$.

Formally:

Definition 5.3: Let A and B be sets. A *function $f : A \rightarrow B$* is subset f of $A \times B$ such that

(F1) If $x \in A$ there exists $y \in B$ such that $(x, y) \in f$.
(F2) Such an element y is unique: in other words, if $x \in A$ and $y, z \in B$ are such that $(x, y) \in f$ and $(x, z) \in f$, it follows that $y = z$.

A function is also called a *map* or *mapping*.

In terms of this definition, the 'square' function with domain $\mathbf{R}$ is the subset

$$\{(x, x^2) \mid x \in \mathbf{R}\}$$

of $\mathbf{R} \times \mathbf{R}$. The curious function above is the set

$$\{(\text{camel, March}), (\text{lion, January}), (\text{elephant, March})\}.$$

We recover the usual notation by defining $f(x)$ to be the unique element $y \in B$ such that $(x, y) \in f$.

The definition of a function in terms of a set of ordered pairs is formally very neat, because it states everything in terms of sets. But it is pedantic and pointless to use ordered pairs when we wish to define a specific function. Instead, we use a form of words along the following lines:

'Define a function $f : A \rightarrow B$ by $f(x) =$ whatever for all $x \in A$.'

An alternative notation, often employed, is

'Define a function $f : A \rightarrow B, x \mapsto$ whatever.'

In any particular case, 'whatever' is replaced by a specific prescription to find $f(x)$ given x. These statements are interpreted formally as:

'f is the subset of $A \times B$ consisting of $(x, f(x)) \in A$.'

Then we must check that the prescription defines $f(x)$ uniquely, and that $f(x) \in B$, for all $x \in A$.

Example 5.4: Define the function $f : \mathbf{N} \to \mathbf{Q}$ by

$$f(n) = \sqrt{2} \text{ to } n \text{ decimal places.}$$

Then

$$f(1) = 1{\cdot}4,$$
$$f(2) = 1{\cdot}41,$$
$$f(3) = 1{\cdot}414,$$

and so on. This rule defines $f(n)$ uniquely because $\sqrt{2}$ is irrational, so it is not one of the decimals that can either end in lots of 0s or lots of 9s. Also, $f(n)$ is always rational since it is a terminating decimal.

The formal function is the subset of $\mathbf{N} \times \mathbf{Q}$ comprising all ordered pairs

$$(n, \sqrt{2} \text{ to } n \text{ decimal places}),$$

namely

$$\{(1, 1{\cdot}4), (2, 1{\cdot}41), (3, 1{\cdot}414), \ldots\}.$$

The advantage of the informal usage is manifest. But knowing how to translate it into formal terms means that the informality is safe.

Non-Examples 5.5: Here are a number of statements that look as if they define functions, but on closer inspection fail in some respect.

(1) Define $f : \mathbf{R} \to \mathbf{R}$ by $f(x) = \dfrac{x^2 + 17x + 93}{x + 1}$.
This does not define a function since when $x = -1$, $1/(x+1)$ is not defined, so $f(-1)$ has not been specified as a real number. If we change the definition to start '$f : \mathbf{R} \backslash \{-1\} \ldots$' then we're all right.

(2) Define $f : \mathbf{Q} \to \mathbf{Q}$ by $f(x) = \sqrt{x}$ (positive square root).
This does not define a function because for some x, for instance $x = 2$, the value $f(x) = \sqrt{2}$ does not belong to $\mathbf{Q}$. If we change the second $\mathbf{Q}$ to an $\mathbf{R}$ then all will be well.

(3) Define $f : \mathbf{R} \to \mathbf{Q}$ by $f(x) =$ the rational number nearest x.
This does not define a function: the supposed $f(x)$ does not exist.

(4) Define $f : \mathbf{R} \to \mathbf{N}$ by $f(x) =$ the integer number nearest x.
This almost works: the trouble is that both 0 and 1 are equidistant from $\frac{1}{2}$, so $f\left(\frac{1}{2}\right)$ is not defined *uniquely*.

General Properties of Functions

Next we introduce an important subset associated with a function.

Definition 5.6: If $f : A \to B$ is a function, then the *image* of f is the subset

$$f(A) = \{f(x) \,|\, x \in A\}$$

of B. Another common notation is im(f).

The image of f is the set of values obtained by working out $f(x)$ for all x in the domain. It need not be the whole codomain; for example if $f : \mathbf{R} \to \mathbf{R}$ has $f(x) = x^2$ then the image is the set of positive reals, and is not the whole codomain $\mathbf{R}$.

The lack of symmetry in the definition of a function may seem disturbing. We require $f(x)$ to be defined for all $x \in A$, yet we do not require every $b \in B$ to be of the form $f(x)$. The reason is pragmatic. When we *use* a function, we want to be sure that it is defined, so knowledge of the precise domain is essential. However, it is less crucial to know exactly where the values $f(x)$ lie, so we can choose the codomain to be whichever set is convenient.

For instance, if we define

$$f : \mathbf{N} \to \mathbf{R}$$

by

$$f(n) = \sqrt[3]{n!}$$

then the image of f is the set of cube roots of factorials

$$\{1, \sqrt[3]{2}, \sqrt[3]{6}, \sqrt[3]{24}, \sqrt[3]{120}, \ldots\}$$

which is not a very nice set. Images in general can be pretty revolting. So it makes sense to define a function in terms of a codomain, and to leave aside the calculation of exactly which part of the codomain we really require, in the hope (often fulfilled) that it is not needed. If it is, we can work it out.

This brings us to another minor point. Strictly speaking, we cannot talk of 'the' codomain of a function. Consider

$$f : \mathbf{R} \to \mathbf{R} \quad f(x) = x^2,$$
$$g : \mathbf{R} \to \mathbf{R}^+ \quad g(x) = x^2,$$

where $\mathbf{R}^+ = \{x \in \mathbf{R} \,|\, x \geq 0\}$. The first has codomain $\mathbf{R}$ and the second $\mathbf{R}^+$, yet the formal definition of a function as a set of ordered pairs leads to the same set $\{(x, x^2) \,|\, x \in \mathbf{R}\}$ in both cases. The functions f and g are *equal*.

So 'the' codomain of a function is ambiguous. *Any* set that includes the range of the function will do as a codomain. *The* domain, on the other hand, is unique.

We can remove this ambiguity by being more pedantic, and defining a function to be a triple (f, A, B) rather than just a set of ordered pairs f. But at this stage, the definition as a triple is off-putting and not worth the effort, and in any case the notation $f : A \rightarrow B$ tells us which of the possible codomains is intended in any particular instance.

A familiar way to picture a function $f : \mathbf{R} \rightarrow \mathbf{R}$ is to draw its graph; we say more about this topic in the next section. For sets other than $\mathbf{R}$ it is often better to think of a function in terms of a picture like this:

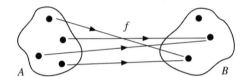

Fig. 5.1 A picture of a function

For each $x \in A$ the value of $f(x)$ is to be found at the far end of the corresponding arrow.

The definition of a function, expressed in pictorial terms, becomes:

(F1′) Every element of A is at the tail end of a unique arrow.
(F2′) All the arrowheads end up in B.

This type of diagram is a pictorial device, on a par with Venn diagrams, but it is useful as a source of motivation and simple examples.

Using such a picture, the range of f is the set of elements of B that lie on the sharp ends of the arrows:

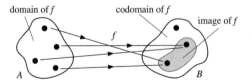

Fig. 5.2 Domain, codomain, and image

The range of f is the whole codomain B if every element of B is at the end of some arrow. This motivates a more formal definition:

Definition 5.7: A function $f : A \rightarrow B$ is a *surjection* (to B) or is *onto* B if each element of B is of the form $f(x)$ for some $x \in A$.

Whether a function is a surjection depends on the choice of codomain. So the statement 'f is a surjection' can be made only if it is clear from the context which codomain is intended—as will be the case in a phrase such as '$f : A \rightarrow B$ is a surjection', where the codomain is B. The next examples clarify this.

Examples 5.8:

(1) $f : \mathbf{R} \rightarrow \mathbf{R}$ where $f(x) = x^2$. This is not a surjection to $\mathbf{R}$, since no negative real number is the square of a real number; in particular, $-1 \in \mathbf{R}$ but is not of the form x^2 for any $x \in \mathbf{R}$.

(2) $f : \mathbf{R} \rightarrow \mathbf{R}^+$ where $f(x) = x^2$. This is a surjection to $\mathbf{R}^+$, since every positive real number has a square root, which is real.

(3) $f : A \rightarrow B$ where $A = \{$all circles in the plane$\}$, $B = \mathbf{R}^+$, and $f(x) =$ the radius of x. This is a surjection to $\mathbf{R}^+$, since given any positive real number we can find a circle with that number as radius.

If no element of B lies at the end of two different arrows, we have another important type of function:

Definition 5.9: A function $f : A \rightarrow B$ is an *injection*, *one-one*, or *one-to-one*, if for all $x, y \in A, f(x) = f(y)$ implies $x = y$.

This time the precise choice of codomain does not lead to any problems. If f is an injection for one choice of codomain, it is also an injection for any other choice. Here are some examples.

Examples 5.10:

(1) $f : \mathbf{R} \rightarrow \mathbf{R}$ where $f(x) = x^2$. This is not an injection, since $f(1) = f(-1)$ but $1 \neq -1$.

(2) $f : \mathbf{R}^+ \rightarrow \mathbf{R}$ where $f(x) = x^2$. This is an injection: if x and y are positive real numbers and $x^2 = y^2$, then $0 = x^2 - y^2 = (x - y)(x + y)$. Therefore either $x - y = 0$ and $x = y$, or $x + y = 0$ which is impossible with both x and y positive unless $x = y = 0$. Either way, $x = y$.

(3) $f : \mathbf{R} \backslash \{0\} \rightarrow \mathbf{R}$ where $f(x) = 1/x$. This is an injection, since if $1/x = 1/y$ then $x = y$.

Nicest of all are functions with both of these properties:

Definition 5.11: A function $f : A \rightarrow B$ is a *bijection* if it is both an injection and a surjection (to B).

Again, this property depends on the choice of codomain. Another common term for the same property is *one-to-one correspondence*. However, it is easy to confuse this with 'one-to-one', so we shall avoid them both. Instead of saying that f is a bijection (injection, surjection), we often say that f is *bijective* (*injective*, *surjective*). Clearly $f : A \rightarrow B$ is a bijection if and only if every $b \in B$ is of the form $b = f(x)$ for a *unique* $x \in A$.

All combinations of injectivity and surjectivity can occur, as the following pictures illustrate:

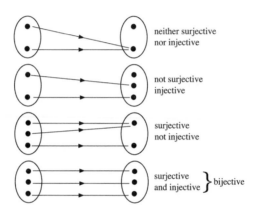

Fig. 5.3 Various kinds of function

One very important, though trivial, function can be defined on each set A.

Definition 5.12: The *identity* function $i_A : A \rightarrow A$ is defined by $i_A(a) = a$ for all $a \in A$.

This is obviously a bijection.

The Graph of a Function

There are two competing ways to picture a *real* function, by which we mean a function whose domain and range are subsets of **R**: the graph, and the blobs-and-arrows diagram. There are interesting connections between the two. A blobs-and-arrows picture of the function $f(x) = x^2$ ($x \in$ **R**) looks something like this:

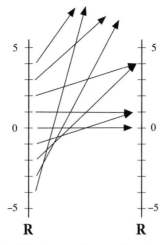

Fig. 5.4 An arrow diagram from **R** to **R**

We can disentangle the arrows better if we place *A* horizontally and let it overlap *B* at 0:

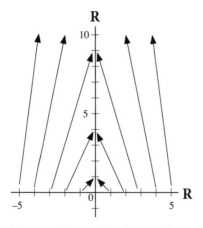

Fig. 5.5 Arrow diagram from horizontal to vertical

However, it is more interesting to use arrows that run only vertically or horizontally:

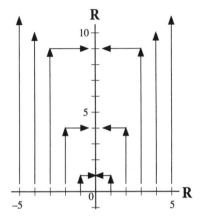

Fig. 5.6 Same diagram with vertical and horizontal arrows

This makes it clear that the important thing is where the corner occurs. If we vary *x*, all the corners lie on a curve:

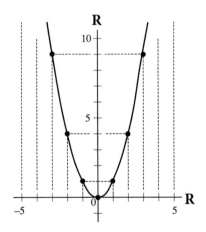

Fig. 5.7 The corners of the arrows are on a curve

Now we can eliminate the arrows, and what we get is the usual graph:

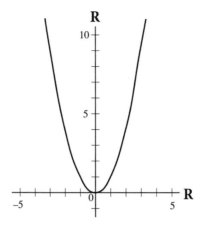

Fig. 5.8 Eliminate the arrows to get the graph

Conversely, given this graph, we can put the arrows back. Starting at $x \in A$ we move vertically until we hit the graph, then horizontally until we hit B. This point will be $f(x)$.

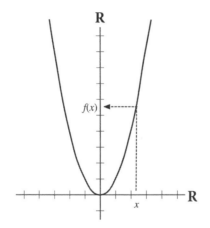

Fig. 5.9 Recovering the arrows from the graph

What is the graph set-theoretically? The plane is $\mathbf{R} \times \mathbf{R}$, and the corner in the arrow from x to $f(x)$ occurs at $(x, f(x))$. So the graph of f is the set

$$\big\{(x, f(x)) \mid x \in \mathbf{R}\big\}.$$

But this, in formal terms, is the *same* as f. By drawing $\mathbf{R} \times \mathbf{R}$ as a plane, we are led to the graph as the natural picture for f.

For a general function $f : A \to B$ we need a corresponding picture. Now we have a way to draw $A \times B$, and we use this instead of the plane. Thus the camel-lion-elephant function we met earlier has the 'graph':

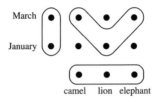

Fig. 5.10 Picturing a more general situation

By analogy with the previous function, suppose that we draw vertical arrows from elements of A, until we hit the graph, and then horizontal arrows until we hit B. The result is:

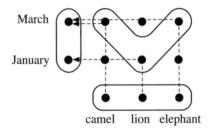

Fig. 5.11 Recovering the arrows

A little distortion recovers the blobs-and-arrows picture:

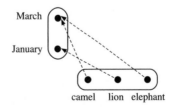

Fig. 5.12 The blobs-and-arrows picture

Strictly speaking, the graph picture for a function $f : \mathbf{R} \to \mathbf{R}$ should look like:

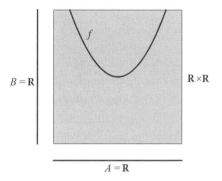

Fig. 5.13 The picture from **R** to **R**

The traditional picture, with A and B drawn *on* the plane as 'axes', is more familiar and convenient:

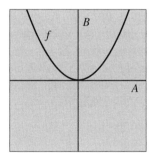

Fig. 5.14 Drawing the axes on the picture

But it should be remembered that these 'axes' are not *part* of the graph, but act as labels for the points (x, y) of the plane.

Composition of Functions

If $f : A \to B$ and $g : C \to D$ are two functions, and the image of f is a subset of C, then we can *compose* f and g by 'first doing f, then g'.

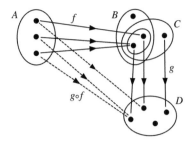

Fig. 5.15 The composition of functions

In formal terms, we define:

Definition 5.13: If $f : A \to B$ and $g : C \to D$ are functions and $f(A) \subseteq C$, the *composition* $g \circ f$ is the function

$$g \circ f : A \to D$$

for which

$$g \circ f(x) = g(f(x)).$$

Of course we must verify that $g \circ f$ *is* a function from A to D, but this is easy.

It is a pity that $g \circ f$ corresponds to 'first f, then g', since a more natural notation would seem to be $f \circ g$. But the latter would make the definition read $f \circ g(x) = g(f(x))$, which looks wrong. One way out is to write $(x)f$ instead of $f(x)$ and let composition be given by $(x)f \circ g = ((x)f)g$. But this looks odd too!

Composition of functions has a very useful property: it is associative, in the following sense:

Proposition 5.14: Let $f : A \to B$, $g : C \to D$, $h : E \to F$ be functions such that the image of f is a subset of C and the image of g is a subset of E. Then the two functions

$$h \circ (g \circ f) : A \to F$$
$$(h \circ g) \circ f : A \to F$$

are equal.

Proof: By 'equal' here we mean that the two subsets of $A \times F$ that define the functions are equal; this in turn means that for each $x \in A$ the two functions take the same value. Now

$$h \circ (g \circ f)(x) = h(g \circ f(x)) = h(g(f(x)));$$
$$(h \circ g) \circ f(x) = h \circ g(f(x)) = h(g(f(x))),$$

which proves the theorem. $\square$

Identity functions have nice properties under composition:

Proposition 5.15: If $f : A \rightarrow B$ is a function, then
$$f \circ i_A = f, \quad i_B \circ f = f.$$

Proof: This is a routine verification from the definitions. $\square$

Inverse Functions

We think of a function $f : A \rightarrow B$ as a rule that takes $x \in A$ and *does something* to it; namely, the rule produces $f(x) \in B$. Sometimes we can find a function g that 'undoes' f. We call g an inverse function to f. However, some care is needed because inverse functions need not exist, and the order in which we perform f and g sometimes matters.

Definition 5.16: Let $f : A \rightarrow B$ be a function. Then a function $g : B \rightarrow A$ is called a

left inverse for f if $g(f(x)) = x$ for all $x \in A$,
right inverse for f if $f(g(y)) = y$ for all $y \in B$,
inverse for f if it is both a left and a right inverse for f.

The three conditions may be stated in equivalent terms:
$$g \circ f = i_A,$$
$$f \circ g = i_B,$$
$$g \circ f = i_A \text{ and } f \circ g = i_B.$$

Here are some illustrations, using single arrows $\rightarrow$ for f and double arrows $\twoheadrightarrow$ for g.

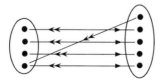

Fig. 5.16 Left inverse

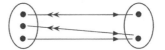

Fig. 5.17 Right inverse

Fig. 5.18 Inverse

These pictures suggest a useful criterion:

Theorem 5.17: A function $f : A \to B$ has a:

(a) left inverse if and only if it is injective
(b) right inverse if and only if it is surjective
(c) inverse if and only if it is bijective.

Proof: (a) Suppose f has a left inverse g. To prove f injective, suppose that $f(x) = f(y)$. Then $x = g(f(x)) = g(f(y)) = y$, so f is an injection. Conversely, suppose that f is injective. If $y \in B$ and $y = f(x)$, define $g(y) = x$. By injectivity, this x is unique. If y is not an element of the range of f, no such x exists; we then pick *any* $a \in A$ and define $g(y) = a$. Now $g(y)$ is defined for all $y \in B$ and $g : B \to A$ is a function. But $g(f(x)) = x$ by the definition of g, so g is a left inverse.

(b) Suppose that f has a right inverse g. If $y \in B$ then $y = f(g(y))$, so is of the form $f(x)$ for $x = g(y)$. Hence f is a surjection to B. Conversely, suppose that f is surjective. Let $y \in B$. Then $y = f(x)$ for some $x \in A$, not necessarily unique. For each $y \in B$ define $g(y)$ to be *one particular choice* of an element in A for which $f(g(y)) = y$. Then g is a function, and a right inverse to f.

(c) The function f has an inverse if and only if it has a left inverse g that is also a right inverse. Therefore f is both injective and surjective, hence bijective. If f is bijective, it has a left inverse g, and it is easy to verify that this g is also a right inverse. Hence f has an inverse. □

Example 5.18:

(1) $f : \mathbf{R} \to \mathbf{R}, f(x) = x^3$. This is bijective and has inverse $g : \mathbf{R} \to \mathbf{R}$, $g(x) = \sqrt[3]{x}$.

(2) $f : \mathbf{R} \to \mathbf{R}, f(x) = x^2$. As it stands this is neither injective nor surjective, so it has neither kind of inverse. So what has happened to square roots? We can make f surjective by taking $\mathbf{R}^+ = \{x \in \mathbf{R} \mid x \geq 0\}$ as codomain. Now $f : \mathbf{R} \to \mathbf{R}^+$ is surjective, and $g(x) = \sqrt{x}$ (positive square root) is a *right* inverse since $f(g(x)) = (\sqrt{x})^2 = x$. But it is *not* a left inverse, since

$$\sqrt{x^2} = x \quad \text{if } x \geq 0,$$
$$\sqrt{x^2} = -x \quad \text{if } x < 0.$$

(3) We assume for this example properties of exponentials and logarithms that we have not proved rigorously in this book. Let $f : \mathbf{R} \to \mathbf{R}$ be given by $f(x) = e^x$. Then f is injective, for if $e^x = e^y$ then $e^{x-y} = 1$ so $x - y = 0$ so $x = y$. Moreover, the function f has a right inverse, defined by (say)

$$g(y) = \log y (y > 0)$$
$$g(y) = 273 (y \leq 0).$$

For $g(f(x))$ is calculated as follows: $f(x) = e^x$ which is positive, so $g(f(x)) = g(e^x) = \log e^x = x$. The arbitrary 273 does not enter into this calculation—it is there merely to define g on the whole of $\mathbf{R}$. Any other definition would do for negative real numbers, because of the way the calculation works.

(4) More sensibly, consider $f : \mathbf{R} \to \mathbf{R}^\#$ where $\mathbf{R}^\# = \{x \in \mathbf{R} \mid x > 0\}$ and $f(z) = e^x$. Now f is a *bijection*, and $g : \mathbf{R}^\# \to \mathbf{R}$, with $g(y) = \log y$, is an inverse:

$$e^{\log y} = y,$$
$$\log e^x = x.$$

(5) In this example we assume properties of trigonometric functions. Consider $f : \mathbf{R} \to \mathbf{R}, f(x) = \sin x$. This is neither injective nor surjective, so it has neither kind of inverse. But what about $\sin^{-1} x$ (or arcsin x), as found in trigonometric tables?

The answer depends on exactly what we are trying to achieve. If $\sin^{-1}(x)$ is defined to be the unique y with $-\pi/2 \leq y \leq \pi/2$ such that $\sin y = x$, then this is a right (but not left) inverse to $f : \mathbf{R} \to \{x \in \mathbf{R} \mid -1 \leq x \leq 1\}$. However, it is not a left inverse; for instance

$$\sin^{-1}(\sin 6\pi) = \sin^{-1} 0 = 0 \neq 6\pi.$$

Sometimes it is said that $\sin^{-1}$ is 'multivalued'. According to our definition, it cannot then be a function in the legal sense of the term.

(6) The most satisfactory procedure is as follows. Let

$$f : \{x \in \mathbf{R} \mid -\pi/2 \le x \le \pi/2\} \to \{x \in \mathbf{R} \mid -1 \le x \le 1\}$$

be defined by $f(x) = \sin x$. Then f is a bijection, and $\sin^{-1}$ is an inverse function for *this f*.

Left and right inverses need not be unique—one reason why their construction involves arbitrary choices. But inverses *are* unique.

Proposition 5.19: If a function has both a left inverse and a right inverse, then it has an inverse. This inverse function is unique, and every left or right inverse is equal to it.

Proof: If $f: A \to B$ have both a left and a right inverse. Then by theorem 5.17, f is a bijection, and has an inverse F. If g is any left inverse, then

$$g = g \circ i_B = g \circ (f \circ F) = (g \circ f) \circ F = i_A \circ F = F.$$

Similarly, if h is any right inverse then $h = F$. Since an inverse is in particular a left inverse, this also proves F unique. $\qquad \square$

The notation for an inverse function to $f : A \to B$, provided it exists, which occurs precisely when f is a bijection, is

$$f^{-1} : B \to A.$$

WARNING. Don't confuse $f^{-1}(x)$ with the reciprocal $1/f(x)$. (For example, if $f(x) = x^2$, then $f^{-1}(x) = \sqrt{x}$ and $1/f(x) = 1/x^2$.)

Proposition 5.20: If $f : A \to B$ and $g: B \to C$ are bijections, then $g \circ f : A \to C$ is a bijection, and

$$(g \circ f)^{-1} = f^{-1} \circ g^{-1}.$$

Proof: It is clear that $g \circ f$ is a bijection. It is also easy to verify directly that $f^{-1} \circ g^{-1}$ is a left inverse, since

$$\begin{aligned}
(f^{-1} \circ g^{-1}) \circ (g \circ f) &= f^{-1} \circ (g^{-1} \circ (g \circ f)) \\
&= f^{-1} \circ ((g^{-1} \circ g) \circ f) \\
&= f^{-1} \circ (i_B \circ f) \\
&= f^{-1} \circ f \\
&= i_A.
\end{aligned}$$

Hence, by theorem 5.17, it is an inverse. $\qquad \square$

This calculation is illustrated below: it is much less horrendous than it may appear.

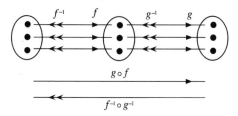

Fig. 5.19 Composing inverse functions

Restriction

Definition 5.21: If $f : A \to B$ is a function and $X \subseteq A$, the *restriction* of f to X is the function

$$f\,|_X : X \to B$$

for which

$$f\,|_X(x) = f(x) \text{ (for } x \in X).$$

This function differs from f only in that we forget about those x that do not lie in X.

For example, if $f : \mathbf{R} \to \mathbf{R}$ has $f(x) = \sin x$, and $X = \{x \in \mathbf{R} \,|\, 0 \le x \le 6\pi\}$, then the graphs of f and $f\,|_X$ are like this:

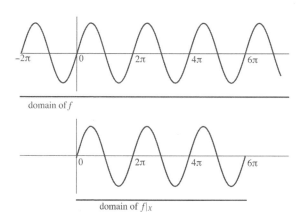

Fig. 5.20 The restriction of a function

Restriction is a relatively trivial operation. Its main use is to concentrate attention on how f behaves on the subset X. Sometimes this is useful: we noted above that $\sin : \mathbf{R} \to I$ is not a bijection when $I = \{x \in \mathbf{R} \,|\, -1 \le x \le 1\}$. If

$$I = \{x \in \mathbf{R} \,|\, -\pi/2 \le x \le \pi/2\},$$

however, then $\sin|_X : X \to I$ *is* a bijection.

Definition 5.22: Restricting the identity function $i_A : A \to A$ to a subset $X \subseteq A$ gives the *inclusion* function

$$i_A |_X : X \to A$$

for which $i_A |_X(x) = x \, (x \in X)$.

This is therefore the same function as i_X, but with a different choice of codomain which leads to a different emphasis: it provides a formal statement of how X sits inside A.

Sequences and n-tuples

We can now use functions to tidy up some questions that arose earlier. In particular we can give precise definitions of sequences and n-tuples. Earlier we gave definitions of ordered pairs, triples, quadruples, and so on, but no general prescription.

Definition 5.23: Let X_n be the set $\{1, 2, 3, \ldots, n\} = \{x \in \mathbf{N} \,|\, 1 \le x \le n\}$. If S is a set then an *n-tuple* of elements of S is defined to be a function

$$f : X_n \to S.$$

This function specifies elements $f(1), f(2), \ldots, f(n)$ of S. If we change notation to $(f_1, f_2, \ldots, f_n)$ we see that two n-tuples $(f_1, \ldots, f_n)$ and $(g_1, \ldots, g_n)$ are equal if and only if $f_1 = g_1, f_2 = g_2, \ldots, f_n = g_n$. This is what an n-tuple should look like.

Similarly, a sequence $a_1, a_2, \ldots$, described earlier as an 'endless list', may be rigorously defined as a function

$$f : \mathbf{N} \to S$$

where now we think of $f(n)$ as a_n.

In the case of ordered pairs, the new definition of (f_1, f_2) turns out *not* to be the same as the one given by Kuratowski, chapter 4. However, it has the same key property for ordered pairs: $(f_1, f_2) = (g_1, g_2)$ if and only if $f_1 = g_1$

and $f_2 = g_2$. This is really all we need, so once again we see that if we focus on the *property* in a formal definition, rather than the object defined, we get the same fundamental mathematical idea. It is another example of a crystalline concept.

Functions of Several Variables

In calculus we encounter 'functions of two variables', such as

$$f(x, y) = x^2 - 3y^3 + \cos xy \quad (x, y \in \mathbf{R}).$$

It is not necessary to go through the whole rigmarole again to make these functions precise. The notation makes it clear that f is just an ordinary function defined on the set of ordered pairs (x, y), that is,

$$f : \mathbf{R} \times \mathbf{R} \to \mathbf{R}.$$

In general, if A and B are sets then a function of two variables $a \in A, b \in B$ is a function $f : A \times B \to C$. Similarly, functions of n variables are just functions defined on a set of n-tuples.

Binary Operations

In many areas of mathematics, it is common to combine two numbers together to get another number, or two objects of a given kind to get another object of the same kind. This leads to the concept of a *binary operation* on a set A, which can be formally defined as:

Definition 5.24: A *binary operation* on a set A is a function $f : A \times A \to A$.

Examples 5.25: Familiar examples include:

(1) Addition on $\mathbf{N} : \alpha : \mathbf{N} \times \mathbf{N} \to \mathbf{N}$, $\alpha(x, y) = x + y$.
(2) Multiplication on $\mathbf{N} : \mu : \mathbf{N} \times \mathbf{N} \to \mathbf{N}$, $\mu(x, y) = xy$.
(3) Subtraction on $\mathbf{N} : \sigma : \mathbf{N} \times \mathbf{N} \to \mathbf{N}$, $\sigma(x, y) = x - y$.
(4) Division on the non-zero elements of $\mathbf{Q}$. Here we let $\mathbf{Q}^* = \{x \in \mathbf{Q} \mid x \neq 0\}$ and define $\delta : \mathbf{Q}^* \times \mathbf{Q}^* \to \mathbf{Q}^*$ by $\delta(x, y) = x/y$.
(5) Union of sets. Let $A = \mathbb{P}(X)$, the set of all subsets of a given set X. Define $u : A \times A \to A$ by $u(Y_1, Y_2) = Y_1 \cup Y_2$.

(6) Composition of functions. For any set X, let M be the set of all functions from X to X, so that $f \in M$ means $f : X \rightarrow X$. Define

$$c : M \times M \rightarrow M \text{ by } c(g,f) = g \circ f.$$

Most occurrences of binary operations in mathematics do not use function notation $f(x, y)$. Instead, they appear in the form $x * y$ for some symbol $*$. In the above examples we have $x + y, xy, x - y, x/y, x \cup y, x \circ y$. For this reason we usually denote a binary operation by $*: A \times A \rightarrow A$ and the image $*(x, y)$ by $x * y$. After examples (2) and (6), $x * y$ is called the *product*, or *composite*, of x and y. In example (2), there is no intervening symbol at all. This economical notation is used often in other mathematical situations where there is no danger of confusion. For instance the composite of functions in (6) is usually written as gf instead of $g \circ f$. You have already become accustomed to such conventions when you learned to read

$$2\pi \text{ as 2 times } \pi,$$
$$2\tfrac{1}{2} \text{ as 2 plus } \tfrac{1}{2},$$
$$21 \text{ as 2 times ten plus 1.}$$

Using $x * y$ notation, we do not usually expect $x * y$ and $y * x$ to be the same. For example, if $*$ is subtraction, then $2 - 1 \neq 1 - 2$. However, when they are the same, the algebra of $*$ can be simplified, so we are led to:

Definition 5.26: If $x * y = y * x$ for all $x, y \in A$, then $*$ is *commutative*.

In examples (1), (2), and (5), the binary operation is commutative; in examples (3) and (4), it is not. The operation in example (6) is non-commutative if X has more than one element. In fact, if $a, b \in X, a \neq b$, we define $f(a) = f(b) = a, g(a) = g(b) = b$, and $f(x) = g(x) = x$ otherwise. Then $g * f(a) = b$, but $f * g(a) = a$, so $g * f \neq f * g$.

Unless we know that a binary operation is commutative, it is essential to maintain the order of elements in a product. Such a product can be extended to three (and more) elements. Given $x, y, z \in A$, then $x * y \in A$, so we can form the product of this with z. Brackets are introduced at this stage, writing $(x * y) * z$ to denote the result and to distinguish it from $x * (y * z)$. Although the latter has x, y, z taken in the same order, it is the product of x and $y * z$, and might conceivably be different. For instance, $(3 - 2) - 1 \neq 3 - (2 - 1)$.

Definition 5.27: If $(x * y) * z = x * (y * z)$ for all $x, y, z \in A$, then $*$ is *associative*.

The operations in examples (1), (2), (5), and (6) are associative, but those in (3) and (4) are not.

When developing the number concept, commutative and associative binary operations (such as addition and multiplication) are essential building blocks. Elementary algebra makes repeated use of these properties, but when we first encounter algebraic ideas the properties are seldom made explicit, because the algebraic symbols stand for numbers. In more advanced algebra, where the symbols can be far more general, commutativity and associativity may or may not be valid, so we have to pay proper attention to them.

Just as we have binary operations $f : A \times A \to A$, we can go on to define ternary operations $t : A \times A \times A \to A$, and so on. It is even possible to think of a function $g : A \to A$ as a 'unary operation' to begin this hierarchy. Such concepts do arise from time to time, but they do not have the central importance of binary operations in mathematics.

Indexed Families of Sets

At the end of chapter 3 we considered sets S whose elements are themselves sets, such as $S = \{S_1, \ldots, S_n\}$ where each S_r is a set. Using the function concept, we can extend this notation. If $\mathbf{N}_n = \{1, 2, \ldots, n\}$, then there is a bijection $f : \mathbf{N}_n \to S$ given by $f(r) = S_r$. There is no reason to restrict attention to $\mathbf{N}_n$ here.

Definition 5.28: If A is *any* set, every element of S is a set, and $f : A \to S$ is a bijection, then we say that S is an *indexed family of sets*, and write

$$S = \{S_\alpha \mid \alpha \in A\}.$$

In this situation A is called the *index set*.

The union

$$\bigcup S = \{x \mid x \in S_\alpha \text{ for some } \alpha \in A\}$$

of such an indexed family is alternatively denoted by

$$\bigcup_{\alpha \in A} S_\alpha,$$

and the intersection

$$\bigcap S = \{x \mid x \in S_\alpha \text{ for all } \alpha \in A\}$$

by

$$\bigcap_{\alpha \in A} S_\alpha.$$

When $A = \mathbf{N}_n$, these are often denoted by $\bigcup\limits_{r=1}^{n} S_r$ and $\bigcap\limits_{r=1}^{n} S_r$. When $A = \mathbf{N}$, the notation is sometimes written $\bigcup\limits_{r=1}^{\infty} S_r$, $\bigcap\limits_{r=1}^{\infty} S_r$. The '$\infty$' symbol in these expressions is part of the historical development of the subject; using modern set-theoretic notation, these are written as $\bigcup\limits_{r \in \mathbf{N}} S_r$ and $\bigcap\limits_{r \in \mathbf{N}} S_r$.

Exercises

In these exercises, any required properties of exponential, logarithmic, and trigonometric functions may be assumed without proof.

1. Find the images of the following functions $f : \mathbf{R} \to \mathbf{R}$:
   (a) $f(x) = x^3$
   (b) $f(x) = x - 4$
   (c) $f(x) = x^2 + 2x + 2$
   (d) $f(x) = x^3 + \cos x$
   (e) $f(x) = 1/x$ if $x \neq 0, f(0) = 1$
   (f) $f(x) = |x|$
   (g) $f(x) = x^2 + x - |x|^2$
   (h) $f(x) = x^{16} + x$.

2. For each of the functions $f : \mathbf{R} \to \mathbf{R}$ defined above, state whether it is
   (a) injective, (b) surjective, (c) bijective.

3. The following functions are to be defined so that their codomain is $\mathbf{R}$, and their domains are certain subsets of $\mathbf{R}$. Say in each case what the largest possible domain is.
   (a) $f(x) = \log x$.
   (b) $f(x) = \log \log \cos x$
   (c) $f(x) = -x$
   (d) $f(x) = \log (1 - x^2)$
   (e) $f(x) = \log(\sin^2 (x))$
   (f) $f(x) = e^{x^2}$
   (g) $f(x) = 1/ (x^2 - 1)$
   (h) $f(x) = \sqrt{(x - 1)(x - 2)(x - 3)(x - 4)(x - 5)(x - 6)}$ (positive square root).

   Find the image of f in each case.

4. Let S be the set of circles in the plane, and let $f : S \to \mathbf{R}$ be defined by

$$f(s) = \text{the area of } S.$$

Is f injective? Surjective? Bijective?

Now let T be the set of circles in the plane whose centre is the origin, $\mathbf{R}^+ = \{x \in \mathbf{R} \mid x \geq 0\}$, and define $g : T \to \mathbf{R}^+$ by

$$g(T) = \text{the length of the circumference of } T.$$

Is g injective? Surjective? Bijective?

5. If A has two elements and B three, how many different functions are there from A to B? From B to A? How many, in each case, are injective? Surjective? Bijective?

6. If A has n elements and B has m elements ($n, m \in \mathbf{N}$), find the number of functions from A to B.

7. Show that if $A = \varnothing, B \neq \varnothing$, then, according to the set-theoretic definition, there is precisely one function from A to B. Show that if $A \neq \varnothing, B = \varnothing$, there are none. How many functions are there from $\varnothing$ to $\varnothing$?

8. Give examples of functions $f : \mathbf{Z} \to \mathbf{Z}$ that are:
   (a) neither injective nor surjective
   (b) injective but not surjective
   (c) surjective but not injective
   (d) surjective and injective.

9. If $f : A \to B$ show that, for $X \subseteq A$ and $Y \subseteq B$, the formulas
$$\hat{f}(X) = \{f(x) \mid x \in X\},$$
$$\hat{f}(Y) = \{x \in A \mid f(x) \in Y\}$$

define two functions $\hat{f} : \mathbb{P}(A) \to \mathbb{P}(B)$, $\tilde{f} : \mathbb{P}(B) \to \mathbb{P}(A)$, where $\mathbb{P}(X)$ denotes the set of all subsets of X. Prove that for all $X_1, X_2 \subseteq A$ and $Y_1, Y_2 \subseteq B$,
   (a) $\hat{f}(X_1 \cup X_2) = \hat{f}(X_1) \cup \hat{f}(X_2)$
   (b) $\hat{f}(X_1 \cap X_2) \subseteq \hat{f}(X_1) \cap \hat{f}(X_2)$, but equality need not hold
   (c) $\tilde{f}(Y_1 \cup Y_2) = \tilde{f}(Y_1) \cup \tilde{f}(Y_2)$
   (d) $\tilde{f}(Y_1 \cap Y_2) = \tilde{f}(Y_1) \cap \tilde{f}(Y_2)$.

Can we improve (b) to equality if f is known to be surjective? Injective? Bijective?

In textbooks the usual notation is $\hat{f}(X) = f(X), \tilde{f}(Y) = f^{-1}(Y)$; for clarity we have used the notation above.

10. Define binary operations $*$ on $\mathbf{Z}$ by
    (a) $x * y = x - y$
    (b) $x * y = |x - y|$
    (c) $x * y = x + y + xy$
    (d) $x * y = \frac{1}{2}(x + y + \frac{1}{2}((-1)^{x+y} + 1) + 1)$.

    Verify that these are binary operations. Which are commutative? Associative?

CHAPTER 6

Mathematical Logic

The essential quality of mathematics that binds it together in a coherent way is the use of mathematical proof to deduce new results from known ones, building up a strong and consistent theory. These techniques include some that are unusual in everyday life. Perhaps the most interesting of them is the method of proof by contradiction (or 'reductio ad absurdum' as it was called in more classically oriented times). To show something is true by this method, we assume that it is false and then demonstrate that this assumption leads to a contradiction. For example:

Proposition 6.1: The least upper bound of $s = \{x \in \mathbf{R} \mid x < 1\}$ is 1.

Proof: Certainly 1 is an upper bound. Let K be another upper bound. Suppose $K < 1$; then, by simple arithmetic, $K < \frac{1}{2}(K + 1) < 1$. This means that $K < \frac{1}{2}(K+1) \in S$, contradicting the fact that K is an upper bound. Thus the assumption $K < 1$ must be false, so $K \leq 1$, and 1 is the least upper bound.

$\square$

This is a typical case of this kind of reasoning. To analyse it more closely, let P stand for the statement 'If K is an upper bound for S then $K \geq 1$'. The major part of the given proof is to establish the truth of P. We assumed P false (that is, there *does* exist an upper bound $K < 1$ for S) and a simple argument led to a contradiction. If the argument is correct, then P cannot be false—so it must be true.

To carry through a proof of this nature, and to be certain of its validity, we must make sure of two vital ingredients.

In the first place, the statement P (and all other statements in the course of the proof, for that matter) must be clearly true or clearly false, although at the time we may not always know which. In everyday conversation we meet comments like 'Almost all drivers exceed the speed limit at some time or other.' This sort of remark would be useless for a contradiction argument. To refute it, is it enough to find just one person who always obeys the speed limit? Do we need to find a 'substantial number' (whatever that means) or even a majority? Everyday language is full of generalities that are vaguely true in most cases, but perhaps not all. Mathematical proof is made of sterner stuff. No such generalities are allowed; all the statements involved must be clearly true or false.

The second essential factor in a proof by contradiction is that the arguments used in the course of a proof must have no flaws. Only if this is so can we be sure in a proof by contradiction that the false link in the chain of argument is the initial assumption: P is false.

An old music hall joke goes something like this:

COMEDIAN: You're not here.

STRAIGHT MAN: Don't be silly, of course I am.

COMEDIAN: You're not, and I'll prove it to you . . . Look, you're not in Timbuktu.

STRAIGHT MAN: No.

COMEDIAN: You're not at the South Pole.

STRAIGHT MAN: Of course I'm not.

COMEDIAN: If you're not in Timbuktu or at the South Pole, you must be somewhere else.

STRAIGHT MAN: Of course I'm somewhere else!

COMEDIAN: Well, if you're somewhere else, you can't be here!

We are amused by this sort of thing, and we all see the logical flaw. But for beginners in mathematical proof techniques it exposes a deep-seated distrust of proof by contradiction. What if some similar ambiguity of terminology happens by accident or sleight of hand in the middle of the proof? When you were first confronted with a proof by contradiction that $\sqrt{2}$ is irrational, were you convinced straight away that it was correct, without any degree of suspicion? Such distrust is fully justified, and the only way to allay it is to make sure that our mathematical logic is flawless.

In the rest of this chapter we concentrate on the precise use of mathematical language and basic terminology in logic. In the following chapter we return to techniques of mathematical proof.

Statements

As we have just seen, it is essential that every statement in a mathematical proof is clearly either true or false. Typical instances are:

Examples 6.2:

   (i) $2 + 3 = 5$.

   (ii) The least upper bound of a bounded non-empty subset of **R** is unique.

   (iii) There is an upper bound K for $S = \{x \in \mathbf{R} \mid x < 1\}$ such that $K < 1$.

   (iv) $\sqrt{2}$ is irrational.

Here (i), (ii), and (iv) are true, but (iii) is false. In mathematics we are naturally more interested in true statements than false ones, but contradiction arguments make it convenient to allow both types of statement.

To distinguish between true and false statements we say that each statement has a *truth value* denoted by the letters t or f, with the obvious interpretation of these symbols: t = true, f = false. Saying that a statement has truth value t is just a fancy way of saying that it is true.

Given a statement P, the sentence 'P is false' is also a statement, and it has the opposite truth value to P. For example, if P is the false statement '$2 + 2 = 5$', then '$2 + 2 = 5$ is false' is a true statement. In logical terminology 'P is false' is usually written

$$\neg P.$$

This is also called 'the negation of P' and may be read simply as 'not P'. It is a convenient shorthand notation; however, when an actual statement is substituted for P it may not read grammatically. In the above example, 'not P' would read 'not $2 + 2 = 5$', which sounds peculiar. The equivalent statements '$2 + 2 = 5$ is false' or '$2 + 2 \neq 5$' are more euphonious. When translating 'not P' into words, it is customary to rephrase it in a suitable way to make it read smoothly.

Predicates

A particularly important type of assertion in mathematics is the predicate, introduced in chapter 3. Recall that a predicate is a sentence involving a symbol, such as x, which is either clearly true or clearly false when we replace x by any element of a set X. For instance, a typical mathematical predicate is 'the real number x is not less than 1'. If we denote this by $P(x)$, then $P(2)$ is true, $P(0)$ is false, $P(\pi/4)$ is false, and so on. If we find the truth value of $P(a)$ for every $a \in \mathbf{R}$, we get a *truth function* $T_P : \mathbf{R} \to \{t, f\}$ for which $T_P(a) = t$ if $P(a)$ is true, and $T_P(a) = f$ if $P(a)$ is false.

This concept dovetails very nicely with our ideas about set theory. The predicate $P(x)$ partitions **R** into two non-overlapping subsets, one containing the elements for which $P(x)$ is true, the other containing the elements for which $P(x)$ is false. The first of these is denoted $\{x \in \mathbf{R} \,|\, P(x)\}$. For example $\{x \in \mathbf{R} \,|\, x \geq 1\}$ is the set just described. The other is written $\{x \in \mathbf{R} \,|\, \neg P(x)\}$, which in the example becomes $\{x \in \mathbf{R} \,|\, x < 1\}$.

This situation mirrors what happens in general. For any predicate $P(x)$ we get a truth function as above. Then, for $a \in S$, we have

$$a \in \{x \in S \,|\, P(x)\} \qquad \text{if and only if } P(a) \text{ is true,}$$
$$a \in \{x \in S \,|\, \neg P(x)\} \quad \text{if and only if } P(a) \text{ is false.}$$

Rather than using vague remarks like 'a predicate is some sort of statement ...' we could use truth functions to give a set-theoretic definition. Suppose we define a truth function T_P on a set S to be any function $T_P : S \to \{t, f\}$. Then we could propose the definition: 'a predicate $P(x)$ associated with T_P is any sentence equivalent to "$T_P(x) = t$"'. The only trouble with this approach is that predicates that appear different may have the same truth function. For example,

$$P_1(x): \text{'}x \text{ is an upper bound for } \{s \in \mathbf{R} \,|\, s < 1\}\text{'},$$
$$P_2(x): \text{'}x \geq 1\text{'}.$$

It is a major part of a mathematician's job to show that such predicates are equivalent, or, more generally, that the truth of one implies the truth of the other. Therefore the predicates dealt with by practising mathematicians have the structure just described. Explaining this is a bit like explaining colour by pointing to something and saying 'that's blue'. A formal definition needs a lot to set it up; this would be appropriate in a formal course on mathematical logic, but it seems pointless here.

If more than one variable occurs in a sentence, we talk about a 'predicate in two variables', or 'three variables', and so on. For example the sentence '$x > y$' is a predicate (which we will denote by $Q(x, y)$) in two variables. If real numbers are substituted for x and y then we get a statement. For instance, $Q(3, 2)$ is true, but $Q\left(7\frac{1}{4}, 10 + \sqrt{2}\right)$ is false. Here the truth function can be considered as

$$T_Q : \mathbf{R} \times \mathbf{R} \to \{t, f\}$$

where

$$T_Q(x, y) = t \text{ if } Q(x, y) \text{ is true and } T_Q(x, y) = f \text{ if } Q(x, y) \text{ is false.}$$

In the same way we can consider '$x^2 + y^2 = z$' as a predicate in three variables $x, y, z \in \mathbf{R}$ which we denote by $P(x, y, z)$. The truth function is $T_P : \mathbf{R} \times \mathbf{R} \times \mathbf{R} \to \{t, f\}$, where

$$T_P(x, y, z) = \begin{cases} t & \text{if } x^2 + y^2 = z \\ f & \text{if } x^2 + y^2 \neq z \end{cases}.$$

In practice, mathematicians do not always mention explicitly the set to which a predicate refers, assuming that it is implied by the context. For example, the predicate '$x > 3$' is evidently meant to apply to real numbers x, whereas '$n > 3$' refers to integers n. This follows from the standard convention that *unless otherwise stated* symbols x, y, z refer to real numbers.

In particular, when we write '$x > 3$' we assume that no one would dream of substituting something for x that doesn't make sense. In the same way, it is a time-honoured convention that certain letters normally stand for elements from a specified set. For example, n is usually used to denote a natural number, or perhaps an integer. In this context the predicate '$n > 3$' would be taken to refer only to natural numbers. We have already seen cases of this earlier in the book, for instance in the definition of convergence (Definition 2.7 on page 35) we wrote:

A sequence (a_n) of real numbers tends to a limit l if, given any $\varepsilon > 0$, there is a natural number N such that $\left| a_n - l \right| < \varepsilon$ for all $n > N$.

Nowhere in this definition do we actually mention that n is a natural number, but it is clearly implied by the context. In fact, since (a_n) is a sequence, n must be a natural number.

There is a good reason for conventions of this kind, although at first sight they may seem a little sloppy stylistically. The more explicit we are in mathematics, the more symbols we need. If making everything explicit is taken to ridiculous lengths, the page gets so cluttered with symbols that it gets difficult to read the overall meaning because of the mass of detail. It then becomes a question of judgement and mathematical style to select symbols that express the ideas as clearly and succinctly as possible. On some occasions it may be appropriate to ignore standard conventions. For example, in a given context it may be appropriate to use the letter x for an integer.

All and Some

Given a predicate $P(x)$ that makes sense for elements in a set S, we can ask whether it is true for all elements in S, or whether it is true for at least some elements in S. We can then make the statements 'for all $x \in S, P(x)$ is true' or 'for some $x \in S, P(x)$ is true'. These statements can, of course, themselves be true or false. We write them in symbols using the 'universal quantifier' $\forall$ and the 'existential quantifier' $\exists$.

$\forall x \in S : P(x)$ is read 'for all $x \in S$, $P(x)$'

$\exists x \in S : P(x)$ is read 'there exists (at least) one $x \in S$ such that $P(x)$'.

If the predicate $P(x)$ is true for all $x \in S$, then the statement $\forall x \in S : P(x)$ is true; otherwise it is false. On the other hand, when $P(x)$ is true for at least one $x \in S$, then the statement $\exists x \in S : P(x)$ is true, otherwise it is false.

The symbols $\forall x \in S : P(x)$ can be read as 'for every $x \in S$, $P(x)$' or 'for each $x \in S : P(x)$', or any grammatically equivalent way. Similarly, $\exists x \in S : P(x)$ can be translated as 'there is an $x \in S$ such that $P(x)$', 'for some $x \in S : P(x)$' and so on.

In ordinary language there are subtle overtones in a statement like 'some politicians are honest'. We get the message that *some* are honest, but we also tend to assume that some are not, because otherwise the statement would have been 'all politicians are honest'. Mathematical usage carries no such implication. The statement 'for some $x \in S$, $P(x)$' does not have the connotation that there exist certain other $x \in S$ for which $P(x)$ is false. Consider the statement:

'some of the numbers 3677, 601, 19, 257, 11119, are prime'.

Since 19 is prime, the statement is true. The other numbers are *also* prime, but this does not invalidate the conclusion. At the other end of the scale, 'some' may mean only one; for instance

'some of the numbers 2, 3, 5, 7, 11 are even'

is also true, because 2 is even. This convention greatly simplifies the task of verifying the truth of '$\exists x \in S : P(x)$'. We need only find a single value of x for which $P(x)$ is true.

Examples 6.3:

(i) $\forall x \in \mathbf{R} : x^2 \geq 0$ means 'for every $x \in \mathbf{R}$, $x^2 \geq 0$' or 'the square of any real number is non-negative', or some grammatical equivalent. This is a true statement.

(ii) $\exists x \in \mathbf{R} : x^2 \geq 0$ reads 'for some $x \in \mathbf{R}$, $x^2 \geq 0$' or 'there exists a real number whose square is non-negative'. This is also true.

(iii) $\forall x \in \mathbf{R} : x^2 \geq 0$ is false (since $0^2 \not> 0$).

(iv) $\exists x \in \mathbf{R} : x^2 \geq 0$ is true (since $1^2 > 0$. In this case there are a lot of other elements of $\mathbf{R}$ besides 1 which would do just as well.)

(v) $\exists x \in \mathbf{R} : x^2 < 0$ is false.

If the symbol x is replaced throughout a quantified statement by another symbol, then we regard the new statement as being equivalent to the old.

$$\exists x \in S : P(x) \text{ means the same as } \exists y \in S : P(y).$$

For instance, $\exists x \in \mathbf{R} : x^2 > 0$ is equivalent to $\exists y \in \mathbf{R} : y^2 > 0$. Both statements say 'there exists a real number whose square is positive'.

More Than One Quantifier

Given a predicate in two or more variables, we can use a quantifier for each variable. For example, if $P(x, y)$ is the predicate '$x + y = 0$', then the statement $\forall x \in \mathbf{R} \ \exists y \in \mathbf{R} : P(x, y)$ is read as 'for every $x \in \mathbf{R}$ there is a $y \in \mathbf{R}$ such that $x + y = 0$'. It is standard logical practice to put all the quantifiers at the front of the predicate and read them in order; for instance, $\exists y \in \mathbf{R} \ \forall x \in \mathbf{R} \ : P(x, y)$ reads as 'there is a $y \in \mathbf{R}$ such that for all $x \in \mathbf{R}, x + y = 0$'.

The order matters. Of the two statements given, $\forall x \in \mathbf{R} \ \exists y \in \mathbf{R} : P(x, y)$ is true, because for each $x \in \mathbf{R}$, we can take $y = -x$ to get $x + y = 0$. However, $\exists y \in \mathbf{R} \ \forall x \in \mathbf{R} : P(x, y)$ is false, because it asserts the existence of $y \in \mathbf{R}$ that satisfies $x + y = 0$ for every $x \in \mathbf{R}$. No single value of y will do.

Getting the order of the quantifiers right in such a statement is a vital part of clear mathematical thinking. It is a common error to get it wrong (and not just among beginners). This problem can arise when we try to write a clear but formal logical statement in flowing prose. The word order may be changed around to give a more euphonious sound to the language, some-times at the expense of logical clarity. In particular the quantifiers may be embedded in the middle of the sentence instead of all coming at the be-ginning. We have already done this a few lines above when we wrote '... it asserts *the existence* of $y \in \mathbf{R}$ which satisfies $x + y = 0$ *for every* $x \in \mathbf{R}$.'

Consider the statement 'every non-zero rational number has a rational in-verse'. What we mean here is 'given $x \in \mathbf{Q}$ where $x \neq 0$ there is an element $y \in \mathbf{Q}$ such that $xy = 1$'. This is, of course, true; if $x = p/q$ where p, q are integers with $p \neq 0$, then we can take $y = q/p$. Written in logical language, the statement becomes

$$\forall x \in \mathbf{Q} \ (x \neq 0) \ \exists x \in \mathbf{Q} : xy = 1.$$

A mathematician might change the order and say 'There's a rational inverse for every non-zero rational number' to convey the same idea, even though this kind of statement could be misinterpreted. You can help matters by mak-ing sure that the meaning of your written mathematics is as clear as you can possibly make it.

The ambiguity only arises when the quantifiers involved are different. If they are the same, there is no such problem. For instance, given the predicate $P(x, y)$: '$(x + y)^2 = x^2 + 2xy + y^2$', the two statements

$$\forall x \in \mathbf{R} \ \forall y \in \mathbf{R} : P(x, y)$$

and

$$\forall y \in \mathbf{R} \ \forall x \in \mathbf{R} : P(x, y)$$

both amount to the same thing: 'for all $x, y \in \mathbf{R}$, $(x + y)^2 = x^2 + 2xy + y^2$', which is of course a true statement.

If the variables involved come from the same set, as in this case, we usually simplify the notation, writing $\forall x, y \in \mathbf{R} : P(x, y)$. The same happens with the existential quantifier. For instance if $P(x, y)$ is 'x, y are irrational and $x + y$ is rational', then $\exists x \in \mathbf{R} \backslash \mathbf{Q} \ \exists y \in \mathbf{R} \backslash \mathbf{Q} : P(x, y)$ and $\exists y \in \mathbf{R} \backslash \mathbf{Q} \ \exists x \in \mathbf{R} \backslash \mathbf{Q} : P(x, y)$ both say 'there exist two real numbers x, y which are irrational but whose sum is rational'. (This is a true statement since $\sqrt{2}$ and $-\sqrt{2}$ are irrational, but 0 is rational.) It may also be written as $\exists x, y \in \mathbf{R} \backslash \mathbf{Q} : P(x, y)$.

There is another minor pitfall in written mathematics. The universal quantifier is not always explicitly written; often it is implied by the context. Take another look at the definition of convergence of a sequence on page 34:

A sequence of real numbers tends to a limit l if, given any $\varepsilon > 0$, there is a natural number N such that $|a_n - l| < \varepsilon$ for all $n > N$.

This is quite a mouthful, and is often cut down to make it as brief as possible. A more precise definition should begin '*for all* $\varepsilon \in \mathbf{R}$, $\varepsilon > 0$ ...'. One of the little words that often gets lost is 'all'. A typical shortened statement is:

Given $\varepsilon > 0$, $\exists N$ such that $n > N$ implies $|a_n - 1| < \varepsilon$.

You will find a lot of minor variations on this definition, but in essence they all mean the same thing. If you understand this, you are a long way along the road to understanding the nature of the problem of communicating mathematics with the appropriate degree of precision.

Negation

On page 123 we introduced the negation $\neg P$ of a statement P. The truth value of $\neg P$ can be represented in the following table (called a *truth table*):

P	$\neg P$
t	f
f	t

Reading along the rows, this says that when P is true, $\neg P$ is false, and conversely. The symbol $\neg$ is called a *modifier* because it modifies a statement, changing its meaning and its truth value.

In the same way, a predicate can be modified using $\neg$. If $P(x)$ is '$x > 5$', then $\neg P(x)$ is '$x > 5$ is false' or equivalently, '$x \ngtr 5$'.

The negation of a statement involving quantifiers leads to an interesting situation. It is easy to see that the statement '$\forall x \in S : P(x)$ is false' is the same as '$\exists x \in S : \neg P(x)$'. (If it is false that $P(x)$ is true for all $x \in S$, then there must exist an $x \in S$ for which $P(x)$ is false, in which case $\neg P(x)$ is true.) That is,

(1) $\neg \forall x \in S : P(x)$ means the same as $\exists x \in S : \neg P(x)$.

Similarly,

(2) $\neg \exists x \in S : P(x)$ means the same as $\forall x \in S : \neg P(x)$.

Statement (2) tells us that 'there is no x for which $P(x)$ is true' is the same as 'for every $x \in S, P(x)$ is false'. An example of (2) is:

$\neg \exists x \in \mathbf{R} : x^2 < 0 \ldots$ there is no $x \in \mathbf{R}$ such that $x^2 < 0$.

$\forall x \in \mathbf{R} : \neg(x^2 < 0) \ldots$ every $x \in \mathbf{R}$ satisfies $x^2 \nless 0$.

These two principles are vital in mathematical arguments. Freely translated, (1) says 'to show that a predicate $P(x)$ is not true for all $x \in S$, it is only necessary to exhibit one x for which $P(x)$ is false'. Similarly, (2) asserts 'to show no $x \in S$ exists for which $P(x)$ is true, it is necessary to prove $P(x)$ false for every $x \in S$'.

As rules of thumb for negating statements involving quantifiers, these ideas come into their own when several quantifiers are involved. A typical instance is the definition of convergence of a sequence:

$$\forall \varepsilon > 0 \, \exists N \in \mathbf{N} \, \forall n > N \left(|a_n - l| < \varepsilon \right).$$

To show that (a_n) does *not* tend to the limit l, we have to prove the negation of this statement:

$$\neg \left[\forall \varepsilon > 0 \, \exists N \in \mathbf{N} \, \forall n > N \left(|a_n - l| < \varepsilon \right) \right].$$

Using principles (1) and (2) this becomes

$$\exists \varepsilon > 0 \neg \left[\exists N \in \mathbf{N} \; \forall n > N \left(|a_n - l| < \varepsilon \right) \right],$$

then

$$\exists \varepsilon > 0 \; \forall N \in \mathbf{N} \neg \left[\forall n > N \left(|a_n - l| < \varepsilon \right) \right],$$

then

$$\exists \varepsilon > 0 \; \forall N \in \mathbf{N} \; \exists n > N \neg \left(|a_n - l| < \varepsilon \right),$$

which translates finally into:

$$\exists \varepsilon > 0 \; \forall N \in \mathbf{N} \; \exists n > N \left(|a_n - l| \geq \varepsilon \right).$$

Therefore, to verify that (a_n) does *not* converge to l, we have to prove that there is some specific $\varepsilon > 0$ such that for any natural number N there is always a larger natural number $n > N$ with $|a_n - l| \geq \varepsilon$.

Much of the difficulty in a subject like mathematical analysis is in manipulating statements like this. Doing so becomes much easier with a little experience and practice, keeping the principles for negating quantifiers in mind.

Logical Grammar: Connectives

In mathematics we give standard conjunctions 'and', 'or', and so on very specific meanings. For instance, 'or' is used in the inclusive sense: if P, Q are statements then P or Q is a statement that is regarded as true provided that one *or both* of P, Q is true. We can represent this by a truth table:

P	Q	P or Q
t	t	t
t	f	t
f	t	t
f	f	f

This is read along the horizontal rows. For example, the second row says that if P is true, Q is false, then P or Q is true.

Other conjunctions in regular use in mathematics are 'and', 'implies', and 'if and only if'. The symbols are & (and), $\Rightarrow$ (implies), $\Leftrightarrow$ (if and only if). They have the following truth tables:

P	Q	$P \& Q$
t	t	t
t	f	f
f	t	f
f	f	f

P	Q	$P \Rightarrow Q$
t	t	t
t	f	f
f	t	t
f	f	t

P	Q	$P \Leftrightarrow Q$
t	t	t
t	f	f
f	t	f
f	f	t

These tables are read in the same way as the table for 'or'. The first and last of these are fairly obvious: $P \& Q$ is regarded as true only when *both* P and Q are true; $P \Leftrightarrow Q$ is regarded as true only when P and Q each have the same truth value.

The interesting table is the one for $P \Rightarrow Q$. If P is true, then the first and second lines say that the implication $P \Rightarrow Q$ is true when Q is true and false when Q is false. This shows that the truth of $P \Rightarrow Q$ means that if P is true, then Q must be true. This is the normal interpretation of the implication sign $\Rightarrow$, and for this reason $P \Rightarrow Q$ is often interpreted as 'if P, then Q'.

What of the situation when P is false? The third and fourth lines say that whether Q is true or false, $P \Rightarrow Q$ is always regarded as true. In many places a lot of philosophical nonsense is talked about this. 'How can the falsehood of P imply the truth of Q?'

The reason for this situation can be seen in the standard mathematical practice of using connectives with predicates rather than statements. If $P(x)$ and $Q(x)$ are predicates both valid for $x \in S$, then we can use the connectives in the manner above to get predicates $P(x)$ or $Q(x)$, $P(x) \& Q(x)$, etc. In particular, the predicate $P(x) \Rightarrow Q(x)$ has the stated truth table. Sometimes $P(x) \Rightarrow Q(x)$ is true for all $x \in S$. This is where the truth table comes into its own. For example when $P(x)$ is '$x > 5$' and $Q(x)$ is '$x > 2$' then every mathematician would agree that $P(x) \Rightarrow Q(x)$ is true, although some would read this as 'If $x > 5$, *then* $x > 2$', they would not be interested in what happens when $x \not> 5$.

Let us substitute some different values for x and see what happens:

If $x = 4$, then $P(4)$ is false and $Q(4)$ is true.
If $x = 1$, then $P(1)$ is false and $Q(1)$ is false.

These are precisely lines three and four of the truth table for '$\Leftrightarrow$' and illustrate how the truth table is arrived at. With this interpretation, the truth table can best be described as follows:

'$P \Rightarrow Q$ is true'

means that

(a) 'If P is true, then Q must be true';

however,

(b) 'If P is false then Q may be either true or false, and no conclusion may be drawn in this case'.

Other connectives are possible; for example the 'exclusive or' (denoted here by OR) with truth table:

P	Q	P OR Q
t	t	f
t	f	t
f	t	t
f	f	f

P OR Q is true when one, but not both, of P, Q is true.

We could write down truth tables for many other connectives, but these can all be manufactured by combining the given ones. For instance, exclusive OR can also be described by $(P$ or $Q)$ & $\neg(P$ & $Q)$. We discuss these ideas in greater detail below in the section on Formulas for Compound Statements.

Mathematicians do not restrict themselves stylistically to the connectives just described. They may also use grammatical connectives like 'but', 'since', or 'because', as fancy takes them. These words are interpreted as grammatical equivalents for the technical words. For instance the truth table for 'P but Q' is the same as that for 'P & Q'. The statement '$\sqrt{2}$ is irrational but $(\sqrt{2})^2$ is rational' means the same thing as $\sqrt{2}$ is irrational and $(\sqrt{2})^2$ is rational'. Similarly, 'P because Q' and 'P since Q' have the same truth table as '$Q \Rightarrow P$'. You can make yourself familiar with these variants by looking at a few examples. (See the exercises at the end of the chapter.)

The Link with Set Theory

If we apply connectives and the modifier $\neg$ to predicates in one variable, we find a simple relationship with set-theoretic notation. Suppose $P(x)$ and $Q(x)$ are predicates valid on the same set S, and look at the subsets for which the various compound statements are true. For '&' we obtain:

$$\{x \in S \,|\, P(x) \,\&\, Q(x)\} = \{x \in S \,|\, P(x)\} \cap \{x \in S \,|\, Q(x)\}.$$

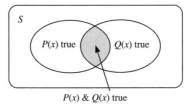

Fig. 6.1 $P(x) \,\&\, Q(x)$

Similarly,

$$\{x \in S \,|\, P(x) \text{ or } Q(x)\} = \{x \in S \,|\, P(x)\} \cup \{x \in S \,|\, Q(x)\}.$$

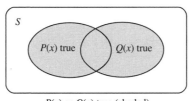

P(x) or Q(x) true (shaded)

Fig. 6.2 Inclusive $P(x)$ or $Q(x)$

This is one reason for using the 'inclusive or', corresponding to set-theoretic union, rather than the 'exclusive OR' which corresponds to the 'symmetric difference' in set theory, represented by the shaded area in the diagram:

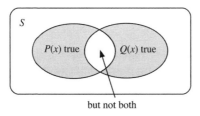

but not both

Fig. 6.3 Exclusive $P(x)$ OR $Q(x)$

The modifier $\neg$ applied to a single predicate $P(x)$ corresponds to the set-theoretic complement:

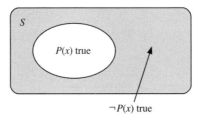

Fig. 6.4 Not $P(x)$

When we consider the implication $P(x) \Rightarrow Q(x)$ we look at the situation a little differently. We are really interested only in the case where $P(x) \Rightarrow Q(x)$ is true for all x.

In this case, if $P(x)$ is true, so must $Q(x)$ be; that is, if $a \in \{x \in S \mid P(x)\}$ then $a \in \{x \in S \mid Q(x)\}$ which means $\{x \in S \mid P(x)\} \subseteq \{x \in S \mid Q(x)\}$. The truth of the statement $P(x) \Rightarrow Q(x)$ for all $x \in S$ corresponds to set-theoretic inclusion:

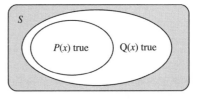

Fig. 6.5 $P(x) \Rightarrow Q(x)$

In the same way, $P(x) \Leftrightarrow Q(x)$ is true for all $x \in S$ if and only if

$$\{x \in S \mid P(x)\} = \{x \in S \mid Q(x)\}.$$

Formulas for Compound Statements

Using connectives and modifiers we can form more complex statements and predicates from given ones, for instance $(P \,\&\, Q)$ or R. This involves three statements, so the truth table has $2^3 = 8$ lines:

P	Q	R	Intermediate calculation P & Q	(P & Q) or R
t	t	t	t	t
t	t	f	t	t
t	f	t	f	t
t	f	f	f	f
f	t	t	f	t
f	t	f	f	f
f	f	t	f	t
f	f	f	f	f

The symbol '$(P \& Q)$ or R' is really a recipe for forming a new statement or predicate out of three given ones P, Q, R. To emphasise this, we call it a *compound statement formula* when P, Q, R stand for unspecified statements or predicates. When we replace P, Q, R by specific statements, for instance

$$(2 > 3 \& 2 > 6) \text{ or } 2 > 1,$$

we call it a *compound statement*. If instead we use specific predicates, we call the result a *compound predicate*. For example,

$$(x > 3 \& x > 6) \text{ or } x > 1$$

is a compound predicate.

Most mathematical proofs involve manipulating compound statements and predicates. Brackets are often essential to show how these statements and predicates are constructed. For instance $P \& (Q \text{ or } R)$ is different from $(P \& Q)$ or R. In fact, looking at the seventh line in the above truth table, if P is false, Q is false, and R is true, then $(P \& Q)$ or R is true; but a calculation shows that in this case $P \& (Q \text{ or } R)$ is false. The same goes for predicates. So we must take care to put the brackets in the right places whenever ambiguities would arise.

Sometimes, however, it is permissible to omit brackets. For instance, $(P \& Q) \& R$ has the same truth table as $P \& (Q \& R)$, so it would not cause any problem to write just $P \& Q \& R$.

When we build up compound statement formulas using connectives and modifiers, we often find formulas that look different but have the same truth table. An example is the two statements $P \Rightarrow Q$ and $(\neg Q) \Rightarrow (\neg P)$:

P	Q	$P \Rightarrow Q$
t	t	t
t	f	f
f	t	t
f	f	t

		Intermediate calculations		
P	Q	$\neg P$	$\neg Q$	$(\neg Q) \Rightarrow (\neg P)$
t	t	f	f	t
t	f	f	t	f
f	t	t	f	t
f	f	t	t	t

We can summarise the result as

P	Q	$(\neg Q) \Rightarrow (\neg P)$
t	t	t
t	f	f
f	t	t
f	f	t

and the final column for both formulas is $t\,f\,t\,t$.

In this case the compound statement formulas are said to be *logically equivalent*. Denoting two compound statement formulas by S_1, S_2, we write

$$S_1 \equiv S_2$$

for logical equivalence. For instance, our result above can be expressed as

$$P \Rightarrow Q \equiv (\neg Q) \Rightarrow (\neg P).$$

Sometimes two compound statement formulas can be considered to be logically equivalent, even though they are composed of different symbols. This happens when changing the truth value of a particular symbol does not affect the final result. For instance, $P \,\&\, (\neg P)$ is always false. The truth table for $(P \,\&\, (\neg P))$ or $(\neg Q)$ has the same truth value as $(\neg Q)$, regardless of the truth value of P. One way of looking at this, typical of the mathematical fraternity, is to think of $(\neg Q)$ as a function of both P and Q, so that its truth table becomes

P	Q	$(P \,\&\, (\neg P))$ or $(\neg Q)$
t	t	f
t	f	t
f	t	f
f	f	t

P	Q	$\neg Q$
t	t	f
t	f	t
f	t	f
f	f	t

By this formal device we can legitimately write

$$\neg Q \equiv (P \,\&\, (\neg P)) \text{ or } (\neg Q).$$

Definition 6.4: A compound statement formula is a *tautology* if it is true regardless of the truth values of its constituent statement symbols.

Typical examples of tautologies are:

(i) P or $(\neg P)$
(ii) $P \Rightarrow (P \text{ or } Q)$
(iii) $(P \,\&\, Q) \Rightarrow P$
(iv) $(P \Rightarrow Q) \Leftrightarrow ((\neg Q) \Rightarrow (\neg P))$.

Check that the truth tables for these always yield the value t.

Definition 6.5: If a compound statement formula always takes truth value f, regardless of the truth values of its constituent statement symbols, then it is a *contradiction*.

For example, $P \,\&\, (\neg P)$ is a contradiction.

Any two tautologies are logically equivalent, and any two contradictions are logically equivalent. Moreover, a compound statement formula S is a tautology if and only if $\neg S$ is a contradiction.

It is useful to use the symbol T for a tautology and C for a contradiction. We then get some interesting results. For example, $(\neg P) \Rightarrow C$ is logically equivalent to P. The truth tables are

P	$(\neg P) \Rightarrow C$
t	t
f	f

P	P
t	t
f	f

When calculating the first table, remember that C always has truth value f.

If we replace C by a specific contradiction, say $Q \,\&\, (\neg Q)$, we will still get the same result:

P	Q	$(\neg P) \Rightarrow (Q \,\&\, (\neg Q))$
t	t	t
t	f	t
f	t	f
f	f	f

P	Q	P
t	t	t
t	f	t
f	t	f
f	f	f

You should check all the intermediate calculations in the first table to get a feeling for what is happening.

Instead of comparing the truth tables of two compound statement formulas S_1 and S_2 to check logical equivalence, we can look at the single table for $S_1 \Leftrightarrow S_2$. If S_1 is logically equivalent to S_2, then $S_1 \Leftrightarrow S_2$ is a tautology, and vice versa. For example, the logical equivalence of $P \Rightarrow Q$ and $(\neg Q) \Rightarrow (\neg P)$ corresponds to the fact that $[P \Rightarrow Q] \Leftrightarrow [(\neg Q) \Rightarrow (\neg P)]$ is a tautology.

Logical Deductions

The overall strategy behind a proof often arises by proving not the truth of a given statement, but the truth of a logically equivalent one. Important examples are as follows.

Examples 6.6:

(1) *The contrapositive* $P \Rightarrow Q \equiv (\neg Q) \Rightarrow (\neg P)$. To prove $P \Rightarrow Q$, we establish the truth of $(\neg Q) \Rightarrow (\neg P)$.

(2) *Proof by contradiction* $P \equiv (\neg P) \Rightarrow C$ where C is a contradiction. To prove P, we establish the truth of $(\neg P) \Rightarrow C$.

(3) *'If and only if' proof* $P \Leftrightarrow Q \equiv (P \Rightarrow Q) \& (Q \Rightarrow P)$. To prove $P \Leftrightarrow Q$, we prove both $P \Rightarrow Q$ and $Q \Rightarrow P$.

(4) *'If and only if' version II* $P \Leftrightarrow Q \equiv (P \Rightarrow Q) \& ((\neg P) \Rightarrow (\neg Q))$. To prove $P \Leftrightarrow Q$, we establish the truth of both $P \Rightarrow Q$ and $(\neg P) \Rightarrow (\neg Q)$.

We establish the truth of a given statement from known ones by seeing how the new statement is made up, and using truth tables. For instance, we might know that P is true and that $(\neg Q) \Rightarrow (\neg P)$ is true. From these facts we can deduce that Q must be true. The given statements might be compound ones, like $(\neg Q) \Rightarrow (\neg P)$, and although we know that the total statement is true, we may have no information on the truth of its constituents. Thus we might know that $(\neg Q) \Rightarrow (\neg P)$ is true, but have no knowledge of the truth values of P or of Q. This still allows us to make some deductions; for example if $(\neg Q) \Rightarrow (\neg P)$ is true, then we know that the equivalent statement $P \Rightarrow Q$ is true.

Here are a few situations in which we can deduce the truth of the statement in the second column from those in the first.

If these statements are true . . .	*. . . then this must be true*
$P, (\neg Q) \Rightarrow (\neg P)$	Q
$(\neg P) \Rightarrow C$ (contradication)	P
$P, P \Rightarrow Q$	Q
$P \Rightarrow Q, Q \Rightarrow R$	$P \Rightarrow R$
P or $Q, \neg P$	Q
$P \& Q$	P or Q
$P \Rightarrow Q, Q \Rightarrow P$	$P \Leftrightarrow Q$
$P_1, \ldots, P_n$	$P_1 \& \ldots \& P_n$
$P_1, \ldots, P_n, (P_1 \& \ldots \& P_n) \Rightarrow Q$	Q

This table can be continued indefinitely. To obtain a new entry, write a number of compound statement formulas $S_1, \ldots, S_n$ in the left-hand column. In the corresponding position in the right-hand column, put any compound statement formula D whose truth is ensured when $S_1, \ldots, S_n$ are true. This involves looking at the truth tables for $S_1, \ldots, S_n, D$, but we can formulate the condition in one composite table by considering the formula $(S_1 \& \ldots \& S_n) \Rightarrow D$. If this is a *tautology* then the truth of $S_1, \ldots, S_n$ ensures the truth of D, as required.

A tautology of the form $(S_1 \& \ldots \& S_n) \Rightarrow D$ is called a *rule of inference*. Given such a rule of inference, and substituting actual statements into the compound statement formulas involved, if $S_1, \ldots, S_n$ are true then we may infer that D is true.

When the statements concerned involve quantified predicates we must look at how they are composed to see if the truth of one statement is a natural consequence of given ones. In a simple case we might know that $\forall x \in S : P(x)$ is true, and infer that $\exists x \in S : P(x)$ is also true. Given the truth of $\forall x \in S : P(x)$ and $\forall x \in S : Q(x)$ we can deduce a whole host of statements, including

$$\forall x \in S : P(x) \& Q(x),$$
$$[\forall x \in S : P(x)] \text{ or } [\forall x \in S : Q(x)],$$
$$P(a) \& Q(b) \text{ where } a, b \in S,$$

and so on. Again, we can make a list of deductions that can be made from statements involving quantified predicates.

If these statements are true . . .	*. . . then this must be true*
$\forall x \in S : P(x), \forall x \in S : Q(x)$	$\forall x \in S : [P(x) \& Q(x)]$
$\forall x \in S : P(x)$	$\exists x \in S : P(x)$
$\forall x \in S : P(x)$	$P(a) \ (a \in S)$
$P(a) \ (a \in S)$	$\exists x \in S : P(x)$
$\neg [\forall x \in S : P(x)]$	$\exists x \in S : [\neg P(x)]$
$\neg [\exists x \in S : P(x)]$	$\forall x \in S : [\neg P(x)]$
$\exists x \in S \, \forall y \in S \, \forall z \in M : \neg [P(x, y, z)]$	$\neg [\forall x \in S \, \exists y \in S \, \exists z \in M : P(x, y, z)]$

Again, this list could easily be extended. In the left-hand column we put statements $S_1, \ldots, S_n$, which may involve quantified predicates; in the right-hand column we put a statement D whose truth follows when all of $S_1, \ldots, S_n$ are true. Again this can be formulated as the requirement that the single statement $(S_1 \& \ldots \& S_n) \Rightarrow D$ must be true.

In this book our main aim is not to proliferate more and more complex logical statements, it is to seek simpler ways of expressing complicated ideas to make proofs easier to read and write.

Proof

In practice, when seeking a proof of some mathematical statement, we start from a number of statements $H_1, \ldots, H_r$ (called the *hypotheses*) and attempt to deduce the truth of a statement D. The process may become quite involved, with the introduction of other subsidiary statements. For this reason we perform the process in a number of steps by writing down a sequence of statements $L_1, \ldots, L_n$, where $L_n = D$ and each L_m is either one of the hypotheses $H_1, \ldots, H_r$, or a statement whose truth can be deduced from the truth of $L_1, \ldots, L_{m-1}$ for each $m = 1, 2, \ldots, n$. Therefore L_1 must be one of the hypotheses, and each successive statement $L_2, \ldots, L_n$ must either be a true deduction from previous L_j or a hypothesis. These conditions clearly imply that the last line D is true.

The truth of the deduction of L_m from previous L_j is checked, as before, by verifying the truth of the statement $(L_1 \& \ldots \& L_{m-1}) \Rightarrow L_m$. If L_m is a hypothesis, it follows immediately from the truth table for $\Rightarrow$ that $(L_1 \& \ldots \& L_{m-1}) \Rightarrow L_m$ is true; but when L_m is not a hypothesis we need to check more fully.

When the final deduction D is of the form $P \Rightarrow Q$, mathematicians often vary the prescription by writing down lines $L_1, \ldots, L_n$ with P as the first line L_1 and Q as the last line L_n. Here each intermediate line must either be a hypothesis, or its truth must follow from previous lines, as before. Some lines may well be predicates; again the important thing is to ensure that $(L_1 \& \ldots \& L_{m-1}) \Rightarrow L_m$ is always true.

Example 6.7: Given hypotheses

H_1: $5 > 2$
H_2: $\forall x, y, z \in \mathbf{R} : (x > y) \& (y > z) \Rightarrow (x > z)$,

we can write down the proof that $(x > 5) \Rightarrow (x > 2)$ as

L_1: $x > 5$
L_2: $5 > 2$
L_3: $\forall x, y, z \in \mathbf{R} : (x > y) \& (y > z) \Rightarrow (x > z)$
L_4: $x > 2$.

Although this particular deduction is not exactly mind-boggling, it embodies the general prescription for a proof, which we crystallise as follows:

Definition 6.8: Let P and Q be statements or predicates. A *proof* of the statement $P \Rightarrow Q$, given the statements $H_1, \ldots, H_r$, consists of a finite number of statements

$$L_1 = P$$
$$L_2$$
$$\vdots$$
$$L_n = Q$$

where each L_m $(2 \le m \le n)$ is either a hypothesis H_s $(1 \le s \le r)$ or a statement or predicate, such that

$$(L_1 \& \ldots \& L_{m-1}) \Rightarrow L_m$$

is a true statement for all $m \le n$.

We call the L_j the *lines* of the proof.

Under these conditions, if P is true, then each succeeding line must also be true, so in particular Q is true. The truth table for $\Rightarrow$ then shows that $P \Rightarrow Q$ is true.

It is worth looking at what happens when P is false. This could easily occur if P is a predicate, when substituting a particular value for the variable renders the predicate false. Thus in the above example, $x > 5$ is false when $x = 1$, in which case line L_4 becomes $1 > 2$, which is also false. On the other hand, if $x = 3$ then L_1 is false, but L_4 is $3 > 2$ which is true. In short, if P is false we can draw no conclusions about the validity of succeeding statements L_j: the only thing we are certain of is that the *compound* statement $P \Rightarrow Q$ is true. This happens because, although we know that the deduction $(L_1 \& \ldots \& L_{m-1}) \Rightarrow L_m$ is true, the falsity of $L_1 = P$ can lead to the falsity of L_m.

This is the most important factor in proof by contradiction. Such a proof has exactly the same format as the one above. To establish P we prove an equivalent statement $(\neg P) \Rightarrow C$, where C is some contradiction. So we start with the first line L_1 as $(\neg P)$, and end up with the last line L_n being C. On the assumption that $(\neg P)$ is true, each succeeding line must also be true. But L_n is manifestly false, being a contradiction. Hence $(\neg P)$ cannot be true, so P must be true. In this way we establish the truth of P 'by contradiction'.

The proof of $P \Rightarrow Q$ by establishing the logically equivalent statement $(\neg Q) \Rightarrow (\neg P)$ also has the same basic structure, starting with the line $(\neg Q)$, and ending with the line $(\neg P)$.

This is the formal definition of the logical steps in a proof. What do we actually need to write down in practice? The next chapter provides a possible answer.

Exercises

1. Write down truth tables for the following compound statements:
   (a) $P \Rightarrow (\neg P)$
   (b) $((P \Rightarrow R) \& (Q \Rightarrow R)) \Leftrightarrow ((P \& Q) \Rightarrow R)$
   (c) $(P \& Q) \Rightarrow (P \text{ or } Q)$
   (d) $(P \Rightarrow Q) \text{ or } (Q \Rightarrow P) \text{ or } (\neg Q)$.
   Which are tautologies?

2. Write the following statements using quantifiers $\forall$, $\exists$ and state which of them are true:
   (a) For every real number x there exists a real number y such that $y^3 = x$.
   (b) There exists a real number y such that for every real number x, the sum $x + y$ is positive.
   (c) For each irrational number x, there is an integer n satisfying
   $$x < n < x + 1.$$
   (d) The square of every integer leaves remainder 0 or 1 on division by 4.
   (e) The sum of the squares of two prime numbers which are not equal to 2 is an even number.

3. Translate the following statements into prose:
   (a) $\forall x \in \mathbf{R} \; \exists y \in \mathbf{R} : x^2 - 3xy + 2y^2 = 0$.
   (b) $\exists y \in \mathbf{R} \; \forall x \in \mathbf{R} : x^2 - 3xy + 2y^2 = 0$.
   (c) $\exists N \in \mathbf{N} \; \forall \varepsilon \in \mathbf{R} : [(\varepsilon > 0) \& (n > N)] \Rightarrow (1/n < \varepsilon)$.
   (d) $\forall x \in \mathbf{N} \; \forall y \in \mathbf{N} \; \exists z \in \mathbf{N} : x + z = y$.
   (e) $\forall x \in \mathbf{Z} \; \forall y \in \mathbf{Z} \; \exists z \in \mathbf{Z} : x + z = y$.
   Read your translations carefully, and if you think they sound stilted, rewrite them in a more flowing style (but don't change their meaning!). State which of (a)–(e) are true and which are false, giving a reason in each case.

4. In each of the following cases, write out truth tables and say whether the two statements are equivalent or not.
   (a) $\neg [P \& (\neg P)]$, $P \text{ or } (\neg P)$
   (b) $P \Rightarrow Q$, $(\neg P) \& Q$

(c) $P \Rightarrow Q$, $(\neg Q) \& P$

(d) $(P \Rightarrow Q) \& R$, $P \Rightarrow (Q \& R)$

(e) $[P \& (\neg Q)] \Rightarrow [R \& (\neg R)]$, $P \Rightarrow Q$.

5. Use truth tables (where possible) to verify the rules of inference listed in the section 'logical deductions'.

6. Which of the following are logically correct deductions?

   (a) If an International Weapons Limitation Agreement is signed, or the United Nations approve a disarmament plan, then shares in the arms industry will slump. But armament shares will not slump, so an International Weapons Limitation Agreement will not be signed.

   (b) If Britain leaves the European Union or if the trade deficit is reduced, the price of butter will fall. If Britain stays in the European Union exports will not increase. The trade deficit will increase unless exports are increased. Therefore the price of butter will not fall.

   (c) Some politicians are honest. Some women are politicians. Therefore some women politicians are honest.

   (d) If I do not work hard I will sleep. If I am worried I will not sleep. Therefore if I am worried I will work hard.

7. Consider the statement

$$x \leq y \text{ but } y > z$$

in the following cases:

   (a) $x = 1, y = 2, z = 0$

   (b) $x = 1, y = 2, z = 3$

   (c) $x = 2, y = 1, z = 0$

   (d) $x = 2, y = 1, z = 3$.

Which cases yield a true statement? Use this information to draw up a truth table for 'but' and check that

$$(P \text{ but } Q) \Leftrightarrow (P \& Q)$$

is a tautology.

   Do the same for 'since' and 'therefore' and compare with 'implies'. What happens for 'unless'?

8. What are the negations of the following statements?

   (a) $\forall x : (P(x) \& Q(x))$

   (b) $\exists x : (P(x) \Rightarrow Q(x))$

   (c) $\forall x \in \mathbf{R} \, \exists y \in \mathbf{R} : x \geq y$

   (d) $\forall x \in \mathbf{R} \, \forall y \in \mathbf{R} \, \exists z \in \mathbf{Q} : x + y \geq z$.

   Are (c) and (d) true or false?

9. Prove by contradiction the following theorems:
   (a) If $x, y \in \mathbf{R}$ and $y \leq x + \varepsilon$ for all $\varepsilon > 0$ $(\varepsilon \in \mathbf{R})$, then $y \leq x$.
   (b) For all real numbers x, either $\sqrt{3} + x$ is irrational or $\sqrt{3} - x$ is irrational.
   (c) There is no smallest rational number greater than $\sqrt{2}$.

10. Consider the connectives $\neg$, &, or, $\Rightarrow$. Show that

$$P \Rightarrow Q \equiv (\neg P) \text{ or } Q$$
$$P \text{ or } Q \equiv \neg[(\neg P) \& (\neg Q)]$$

and hence that any compound statement can be written in terms of the connectives $\neg$, & alone. Is it possible to write every compound statement in terms of just one of the connectives $\neg$, &, or, $\Rightarrow$?

Define the *stroke* connective | by the truth table

P	Q	$P \mid Q$
t	t	f
t	f	t
f	t	t
f	f	t

and show that

$$P \mid P \equiv (\neg P) \text{ or } (\neg Q).$$

Show further that
(a) $(\neg P) \equiv P \mid P$
(b) $(P \& Q) \equiv (P \mid Q) \mid (Q \mid P)$
(c) $(P \text{ or } Q) \equiv (P \mid P) \mid (Q \mid Q)$
(d) $(P \Rightarrow Q) \equiv P \mid (Q \mid Q)$.
Hence deduce that any compound statement may be written using only the stroke connective.

   Remark: this may be economical in terms of connectives, but is

$$(((P \mid P) \mid Q) \mid ((P \mid P) \mid Q)) \mid (Q \mid Q)$$

easier to read than $((\neg P) \& Q) \Rightarrow Q$? They are equivalent . . .

11. Look back at your responses to the questions at the end of the first chapter and reflect on any change in sophistication that has occurred.

CHAPTER 7

Mathematical Proof

In the last chapter we looked at the logical use of language in mathematics and how the truth of a statement can be deduced from given ones. We showed how a proof may be thought of as a sequence of logical deductions. In practice this formal definition of 'proof' does not provide a satisfactory way of *writing* proofs: to include every single step leads to a stereotyped format and is usually unbearably long-winded (see [8]). In this chapter we look at how proofs are actually written by practising mathematicians. In addition to the underlying logical skeleton, the writing of mathematical proofs needs a sense of judgement about how much detail is appropriate: what must be included and what may safely be left out. Too little detail may omit vital portions of the argument; too much may obscure the overall story.

We begin by taking an actual proof, written in normal mathematical style, and comparing it with the formal structure of the previous chapter.

Theorem 7.1: If (a_n), (b_n) are sequences of real numbers such that $a_n \to a$ and $b_n \to b$ as $n \to \infty$, then $a_n + b_n \to a + b$.

Proof: Let $\varepsilon > 0$. Since $a_n \to a$ there exists N_1 such that

$$n > N_1 \Rightarrow |a_n - a| < \tfrac{1}{2}\varepsilon.$$

Since $b_n \to b$ there exists N_2 such that

$$n > N_2 \Rightarrow |b_n - b| < \tfrac{1}{2}\varepsilon.$$

Let $N = \max(N_1, N_2)$. If $n > N$ then $|a_n - a| < \tfrac{1}{2}\varepsilon$ and $|b_n - b| < \tfrac{1}{2}\varepsilon$, so, by the triangle inequality,

$$|(a_n + b_n) - (a + b)| \le |a_n - a| + |b_n - b|$$
$$\le \tfrac{1}{2}\varepsilon + \tfrac{1}{2}\varepsilon$$
$$= \varepsilon.$$

Hence $a_n + b_n \to a + b$, as required. $\qquad\qquad\square$

To analyse the structure of this proof, let us break it down line by line, adding a few words here and there to make the construction clearer.

First look carefully at the statement of the proof and note the hypotheses that are given and the consequence that is to be proved. The theorem is in the form $P \Rightarrow Q$ where the P involves two given hypotheses:

Hypotheses:

H_1. (a_n) is a sequence of real numbers and $a_n \to a$.
H_2. (b_n) is a sequence of real numbers and $b_n \to b$.

The consequence Q to be proved states that:

Consequence to be Proved: $Q : a_n + b_n \to a + b$.

The proof, as given, consists of the following lines:

Proof:

L_1. Let $\varepsilon > 0$.
L_2. Since $a_n \to a$ there exists N_1 such that $n > N_1 \Rightarrow |a_n - a| < \frac{1}{2}\varepsilon$.
L_3. Since $b_n \to b$ there exists N_2 such that $n > N_2 \Rightarrow |b_n - b| < \frac{1}{2}\varepsilon$.
L_4. Let $N = \max(N_1, N_2)$.
L_5. If $n > N$ then $|a_n - a| < \frac{1}{2}\varepsilon$ and $|b_n - b| < \frac{1}{2}\varepsilon$.
L_6. So $|(a_n + b_n) - (a + b)| \leq |a_n - a| + |b_n - b|$ by the triangle inequality.
L_7. $|a_n - a| + |b_n - b| < \frac{1}{2}\varepsilon + \frac{1}{2}\varepsilon$.
L_8. $\frac{1}{2}\varepsilon + \frac{1}{2}\varepsilon = \varepsilon$.
L_9. (There exists $N = \max(N_1, N_2)$ such that)
$\quad n > N \Rightarrow |(a_n + b_n) - (a + b)| < \varepsilon$.
L_{10}. $a_n + b_n \to a + b$, as required. $\quad\square$

Lines L_1 and L_2 are the definition of the limit $a_n \to a$ while lines L_1 and L_3 are the definition of the limit $b_n \to b$. These involve the implicit step that if $\varepsilon > 0$ then $\frac{1}{2}\varepsilon > 0$ so $\frac{1}{2}\varepsilon$ can be used in the definition of limit. In principle we ought to write out these short deductions explicitly, but in practice the steps are omitted when they are known parts of our technique.

Line L_4 is something new: a definition of the symbol N in terms of N_1 and N_2. This definition could be omitted if we desired, and each occurrence of N in the proof replaced by $\max(N_1, N_2)$ without any real change in the proof. In practice, however, it is common to use new symbols to stand for complex concepts built up from known ones, in order to keep the notation looking simple.

Line L_5 follows from L_2, L_3, and L_4, although the statement that $n > N$ implies $n > N_1$ and $n > N_2$ is taken for granted.

Line L_6 subsumes some simple algebraic manipulations rearranging $|(a_n + b_n) - (a + b)|$ to give $|(a_n - a) + (b_n - b)|$, before using the triangle inequality to get the final result. This statement looks like a predicate in n, but tacitly we regard it as coming under the implied quantifier $\forall n > N$ in L_5.

Line L_7 follows from L_5 and L_6, again using an implicitly understood algebraic result, this time addition of inequalities.

Line L_8 is trivial algebra.

Line L_9 follows from lines L_2 to L_8. This is precisely the formal definition of the convergence of $a_n + b_n$ to $a + b$, which gives the final conclusion in L_{10}.

This analysis shows that mathematicians do not write out proofs in precisely the manner described in the previous chapter. Steps are omitted, both when hypotheses are introduced and when deductions are made; new definitions are brought in; the whole package is wrapped up in a flowing prose style in total contrast to a formal sequence of statements.

Why is this? In the first place, mathematicians were writing proofs long before they were logically analysed, so the prose style came first and continues to be used. The main reason is that the omission of trivial detail and the use of new symbols for complicated constructions are part of the process of attempting to make the deductions more comprehensible. The human mind builds up theories by recognising familiar patterns and glossing over details that are well understood, so that it can concentrate on the new material. In fact it is limited by the amount of new information it can hold at any one time, and the suppression of familiar detail is often essential for a grasp of the total picture. In a written proof, the step-by-step logical deduction is therefore foreshortened where it is already a part of the reader's basic technique, so that they can comprehend the overall structure more easily.

When working out a new theory, practising mathematicians tend to distinguish between well-established facts that are part of their technique, and those that are in the new material they are developing. They take the established ideas very much for granted, telescoping several steps into a single line where their technique is fluent, often without giving explicit references to where a proof of these results can be found. This is done in the confidence that, if they were challenged, they would be able to fill in the details (though it might tax their memory to recall them straight away!).

Newly established results constitute the heart of the theory that is being developed, and are therefore treated with greater care. They will be stated clearly as hypotheses when they are needed, and references to their proofs will be given.

When to omit logical steps or references in a proof, and when to give them in full, is part of that elusive quality: mathematical style. Different mathematicians will differ in their opinion. The clue is to look at the context of a proof and see for whom it is intended. Thus the present reader is most probably a student whose experience comes mainly from explanatory textbooks and lectures. Here the balance is likely to lean towards fuller exposition. On the other hand communication between two experts might comprise very sketchy outlines concentrating on the important new details. Nevertheless, both extremes have in common the feature of omitting detail when it may be considered as part of the basic technique in the given context.

As a specific example, when studying analysis, the rules of arithmetic might be subsumed as part of the basic toolkit, but new ideas such as limits, continuity, and so on would be treated with greater respect. Theorems about these new ideas would be proved carefully, and later theorems would refer back explicitly to results established earlier on. In the proof above, arithmetical results were used without comment, though the triangle inequality was mentioned because it was felt to be sufficiently new to be worth reminding the reader about. In more advanced work the triangle inequality would become part of the underlying technique, and be used without special reference.

In principle, a proof as used by practising mathematicians has an underlying structure like the one described in the previous chapter, but the proof occurs in a context where certain results have become a standard part of the technique. So a proof of a statement D from explicit hypotheses $H_1, \ldots, H_n$ now consists of a number of statements $L_1, \ldots, L_n$, where L_n is D, and each L_m is either:

(i) a known truth, which is either a simple deduction from the hypotheses or from the contextual technique,

or

(ii) a deduction from the previous lines $L_1, \ldots, L_{m-1}$ using formal logic and known truths from the contextual technique.

The proof is written in a mixture of prose and mathematical symbolism that makes the logical structure clear. Steps are omitted if the deductions are clear from the context, and new symbols may be introduced to simplify notation. Similarly, an actual proof of a statement $P \Rightarrow Q$ will have the same underlying structure as the formal, logical one, but making tacit use of the contextual technique.

Sometimes the context may be so clear that no explicit hypotheses are mentioned. For example:

Theorem 7.2 (Euclid): There exist infinitely many primes.

Proof: Suppose that there exist only a finite number of primes, say $p_1, \ldots,$ p_n. The number

$$N = 1 + p_1 \ldots p_n$$

is divisible by some prime p. But p cannot be any of $p_1, \ldots, p_n$ since the latter all leave remainder 1 on dividing N. This contradicts our assumption that $p_1, \ldots, p_n$ is the complete list of primes. $\qquad \square$

This proof depends on the context of arithmetic of whole numbers, including factorisation of numbers into primes. It is given in the form of a proof by contradiction. Let P be the statement 'there exist infinitely many primes'. The first line is 'suppose $\neg P$', and the proof thereafter follows the usual line of argument in search of a contradiction. Then $\neg P$ must be false, so P must be true.

In a proof like this we must keep a careful eye on our contextual material for the presence of logical flaws in the parts of the proof that have been omitted, as well as on the actual symbols on the paper. For instance, what is wrong with the following 'proof'?

Theorem(?) 7.3: The largest integer is 1.

Proof(?): Suppose not. Let n be the largest integer. Then $n > 1$. Now n^2 is also an integer, and $n^2 > n \times 1 = n$. So $n^2 > n$, which contradicts n being the largest integer. Therefore our initial assumption is false, so 1 is the largest integer. $\qquad \square$

Where is the flaw? Think about it before reading on.

The flaw is in the statement 'Let n be the largest integer'. This is not the correct negation of '1 is the largest integer'. It should be '1 is not the largest integer', which pulls apart into '$n > 1$ is the largest integer *or there is no largest integer*'. With this statement substituted, the contradiction fails to materialise, since $n^2 > n$ does not contradict the phrase italicised above. This logical flaw is disguised by the informality of the proof. It needs a lot of experience to avoid traps like these.

Axiomatic Systems

To provide a firm basis for the contextual material used, we must start some-where. We do this by taking certain explicit statements as *axioms*, which are assumed to be true; all other results in the theory are deduced from these. Ac-cording to taste, these deductions are called theorems, propositions, lemmas, corollaries, and so on. The words 'theorem' and 'proposition' are often re-garded as being interchangeable, some authors sticking exclusively to one or the other. We prefer to use the word 'proposition' to describe an ordinary run-of-the-mill result, reserving 'theorem' for something more important. In this way the structure of the theory can be seen more clearly, with important theorems standing out in relief from the background of propositions.

To give even more shape to the contours of the theory, and to reduce the strain in particularly long proofs, constituent parts of a proof may be separated out and proved before using them in the proof of a theorem or proposition. Such a preliminary result is called a *lemma*. There may be sev-eral lemmas preceding a major theorem, so that when the proof of the main result is reached, all the spadework has been done, and all that is left is to put the pieces together. In this way it is possible to make the proof of the theorem itself a much more streamlined affair, with its salient features clearly delin-eated and not concealed by the details, which have been subordinated to the lemmas.

The complement of a lemma, which precedes a theorem, is a *corollary*, which follows it. A corollary is a result that can be deduced very simply from a theorem (or proposition) and immediately follows it. Sometimes the proof is so obvious, because of the context, that the proof of a corollary is omitted, or 'left to the reader'.

In chapter 2 we looked at intuitive ideas of the real numbers and proved results in that context. To treat the subject formally, we will have to select certain properties of arithmetic as basic axioms and then build logically on these. (If we are sensible, we will use all the guile we have developed in in-tuitive arithmetic to suggest to us which way we should go in our formal development.) In chapter 8 we will look at suitable axioms for the natural numbers, before moving on to other number systems. Once we have a firm foundation for arithmetic, we can use it as contextual material to go on to more advanced theories. When handling vector spaces, or analysis, or geom-etry, arithmetical results can be subsumed and we can concentrate on the next level of deduction. At each stage it must be made clear which type of result may be used without comment, and which must be documented care-fully. Sometimes the author of a textbook, or a lecturer, may fail to mention this explicitly because it is inherent in the context.

Proof Comprehension and Self-Explanation

The question now arises: how can you make sense of a proof when the writer of the proof has made various stylistic decisions to express the essential ideas, but has glossed over details that are assumed to be implicitly well known? The answer is that when you read a proof, it is essential to consider each line in turn and to explain *to yourself* why each successive line is justified. This process is called *self-explanation* and you can practise what it means by working through the appendix at the end of this book. It involves mentally giving a mathematical reason why each successive line follows from earlier ones. You can do this silently in your own mind, or by making written comments on the page. It requires more than simply repeating or paraphrasing what is written in the proof. By making a mental effort to justify the status of each successive line in a proof, you are much more likely to make firmer links in your brain than you would by passively reading one line after another.

To see self-explanation in the context of this chapter, consider the following proof that was given earlier:

Theorem: If (a_n), (b_n) are sequences of real numbers such that $a_n \to a$ and $b_n \to b$ as $n \to \infty$, then $a_n + b_n \to a + b$.

Proof: Let $\varepsilon > 0$. Since $a_n \to a$ there exists N_1 such that

$$n > N_1 \Rightarrow |a_n - a| < \tfrac{1}{2}\varepsilon.$$

Since $b_n \to b$ there exists N_2 such that

$$n > N_2 \Rightarrow |b_n - b| < \tfrac{1}{2}\varepsilon.$$

Let $N = \max(N_1, N_2)$. If $n > N$ then $|a_n - a| < \tfrac{1}{2}\varepsilon$ and $|b_n - b| < \tfrac{1}{2}\varepsilon$, so, by the triangle inequality,

$$|(a_n + b_n) - (a + b)| \leq |a_n - a| + |b_n - b|$$
$$\leq \tfrac{1}{2}\varepsilon + \tfrac{1}{2}\varepsilon$$
$$= \varepsilon.$$

Hence $a_n + b_n \to a + b$, as required. $\qquad\square$

Read each line in turn and explain to yourself why the successive lines can be justified. Why does the first line simply state 'Let $\varepsilon > 0$'? (The reason relates to the fact that you are asked to prove that $a_n + b_n \to a + b$. This is expressed mathematically by the statement that begins 'given $\varepsilon > 0$' and you have to find an N such that if $n > N$, then $|(a_n + b_n) - (a + b)| \leq \varepsilon$.) Having written down 'Let $\varepsilon > 0$', you must use what you are given to find such an N, which must have the properties required to establish that $a_n \to a$ and $b_n \to b$. Now explain to yourself how this information is used in lines 2 and 3 and why you use $\frac{1}{2}\varepsilon$ rather than ε. Then go on to reflect on each line in turn to see where it comes from. Is it a definition? Is it a new symbol introduced to make the argument look simpler? Are there implicit assumptions that are easily justified and can be omitted? Is it a deduction from previous lines? If so, precisely which lines?

Do this now, and do it seriously.

Then read the appendix in this book on How to Read Proofs: The 'Self-Explanation' Strategy. Research using eye-tracking devices to find out what the reader is looking at when they read a proof shows that students using self-explanation techniques are more successful in making sense of the proof and retaining the ideas over a period of time (see [3]). If you make meaningful links between ideas, they are more likely to be manipulated easily in the mind. If you don't, then the ideas are more likely to be diffuse and less likely to form a basis for making sense in the longer term.

Examination Questions

One situation that is of great concern to students is what constitutes a proof acceptable in an examination. To a certain extent, the answer depends on the examiner, but part of the anxiety is due to uncertainty about the context. In a book, the context of a statement is usually clear from its position. A proof in chapter 7 is obviously allowed to assume results from chapters 1–6. But in an examination it may not be clear at which level a proof is required. Do all the steps have to be included? What can safely be missed out?

If the question is well posed, it will make the context clear. The phrase 'show from first principles . . . ' asks for a careful proof from the basic definitions and axioms. A question on the more advanced parts of a subject will not expect this kind of answer, and it is safe to assume any preceding material that is well established as contextual material for that level, never going into greater detail than is appropriate for the concepts used in the question. In particular, if a question is asked in a manner that makes familiar use of certain ideas, then they can be used at the same level of familiarity in the solution. This avoids long-winded answers that include proofs of elementary material that ought to be subsumed into the context.

Levels may vary within a single question, with the first part being elementary and later parts more advanced. The wise student will sensibly increase the power of their reasoning to the appropriate context, freely using ideas commensurate with the new situation.

Exercises

1. Is the following a proof? If not, why not? Read it through and carefully explain to yourself how each line follows (or fails to follow) from the assumptions and the previous lines in the proof.

   Theorem: For all real numbers $x, y, \frac{1}{2}(x + y) \geq \sqrt{xy}$.

   Proof: Squaring and multiplying through by 4,

   $$x^2 + 2xy + y^2 \geq 4xy,$$

   so subtracting $4xy$ from each side,

   $$x^2 - 2xy + y^2 \geq 0.$$

   But $x^2 - 2xy + y^2 = (x - y)^2$ which is always ≥ 0, so the theorem is proved. $\qquad\square$

2. Is the following a proof? If not, why not?

   Theorem: The base angles of an isosceles triangle are equal.

   Proof: Let $\triangle ABC$ be an isosceles triangle with sides $AB = AC$. Then $\triangle ABC$ is congruent to $\triangle ACB$ because the corresponding sides are equal: $AB = AC$, $BC = CB$, $AC = AB$. Here, corresponding angles are equal: in particular $\angle ABC = \angle ACB$.
   (You may assume the usual geometrical properties of congruent triangles.) $\qquad\square$

3. Are the 'proofs' given in chapter 2 of this book genuine proofs within a suitable context? If so, what is the context? If not, what are the proofs?

4. Analyse the proof of proposition 3.10 from chapter 3, showing how each statement follows from previous ones. What must be added to the proof as written to make it fit the logical definition of a proof?
   Repeat the exercise for other proofs from chapter 3.

5. Find a mathematics textbook, select a theorem (whose proof is neither too long nor too short) and analyse its structure. Which results are assumed as contextual background?
   Repeat the exercise for several other theorems, preferably from different texts and in different branches of mathematics.

6. The following are axioms for a (hitherto undefined) mathematical structure known as a *bureaucracy*. This consists of:

> a set B of *bureaucrats*,
> a set C of *committees*,
> a relation S between B and C (read *serves on*),

satisfying the following axioms:

(B1) Every bureaucrat serves on at least three different committees.

(B2) Every committee is served on by at least three different bureaucrats.

(B3) Given two distinct committees, exactly one bureaucrat serves on both.

(B4) Given two distinct bureaucrats, there is exactly one committee on which they both serve.

Prove from these axioms that if the number of bureaucrats is finite, so is the number of committees. Prove that there are always at least seven bureaucrats in a bureaucracy, and find a bureaucracy with exactly seven bureaucrats.

7. The following proof fits the logical definition. Analyse it to find out what is really going on.

Theorem: If A, B, C are sets then $(A \cap B) \cap C = A \cap (B \cap C)$.

Proof:

L_1: Let $a \in (A \cap B) \cap C$.
L_2: $a \in A \cap B$.
L_3: Let $b \in A \cap (B \cap C)$.
L_4: $a \in C$.
L_5: $b \in B \cap C$.
L_6: $b \in B$.
L_7: $a \in B$.
L_8: $b \in C$.
L_9: $\{a, b\} \subseteq B$.
L_{10}: $b \in A$.
L_{11}: $a \in A$.
L_{12}: $b \in A \cap B$.
L_{13}: $a \in A \cap B$.
L_{14}: $\{a, b\} \subseteq A \cap B$.

L_{15}: $a \in B \cap C$.
L_{16}: $a \in A \cap (B \cap C)$.
L_{17}: $(A \cap B) \cap C \subseteq A \cap (B \cap C)$.
L_{18}: $b \in (A \cap B) \cap C$.
L_{19}: $(A \cap B) \cap C \supseteq A \cap (B \cap C)$.
L_{20}: $(A \cap B) \cap C = A \cap (B \cap C)$. □

Rewrite it in a sensible style to reveal the structure of the argument.

PART III
The Development of Axiomatic Systems

Now we turn to the number systems themselves, analysing their structure and aiming to find a formal list of axioms that will describe them precisely. We also show how to construct systems that satisfy these axioms, using the raw materials of set theory. This places our intuitive ideas on a firm basis, and lets us use them without logical qualms.

Metaphorically, we are now constructing our building, or growing our plant: the important thing is to take as much care as is required to make sure that nothing goes wrong. This means a certain amount of attention to detail, and the result can look rather tortuous and pedantic.

The attitude of mind demanded of the reader is now a little different. Although intuitive ideas may be used as a source of inspiration, nothing may be used as part of a proof unless it has been given a rigorous logical demonstration. It therefore becomes necessary to give rigorous proofs, from the axioms, of properties that, on an intuitive level, we already accept. We must do this in order to be sure that, in this axiomatic sense, they really are true and can be proved logically from the axioms. By doing so, we put our ideas on a sound basis.

In chapter 8 we give highly detailed proofs, even of statements that may seem obvious. However, from chapter 9 onwards, having established a rich schema of ideas proven rigorously from the axioms of a theory and from definitions made within that context, we use proven results from the context without revisiting the detail, which you could check for yourself, should that be necessary. This avoids the danger of losing track of the main outline beneath an accumulation of ever more elaborate detail. The step-by-step method, if carried too far, obscures the overall picture.

CHAPTER 8

Natural Numbers and Proof by Induction

W hat is a number? It took mathematicians a long time to get round to wondering what the answer was, and a lot longer to find one. The first step was to characterise natural numbers. It turned out that their most important defining feature wasn't counting, or arithmetic: it was the possibility of proving theorems using mathematical induction. But at first sight, proof by induction does not seem to fit the pattern of proof described in the previous chapter. Look at a typical instance:

Proposition 8.1: The sum of the first n natural numbers is $\frac{1}{2}n(n + 1)$.

Proof: This is trivially true for $n = 1$. If it is true for $n = k$,

$$1 + 2 + \cdots + k = \tfrac{1}{2}k(k + 1),$$

then adding $k + 1$ to each side we obtain

$$1 + 2 + \cdots + (k + 1) = \tfrac{1}{2}k(k + 1) + (k + 1) = \tfrac{1}{2}(k + 1)(k + 2).$$

This is the sum of the first $k + 1$ natural numbers, and the formula is true for $n = k + 1$. By induction, the formula is true for all natural numbers. □

Many people regard this type of proof as an 'and so on . . . ' sort of argument. The truth of the statement is established for $n = 1$; then, having established the general step from $n = k$ to $n = k + 1$, this is applied for $k = 1$ to get us from $n = 1$ to $n = 2$, then used again to go from $n = 2$ to $n = 3$, and so on, as far as we wish to go. For instance, we could reach $n = 593$ after 592 applications of the general step. The only trouble with thinking this way is that reaching large values of n requires a large number of applications of the general step. We can never actually cover *all* natural numbers in a finite number of deductions if we proceed one number at a time. But a proof, by definition, comprises a *finite* number of lines.[1]

The way out of this dilemma is to remove the 'and so on . . .' part from the proof and place it squarely in the *definition* of the natural numbers. Proof by induction then fits naturally into the type of mathematical proof described in the last chapter.

Natural Numbers

The natural numbers form a highly non-trivial set, because we cannot write down a complete list of elements: they go on forever. Describing them satisfactorily needs a different approach. Fortunately, the intuitive idea of counting can easily be modelled in a set-theoretic way. We begin with 1, then comes 2, then 3, and we carry on in this way, naming each successive number as far as we wish.

To grasp the concept of the set of natural numbers 'all in one go', we regard this succession as a *function* on the set N of natural numbers. That is, we seek a function $s : N \rightarrow N$ with suitable properties. Here s stands for 'successor' and $s(1) = 2$, $s(2) = 3$, and so on. Two obvious properties that we need are:

  (i) s is not surjective (because $s(n) \neq 1$ for any $n \in N$),
  (ii) s is injective ($s(m) = s(n)$ implies that $m = n$).

There is a third vital property, giving rise to induction proofs, as follows:

  (iii) Suppose that $S \subseteq N$ is such that $1 \in S$; and for all $n \in N$ if $n \in S$ then $s(n) \in S$. Then $S = N$.

In words, (iii) says that a subset containing 1, which includes $s(n)$ whenever it contains n, exhausts the whole set of natural numbers.

Surprisingly, these three properties are all that are required to describe the natural numbers. An axiomatic basis for arithmetic requires only that we postulate the existence of a set with these three properties.

For technical reasons, it is more profitable to start with 0 rather than 1. Although in counting we usually start with 1, the empty set has 0 elements and it is useful to be able to say so. Again, in arithmetic it is convenient to have the zero element. For these and other reasons we start with 0 in our axiomatic system, and to avoid confusion with our intuitive concept N of the natural numbers we use $\mathbb{N}_0$ to denote the formal system. The 'black-board bold' font $\mathbb{N}$ distinguishes the formal concept of the natural numbers from the informal one, and the subscript 0 reminds us that 0 is included.

---

[1] Textbooks would become very expensive if not, for a start.

We then obtain the *Peano Axioms* for the natural numbers, named after Giuseppe Peano, the Italian mathematician responsible for this approach at the end of the nineteenth century:

Peano Axioms: Suppose that there exists a set $\mathbb{N}_0$ and a function $s : \mathbb{N}_0 \to \mathbb{N}_0$ such that

(N1) s is not surjective: there exists $0 \in \mathbb{N}_0$ such that $s(n) \neq 0$ for any $n \in \mathbb{N}_0$.

(N2) s is injective: if $s(m) = s(n)$ then $m = n$.

(N3) If $S \subseteq \mathbb{N}_0$ is such that $0 \in S$ and $n \in S \Rightarrow s(n) \in S$ for all $n \in \mathbb{N}_0$, then $S = \mathbb{N}_0$.

There is no guarantee that any such set $\mathbb{N}_0$ exists, so we take its existence as a basic axiom for mathematics:

Existence Axiom for Natural Numbers: There exists a set $\mathbb{N}_0$ and a function $s : \mathbb{N}_0 \to \mathbb{N}_0$ satisfying (N1)–(N3).

From these slender beginnings we can develop all the usual properties of arithmetic, then later build up the other number systems including real and complex numbers. We will also see how axiom (N3) enshrines the idea of proof by induction, as in the following simple case:

Proposition 8.2: If $n \in \mathbb{N}_0$, $n \neq 0$, then there exists a unique $m \in \mathbb{N}_0$ such that $n = s(m)$.

Proof: Let $S = \{n \in \mathbb{N}_0 \mid n = 0 \text{ or } n = s(m) \text{ for some } m \in \mathbb{N}_0\}$. Certainly $0 \in S$. If $n \in S$ then either $n = 0$, in which case $s(n) = s(0)$ so $s(n) \in S$; or $n = s(m)$ and $s(n) = s(s(m))$ where $s(m) \in \mathbb{N}_0$, so $s(n) \in S$. Hence, by axiom (N3), $S = \mathbb{N}_0$. This shows that the required m exists. Uniqueness follows from (N2). □

Proposition 8.2 tells us that 0 is the only element that is not a successor, a property that distinguishes it from all other elements. The set $\mathbb{N} = \mathbb{N}_0 \setminus \{0\}$ will be called the *natural numbers*. We shall denote $s(0)$ by 1. This element lies in $\mathbb{N}$ and will prove of paramount importance.

Look at the proof of proposition 8.2 once more. Its essential structure consists of defining a set S, then

(i) showing that $0 \in S$,

(ii) showing that $n \in S \Rightarrow s(n) \in S$,

(iii) invoking axiom (N3) to deduce that $S = \mathbb{N}_0$.

A proof by induction always has this format.

In practice S is of the form

$$S = \{n \in \mathbb{N}_0 \mid P(n)\}$$

where $P(n)$ is a predicate known to be true or false for each $n \in \mathbb{N}_0$. The statements (i), (ii), (iii) translate into

(i)′ showing $P(0)$ is true,
(ii)′ showing that if $P(n)$ is true then $P(s(n))$ is true,
(ii)′ invoking (N3) to deduce that $P(n)$ is true for all $n \in \mathbb{N}_0$.

Axiom (N3) finishes the proof without a breath of an 'and so on . . .' type of argument.

The reader will recognise the basic skeleton of this method in proposition 8.1, except that we began at 1 instead of 0 and wrote $n + 1$ instead of $s(n)$. Later we show that the same method applies starting at any $k \in \mathbb{N}_0$, in particular at $k = 1$, so the proposition at the beginning of the chapter is just a simple example of an induction proof depending on the use of axiom (N3).

In practice, axiom (N3) may not be mentioned explicitly. The proof may be phrased entirely in terms of a predicate $P(n)$, and, when steps (i)′ and (ii)′ are established, the conclusion '$P(n)$ is true for all n' is said to be established 'by induction'. You should always interpret this as an implicit use of axiom (N3), which is referred to as the *induction axiom* for this very reason. During the course of such a proof, the assumption that $P(n)$ is true is called the *induction hypothesis* and the proof that $P(n) \Rightarrow P(s(n))$ is called the *induction step*. For the moment, we make the set S explicit.

Definition by Induction

The most important task is to set up arithmetic in $\mathbb{N}_0$. To get started, we must define the basic operations of addition and multiplication.

We can define addition by setting

$$m + 0 = m \tag{8.1}$$

for all $m \in \mathbb{N}_0$, and then, once we have calculated $m + n$, we can calculate $m + s(n)$ by

$$m + s(n) = s(m + n). \tag{8.2}$$

The induction axiom seems tailor-made for definitions, as well as proofs. If S is the set $n \in \mathbb{N}_0$ for which $m + n$ is defined, then $0 \in S$ (by (8.1)), and, if $n \in S$ then $m + n$ is defined and by (8.2) we can use $s(m + n)$ to define $m + s(n)$ so that $s(n) \in S$.

However, there is a subtle point here, which involves a difference between proof and definition. In an induction proof, the induction step $n \in S \Rightarrow s(n) \in S$ involves only a demonstration that *if* $n \in S$ is true then so is $s(n) \in S$. But when making an inductive definition of addition, in order to be able to *define* the sum $m + s(n)$ as $s(m + n)$, it is essential first to *know* the value of $m + n$.

Our intuitive model $\mathbf{N}_0$ tells us that for any $n \in \mathbb{N}_0$ we can start at 0, count on 1, 2, 3, ..., and eventually reach n. For instance, if $n = 101$ we can start from the definition (8.1) at 0 and use step (8.2) 101 times to find $m + n$. Unfortunately we have not established any such principle for $\mathbb{N}_0$; indeed, given $m \in \mathbb{N}_0$, we don't yet know that if we start with 0 and form successors $1 = s(0)$, $2 = s(1)$, and so on, we eventually reach m. Moreover, our long-term objective is to eliminate 'and so on ...' arguments. To remove this flaw we prove a general principle about the validity of such definitions based only on the Peano axioms. It helps to formulate the theorem in the general case of the repeated composition of any function f in any set X and then apply it to the successor function s to make definitions by recursion. The proof is quite technical (probably one of the most intricate in the whole book). It may help to use the self-explanation technique to slowly consider each step to seek to explain it to yourself.

Theorem 8.3 (Recursion Theorem): If X is a set, $f : X \to X$ a function, and $c \in X$, then there exists a unique function $\phi : \mathbb{N}_0 \to X$ such that

(i) $\phi(0) = c$,
(ii) $\phi(s(n)) = f(\phi(n))$ for all $n \in \mathbb{N}_0$.

PRE-PROOF DISCUSSION. Essentially, we start with a function $f : X \to X$ and $c \in X$ and apply f again and again to get

$$\phi(0) = c, \phi(1) = f^1(c) = f(c), \phi(2) = f^2(c) = f(f(c)), \text{ and so on,}$$

to give the function $\phi(n) = f^n(n)$ (where we can consider $f^0(c)$ to be c). To eliminate the 'and so on' argument, we use the set-theoretic definition of a function $\phi : \mathbb{N}_0 \to X$ as a set of ordered pairs and consider those subsets U of $\mathbb{N}_0 \times X$ satisfying

(a) $(0, c) \in \phi$,
(b) $(n, y) \in \phi \Rightarrow (s(n), f(x)) \in \phi$.

There are many such subsets, including the whole set $U = \mathbb{N}_0 \times X$. We show that the one we require is the intersection of all such subsets.

Proof: Let ϕ be the intersection of all subsets U of $\mathbb{N}_0 \times X$ satisfying

$$(0, c) \in U, \tag{8.3}$$

$$(n, c) \in U \Rightarrow (s(n), f(x)) \in U. \tag{8.4}$$

Let

$$S = \{n \in \mathbb{N}_0 \,|\, (n, x) \in \phi \text{ for some } x \in X\}.$$

Then $0 \in S$ by (8.3). And by (8.4), $n \in S \Rightarrow s(n) \in S$. By induction, $S = \mathbb{N}_0$.

So every $n \in \mathbb{N}$ does have some x such that $(n, x) \in \phi$ for some $x \in S$. However, to show that ϕ is a function, we also need to prove that x is unique. Let

$$T = \{n \in \mathbb{N}_0 \,|\, (n, x) \in \phi \text{ for a unique } x \in X\}.$$

We seek to prove that $T = \mathbb{N}_0$ by induction.

Starting with $n = 0$, we know that $(0, c) \in \phi$. If also $(0, d) \in \phi$ with $c \neq d$, let $\phi^- = \phi \setminus \{(0, d)\}$. Then ϕ^- satisfies (8.3); and if $(n, x) \in \phi^-$ then $(s(n), f(x)) \in \phi$ and is not $(0, d)$ because $s(n) \neq 0$ by axiom (N1). So $(s(n), f(x)) \in \phi^-$ and ϕ^- satisfies (8.4). Since ϕ is the smallest set satisfying (8.3) and (8.4) this is a contradiction, hence no such d exists, so $0 \in T$.

The induction step that $n \in T$ implies $s(n) \in T$ uses a similar argument, as follows.

If $n \in T$ then $(n, x) \in \phi$ for precisely one $x \in X$. From (b) in the pre-proof discussion we have $(s(n), f(x)) \in \phi$, so to establish that $s(n) \in T$ we must show that no other ordered pair $(s(n), y) \in \phi$ with $y \neq f(x)$. If there were such a pair, consider $\phi^* = \phi \setminus \{(s(n), y)\}$. Again, since $0 \neq s(n)$, we know that ϕ^* satisfies (8.3).

To check (8.4) we need to prove that

$$(m, z) \in \phi^* \Rightarrow (s(m), f(z)) \in \phi^* \text{ for all } m \in \mathbb{N}_0.$$

This is true for $m = n$, since there is a unique $x \in X$ such that $(n, x) \in \phi$, and for this x, $(s(n)), f(x)) \in \phi$ by (b) and is not $(s(n), y)$ since $y \neq f(x)$. For $m \neq n$ we have $(s(m), f(z)) \in \phi$ by (b), and $s(m) \neq s(n)$ by (N2). Hence $(s(m), f(z)) \neq (s(n), y)$, so $(s(m), f(z)) \in \phi^*$. Either way, ϕ^* satisfies (8.4) and again we have a contradiction.

By induction, $T = \mathbb{N}_0$. $\qquad\square$

Definitions that employ this theorem are said to be *recursive*. The recursion theorem opens the floodgates to give a wide range of examples. These include:

(1) *Addition.* $\alpha_m : \mathbb{N}_0 \to \mathbb{N}_0$, $\alpha_m(n) = m + n$, defined by

$$\alpha_m(0) = m$$
$$\alpha_m(s(n)) = s(\alpha_m(n)).$$

Here $c = m, f = s$.

(2) *Multiplication.* $\mu_m : \mathbb{N}_0 \to \mathbb{N}_0$, $\mu_m(n) = mn$, defined by

$$\mu_m(0) = 0$$
$$\mu_m(s(n)) = \mu_m(n) + m.$$

Here $c = 0, f(r) = r + m$.

(3) *Powers.* $\pi_m : \mathbb{N}_0 \to \mathbb{N}_0$, $\pi_m(n) = m^n$, defined by

$$\pi_m(0) = 1$$
$$\pi_m(s(n)) = m\pi_m(n).$$

Here $c = 1, f(r) = rm$.

(4) *Repeated composition* of a map $f : X \to X$, defined by

$$f^0(x) = x$$
$$f^{s(n)}(x) = f(f^n(x)) \quad \text{for all } x \in X.$$

Laws of Arithmetic

With addition and multiplication properly defined by recursion, it is now relatively easy to prove the usual laws of arithmetic using induction. The proofs are not always easy to find without guidance, and you are encouraged to follow through the arguments and explain the proof to yourself. By building up your own knowledge schemas, you may be able to find slicker proofs than ours.

For reference, we note the definitions:

$$(\alpha 1)\ m + 0 = m, \quad (\alpha 2)\ m + s(n) = s(m + n),$$

$$(\mu 1)\ m0 = 0, \quad (\mu 2)\ ms(n) = mn + m.$$

Now from $(\alpha 2)$ and $(\alpha 1)$ we see that $m + s(0) = s(m + 0) = s(m)$. We have already denoted $s(0)$ by 1, so $s(m) = m + 1$.

Lemma 8.4: For all $m \in \mathbb{N}_0$,

(a) $0 + m = m$,
(b) $1 + m = s(m)$,

(c) $0m = 0$,

(d) $1m = m$.

Proof: In each case, use induction on m. We verify (a) and leave the rest as an exercise for you to explain for yourself. Let

$$S = \{m \in \mathbb{N}_0 \,|\, 0 + m = m\}.$$

Trivially $0 \in S$ by $(\alpha 1)$. If $m \in S$ then $0 + m = m$, so by $(\alpha 2)$,

$$0 + s(m) = s(0 + m) = s(m).$$

Therefore $s(m) \in S$.

By (N3), $S = \mathbb{N}_0$. $\qquad\qquad\qquad\qquad\qquad\qquad\qquad\qquad\qquad\qquad\qquad\quad\square$

Theorem 8.5: For all $m, n, p \in \mathbb{N}_0$,

(a) $(m + n) + p = m + (n + p)$

(b) $m + n = n + m$

(c) $(mn)p = m(np)$

(d) $mn = nm$

(e) $m(n + p) = mn + mp$.

Proof: (a) is proved by induction on p, using

$$S = \{p \in \mathbb{N}_0 \,|\, (m + n) + p = m + (n + p)\}.$$

First

$$
\begin{aligned}
(m + n) + 0 &= m + n & \text{by } (\alpha 2) \\
&= m + (n + 0) & \text{by } (\alpha 1),
\end{aligned}
$$

so $0 \in S$. Second, if $p \in S$ then

$$(m + n) + p = m + (n + p), \qquad\qquad\qquad\qquad (8.5)$$

so

$$
\begin{aligned}
(m + n) + s(p) &= s((m + n) + p) & \text{by } (\alpha 2) \\
&= s(m + (n + p)) & \text{by } (8.5) \\
&= m + s(n + p) & \text{by } (\alpha 2) \\
&= m + (n + s(p)) & \text{by } (\alpha 2)
\end{aligned}
$$

implying $s(p) \in S$. By induction, $S = \mathbb{N}_0$.

(b) is proved by induction on n using

$$S = \{n \in \mathbb{N}_0 \,|\, m + n = n + m\}.$$

Lemma 8.4(a) shows that $0 \in S$. If $n \in S$ then

$$m + n = n + m \tag{8.6}$$

and then

$$
\begin{aligned}
m + s(n) &= s(m + n) && \text{by } (\alpha 2) \\
&= s(n + m) && \text{by } (8.6) \\
&= n + s(m) && \text{by } (\alpha 2) \\
&= n + (1 + m) && \text{by lemma 8.4(b)} \\
&= (n + 1) + m && \text{by theorem 8.5(a)} \\
&= s(n) + m,
\end{aligned}
$$

hence $s(n) \in S$. By induction $S = \mathbb{N}_0$, establishing (b).

It is convenient to deal with (e) next, using induction on p. Let

$$S = \{p \in \mathbb{N}_0 \mid m(n + p) = mn + mp\}.$$

Then

$$
\begin{aligned}
m(n + 0) &= mn && \text{by } (\alpha 1) \\
&= mn + 0 && \text{by } (\alpha 1) \\
&= mn + m0 && \text{by} (\mu 1),
\end{aligned}
$$

implying $0 \in S$. If $p \in S$, then

$$m(n + p) = mn + mp \tag{8.7}$$

so

$$
\begin{aligned}
m(n + s(p)) &= ms(n + p) && \text{by } (\alpha 2) \\
&= m(n + p) + m && \text{by } (\mu 2) \\
&= (mn + mp) + m && \text{by } (8.7) \\
&= mn + (mp + m) && \text{by theorem 8.5(a)} \\
&= mn + ms(p) && \text{by } (\mu 2).
\end{aligned}
$$

Therefore $s(p) \in S$, and induction gives $S = \mathbb{N}_0$.

The proof of (c) is now relatively straightforward and of the same nature as previous proofs. This leaves (d), which turns out to be a little trickier. Let

$$S = \{n \in \mathbb{N}_0 \mid mn = nm\}.$$

Now $0 \in S$ by lemma 8.4(c). If $n \in S$ then

$$mn = nm \tag{8.8}$$

And

$$
\begin{aligned}
ms(n) &= mn + m && \text{by } (\mu 2) \\
&= nm + m && \text{by } (8.8).
\end{aligned}
$$

If we could show that this equalled $s(n)m$ we would have finished, but unfortunately we don't know this yet. However, we can prove it by a second induction on m. Let

$$T = \{m \in \mathbb{N}_0 \,|\, nm + m = s(n)m\}.$$

Then $0 \in T$, and if $m \in T$ then

$$nm + m = s(n)m \qquad (8.9)$$

So

$$
\begin{aligned}
ns(m) + s(m) &= n(m + 1) + (m + 1) \\
&= (nm + n) + (m + 1) && \text{by (e)} \\
&= nm + (n + (m + 1)) && \text{by (a)} \\
&= nm + ((n + m) + 1) && \text{by (a)} \\
&= nm + ((m + n) + 1) && \text{by (b)} \\
&= nm + (m + (n + 1)) && \text{by (a)} \\
&= (nm + m) + (n + 1) && \text{by (a)} \\
&= s(n)m + s(n) && \text{by (8.9)} \\
&= s(n)s(m) && \text{by } (\mu 2).
\end{aligned}
$$

Hence $s(m) \in T$ and $T = \mathbb{N}_0$. Returning to where we left off, $s(n) \in S$ and $S = \mathbb{N}_0$. This proves (d). $\qquad\qquad\square$

Having performed this massive induction exercise we can now use these arithmetic results freely to provide a coherent context in which we can prove more sophisticated ideas without overburdening the proof with too much detail. To simplify notation and make it look more familiar, we replace $s(n)$ by $n + 1$. The induction axiom now becomes:

If $S \subseteq \mathbb{N}_0$, $0 \in S$, and $n \in S \Rightarrow n + 1 \in S$, then $S = \mathbb{N}_0$.

Axiom (N2) translates into

$$m + 1 = n + 1 \Rightarrow m = n,$$

and this can be extended by induction to give:

Proposition 8.6: For all $m, n, q \in \mathbb{N}_0$,

(a) $m + q = n + q \Rightarrow m = n$
(b) $q \neq 0$, $mq = nq \Rightarrow m = n$.

Proof: (a) Use induction on q. Let

$$S = \{q \in \mathbb{N}_0 \,|\, m + q = n + q \Rightarrow m = n\}.$$

Trivially $0 \in S$. If $q \in S$, suppose that

$$m + (q + 1) = n + (q + 1).$$

By theorem 8.5(a),

$$(m + q) + 1 = (n + q) + 1,$$

hence by (N2)

$$m + q = n + q,$$

and since $q \in S$,

$$m = n.$$

Hence $q + 1 \in S$ and by induction $S = \mathbb{N}_0$.

(b) Let

$$S = \{m \in \mathbb{N}_0 \,|\, q \neq 0, \, mq = nq \Rightarrow m = n\}.$$

To show $0 \in S$, suppose that $q \neq 0$, and

$$nq = 0q = 0.$$

Then $q = p + 1$ for some p. If $n \neq 0$ then $n = r + 1$. Then $nq = (pr + p + r) + 1$ so cannot be 0. Hence $n = 0$, so $0 \in S$.

Now suppose $m \in S$ and $q \neq 0$, with

$$(m + 1)q = nq.$$

As before, $n \neq 0$, so $n = r + 1$ for some $r \in \mathbb{N}_0$. Then $mq + q = rq + q$. By part (a), $mq = rq$; by hypothesis $m = r$. Therefore $m + 1 = n$. $\qquad\square$

We now discuss subtraction. Suppose that $p = r + q$. By proposition 8.6, r is determined uniquely by p and q. We may therefore denote r by $p - q$. For $m, n \in \mathbb{N}_0$ we define a relation $\geq$ by

$$m \geq n \Leftrightarrow \exists r \in \mathbb{N}_0, \, m = r + n.$$

Given $m, n \in \mathbb{N}_0$, the difference $m - n$ is defined only when $m \geq n$. This being so, we can verify various rules of subtraction, such as

$$
\begin{array}{ll}
m - (n - r) = (m - n) + r & \text{for } m \geq n \geq r, \\
m + (n - r) = (m + n) - r & \text{for } n \geq r, \\
m(n - r) = mn - mr & \text{for } n \geq r.
\end{array}
$$

All are routine; for instance the last follows by considering

$$n = s + r \quad (\text{since } n \geq r),$$

whence

$$mn = m(s + r) = ms + mr.$$

Thus by definition,

$$mn - mr = ms = m(n - r)$$

since $s = n - r$.

We may also consider division, and in the case $m = rn$ ($n \neq 0$) we denote r by m/n. We discuss when division is possible in a later section.

Ordering the Natural Numbers

We have already defined the relation $\geq$ on $\mathbb{N}_0$. The other order relations are given by

$$m > n \Leftrightarrow m \geq n \,\&\, m \neq n,$$
$$m \leq n \Leftrightarrow n \geq m,$$
$$m < n \Leftrightarrow n > m.$$

We must prove that these are indeed order relations in the sense of chapter 4. For example:

Proposition 8.7: $m \geq n, n \geq p \Rightarrow m \geq p$ for all m, n, $p \in \mathbb{N}_0$.

Proof: There exist $r, s \in \mathbb{N}_0$ such that $m = r + n, n = s + p$. Hence $m = r + (s + p) = (r + s) + p$, so $m \geq p$. □

A second property of order relations is also easy:

Proposition 8.8: If $m, n \in \mathbb{N}_0$ and $m \geq n, n \geq m$, then $m = n$.

Proof: There exist $r, t \in \mathbb{N}_0$ such that $m = r + n, n = t + m$, so $m = r + t + m$. By proposition 8.5(a), $r + t = 0$. We cannot have $t \neq 0$, since this would imply $t = q + 1$ for some $q \in \mathbb{N}_0$ by lemma 8.4, and then $0 = (r + q) + 1$, contradicting axiom (N1). Therefore $t = 0$, so $n = m$. □

The third property of an order relation requires a more technical proof, which is postponed until proposition 8.13. However, it is a simple matter to see that the relations behave as expected, relative to the arithmetical operations of $\mathbb{N}_0$:

Proposition 8.9: For all m, n, p, $q \in \mathbb{N}_0$,

(a) if $m \geq n, p \geq q$, then $m + p \geq n + q$,
(b) if $m \geq n, p \geq q$, then $mp \geq nq$.

Proof: (a) There exist $r, s \in \mathbb{N}_0$ such that $m = r + n$, $p = s + q$. Hence, after simplification, we find that $m + p = (r + s) + (n + q)$.
(b) Similarly, $mp = nq + (rs + ns + rq)$. $\square$

The zero element 0 is the smallest element of $\mathbb{N}_0$, in the following sense:

Lemma 8.10: If $m \in \mathbb{N}_0$ then $m \geq 0$.

Proof: $m = 0 + m$. $\square$

The element 1 is the next smallest:

Lemma 8.11: If $m \in \mathbb{N}_0$ and $m > 0$ then $m \geq 1$.

Proof: By proposition 8.1, if $m \neq 0$ then $m = q + 1$ for some $q \in \mathbb{N}_0$. Hence $m \geq 1$. $\square$

We could go on to show that $2 = 1 + 1$ is the next smallest after 1, then $3 = 2 + 1$ is the next smallest after 2, and so on. It is more efficient to prove a general proposition:

Proposition 8.12: If $m, n \in \mathbb{N}_0$ and $m > n$ then $m \geq n + 1$.

Proof: We have $m = n + r$ for some $r \in \mathbb{N}_0$, and $r \neq 0$ since $m \neq n$. By proposition 8.2, $r = q + 1$ for some $q \in \mathbb{N}_0$, hence $m = (n + 1) + q$, and $m \geq n + 1$. $\square$

Now we can complete the proof that $\geq$ is an order relation in the sense of chapter 4.

Proposition 8.13: The relation $\geq$ is a (weak) order relation on $\mathbb{N}_0$.

Proof: We must prove that for all $m, n, p \in \mathbb{N}_0$,

(WO1) $m \geq n \,\&\, n \geq p \Rightarrow m \geq p$,
(WO2) Either $m \geq n$ or $n \geq m$,
(WO3) If $m \geq n$ and $n \geq m$ then $m = n$.

We have already established (WO1) and (WO3) in propositions 8.7 and 8.8. To verify (WO2), let

$$S(m) = \{n \in \mathbb{N}_0 \,|\, m \geq n \text{ or } n \geq m\}.$$

We aim to prove that $S(m) = \mathbb{N}_0$ for all $m \in \mathbb{N}_0$. Now for a given m, we have $0 \in S(m)$ since $m \geq 0$. Next suppose that $n \in S(m)$. Either $m \geq n$ or $n \geq m$.

If $n \geq m$ then $n + 1 \geq m$. If $n \geq m$ then either $n = m$, and $m \leq n + 1$, or $m > n$, in which case $m \geq n + 1$ by proposition 8.12. Thus $n + 1 \in S(m)$. By induction, $S(m) = \mathbb{N}_0$. $\qquad\square$

As remarked in chapter 4, it follows that $>$ is a strict order relation. That is, for all $m, n, p \in \mathbb{N}_0$,

$$m > n \,\&\, n > p \Rightarrow m > p.$$

Exactly one of $m > n$, $m = n$, $m < n$ is true (trichotomy law). The next result is almost a converse to proposition 8.6.

Proposition 8.14: For all $m, n, p, q \in \mathbb{N}_0$,

(a) $m + q > n + q \Rightarrow m > n$,
(b) $q \neq 0, mq > nq \Rightarrow m > n$.

Proof: (a) If $m \not> n$ then $m \leq n$ by trichotomy. But $m \leq n$ implies $m + q \leq n + q$ by proposition 8.9(a). This contradicts the hypothesis, so part (a) is proved. Part (b) follows a similar format. $\qquad\square$

Proposition 8.14 is of course valid when $>$ is replaced by $\geq$, and is then an exact converse to proposition 8.6.

Uniqueness of $\mathbb{N}_0$

The set $\mathbb{N}_0$, its arithmetic, and order are essentially unique in a very precise sense. As a down-to-earth illustration, the French counting system 'un, deux, trois, ... ', while undeniably *different* from the English 'one, two, three, ... ', possesses the same arithmetical structure. To see this, we observe that translating from French to English by replacing 'un', by 'one', 'deux' by 'two', and so on, turns valid French arithmetic into valid English arithmetic, and conversely. It is the same with $\mathbb{N}_0$.

Suppose that we can find another set $\mathbb{N}_0'$ with a function $s' : \mathbb{N}_0' \to \mathbb{N}_0'$ satisfying the corresponding axioms (N'1)–(N'3). Then we define $\phi : \mathbb{N}_0 \to \mathbb{N}_0'$ by

$$\phi(0) = 0'$$
$$\phi(s(n)) = s'(\phi(n))$$

for all $n \in \mathbb{N}_0$. This function exists by the recursion theorem, as does $\varphi : \mathbb{N}_0' \to \mathbb{N}_0$ given by

$$\varphi(0') \quad = 0$$
$$\varphi(s'(m)) = s(\varphi(m))$$

for all $m \in \mathbb{N}_0'$. A simple induction proof now shows that ϕ and φ are mutual inverses. Let $S = \{n \in \mathbb{N}_0 \mid \varphi\phi(n) = n\}$ to show that $\varphi\phi = 1_{\mathbb{N}_0}$, and similarly prove that $\phi\varphi = 1_{\mathbb{N}_0'}$. Induction on n also shows that

$$\phi(m + n) = \phi(m) + \phi(n)$$
$$\phi(mn) = \phi(m)\phi(n)$$

and

$$m \geq n \Rightarrow \phi(m) \geq \phi(n).$$

Thus the bijection ϕ between $\mathbb{N}_0$ and $\mathbb{N}_0'$ *preserves* the arithmetic and order: we can use it to 'translate' valid results in one into valid results in the other.

Such a bijection is called an *order isomorphism*. The word 'isomorphism' alone is normally used for a bijection that preserves all relevant arithmetical (algebraic) operations. The word 'order' is used to emphasise that the ordering is also preserved. This usage extends to a variety of mathematical systems.

In this sense there is only one possible structure for a system satisfying (N1)–(N3): all such systems are order isomorphic. The whole ethos of the natural numbers is encapsulated in three simple axioms.

One system that we expect to satisfy these axioms is our intuitive concept $\mathbf{N} \cup \{0\}$, so this ought to correspond in the obvious way to $\mathbb{N}_0$. The vital difference is that the properties we expect of $\mathbf{N} \cup \{0\}$ have been built up by example and experience, whereas those of $\mathbb{N}_0$ have been deduced logically from the axioms. Thus all of the usual properties that we expect of $\mathbf{N} \cup \{0\}$ can be given a rigorous justification in $\mathbb{N}_0$. We could, for example, name the elements of $\mathbb{N}_0$ using decimal notation and calculate addition and multiplication tables. At this stage it is more profitable to omit such technicalities on the understanding that they are routine.

Counting

As in real life, we can count using natural numbers. Let

$$\mathbb{N}(n) = \{m \in \mathbb{N} \mid 1 \leq m \leq n\}$$

for $n \in \mathbb{N}$, and let

$$\mathbb{N}(0) = \varnothing.$$

A set X is said to have n elements ($n \in \mathbb{N}_0$) if there is a bijection

$$f : \mathbb{N}(n) \to X.$$

This models the primitive idea of counting. If we point to the elements $f(1)$, $f(2), \ldots, f(n)$ in turn and call out '1, 2, ..., n' then this is precisely how we count.

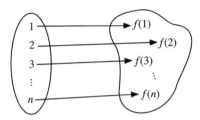

Fig. 8.1 Counting

The useful notational device $\mathbb{N}(0) = \varnothing$ lets us apply the process to the empty set as well. If a set has n elements for some $n \in \mathbb{N}_0$ then it is said to be *finite*; otherwise it is *infinite*.

This manner of counting does not depend on the order in which we count the elements of the set. That is, given a bijection $f : \mathbb{N}(n) \to X$ and a bijection $g : \mathbb{N}(m) \to X$, we always have $m = n$. To see this, let $\varphi = f^{-1}g$. Then $\varphi : \mathbb{N}(m) \to \mathbb{N}(n)$ is a bijection. We prove by induction that if there is a bijection between $\mathbb{N}(n)$ and $\mathbb{N}(m)$, then $m = n$.

This is certainly true for $m = 0$. Suppose it is true for some $m \in \mathbb{N}_0$, and consider a bijection

$$\theta : \mathbb{N}(m+1) \to \mathbb{N}(k).$$

Now $k \neq 0$, or else $m + 1 = 0$ which contradicts (N1). Hence $k = n + 1$ for some $n \in \mathbb{N}_0$. We now construct a bijection $\theta^* : \mathbb{N}(m+1) \to \mathbb{N}(n+1)$ for which $\theta^*(m+1) = n + 1$. If it is already the case that $\theta(m+1) = n + 1$ then we take $\theta^* = \theta$. If not then $\theta(q) = n + 1$ for some $q \leq n$, and we define

$$\theta^*(q) = \theta(m+1)$$
$$\theta^*(m+1) = n+1$$
$$\theta^*(r) = \theta(r) \text{ otherwise.}$$

Restrict θ^* to a map

$$\theta^*|_{\mathbb{N}(m)} : \mathbb{N}(m) \to \mathbb{N}(n).$$

This is clearly a bijection, so by induction $m = n$. Hence $m + 1 = n + 1 = k$, completing the induction step.

This validates the intuitive idea of counting within the formal system.

Von Neumann's Brainwave

As a diversion, we now mention John von Neumann's brilliant method of describing natural numbers, announced in 1923. It is particularly suitable for counting, the number n being defined as a specific *set* with n elements.

To start, there is only one choice for a set with 0 elements, so we put

$$0_v = \varnothing.$$

(Here the suffix v stands for von Neumann.) We now have one object, namely 0_v, so we define

$$1_v = \{0_v\},$$

manifestly a set with 1 element. Now we have two objects 0_v and 1_v, so we define

$$2_v = \{0_v, 1_v\}.$$

It is now clear how to continue. Note that

$$\{0_v, 1_v\} = \{0_v\} \cup \{1_v\} = 1_v \cup \{1_v\}.$$

Having described

$$n_v = \{0_v, 1_v, \ldots, (n-1)_v\}$$

we define

$$\begin{aligned}
(n+1)_v &= n_v \cup \{n_v\} \\
&= \{0_v, \ldots, (n-1)_v\} \cup \{n_v\} \\
&= \{0_v, \ldots, n_v\}.
\end{aligned}$$

This procedure can be made more formal as follows. For any set X we let

$$\sigma(X) = X \cup \{X\}$$

be the *successor* of X. This has the bizarre property

$$X \in \sigma(X) \text{ and } X \subseteq \sigma(X).$$

Now a set Ω whose elements are sets is called *inductive* if

$$\varnothing \in \Omega$$
$$X \in \Omega \Rightarrow \sigma(X) \in \Omega.$$

To avoid an 'and so on . . .' definition, von Neumann postulated:

Axiom of Infinity: There exists an inductive set Ω.

This set Ω may be bigger than we require. But if we let $\mathbb{N}_v$ be the intersection of all inductive subsets of Ω, then it is the smallest inductive subset. Hence if $S \subseteq \mathbb{N}_v$ and S is inductive, it follows that $S = \mathbb{N}_v$.

Since $\mathbb{N}_v$ is inductive, we have $\varnothing \in \mathbb{N}_v$, and $X \in \mathbb{N}_v \Rightarrow \sigma(X) \in \mathbb{N}_v$, so $\sigma : \mathbb{N}_v \to \mathbb{N}_v$ is a function. Also $\varnothing \neq \sigma(n)$ for any $n \in \mathbb{N}_v$, since $n \in \sigma(n)$. We shall prove σ is injective.

First note that if $m, n \in \mathbb{N}_v$ and $m \in n$, then $m \subseteq n$. For let

$$S = \{n \in \mathbb{N}_v \mid m \in n \Rightarrow m \subseteq n\}.$$

Trivially $\varnothing \in S$. Suppose $n \in S$ and $m \in \sigma(n)$. Then either $m \in n$ or $m = n$. In either case $m \subseteq n \cup \{n\} = \sigma(n)$. Hence S is an inductive subset of $\mathbb{N}_v$, so $S = \mathbb{N}_v$.

Now suppose that $\sigma(m) = \sigma(n)$. Then $m \cup \{m\} = n \cup \{n\}$. Thus $m \in n \cup \{n\}$ and either $m \in n$ or $m = n$. By the above remark, $m \subseteq n$. Similarly $n \subseteq m$, hence $m = n$ and σ is injective.

Gathering together these remarks, we find that $\mathbb{N}_v$ is a set, $\sigma : \mathbb{N}_v \to \mathbb{N}_v$ is a function, $\varnothing \in \mathbb{N}_v$, and

  (i) $\varnothing \neq \sigma(n)$ for any $n \in \mathbb{N}_v$,
  (ii) $\sigma(m) = \sigma(n) \Rightarrow m = n$,
  (iii) if $S \subseteq \mathbb{N}_v, \varnothing \in S$, and $n \in S \Rightarrow \sigma(n) \in S$, then $S = \mathbb{N}_v$.

These are the same as the Peano axioms, with $\mathbb{N}_v$ in place of $\mathbb{N}, \sigma$ in place of s, and $\varnothing$ in place of 0. So von Neumann's idea gives an alternative foundation for the natural numbers, and his axiom of infinity acts as a substitute for the existence axiom for the natural numbers. We could have used this approach instead. However, the simplest way to count in von Neumann's system is to say that a set X has n elements if there is a bijection $f : n_v \to X$, that is,

$$f : \{0_v, 1_v, \ldots, (n-1)_v\} \to X.$$

This corresponds to counting '$0_v, 1_v, \ldots, (n-1)_v$' rather than the more primitive '$1, 2, 3, \ldots, n$' to which we are accustomed.

Other Forms of Induction

Sometimes the induction step in a proof by induction needs more than the assumption that $P(n)$ is true in order to deduce $P(n+1)$. For example, we may need to know the truth of $P(1), P(2), \ldots, P(n)$ before being able to pass to $P(n + 1)$. This situation is governed by the so-called General Principle of Induction. If

(GP1) $P(0)$ is true,

(GP2) the truth of $P(m)$ for all $m \in \mathbb{N}_0$ with $m \leq n$ implies the truth of
$P(n + 1)$,

then $P(n)$ is true for all $n \in \mathbb{N}_0$.

At first sight this seems to be a genuine extension of the induction principle, because the second statement seems to use more information. But if we let $Q(n)$ be the predicate

$$P(0)\ \&\ P(1)\ \&\ \ldots\ \&\ P(n),$$

or more formally,

'for all $m \in \mathbb{N}_0$, $m \leq n$, $P(m)$ is true',

then we find that (GP1) and (GP2) become

(i) $Q(0)$ is true,

(ii) the truth of $Q(n)$ implies the truth of $Q(n + 1)$.

Thus the disguise of the 'general' principle is exposed: it is just the ordinary principle for $Q(n)$, and in theory it is no more general than the usual principle of induction. In practice, of course, it sometimes leads to simpler proofs. With it we can prove a highly useful variant of the induction principle. First, we say that a set S has a *least element a* if $a \in S$ and $a \leq s$ for all $s \in S$. Then we can state:

Theorem 8.15 (Well Ordering Principle): Every non-empty subset $S \subseteq \mathbb{N}_0$ has a least element.

Proof: We have to show that if $\varnothing \neq S \subseteq \mathbb{N}_0$ then there exists $a \in S$ such that for all $s \in S$ we have $a \leq s$. For a contradiction, suppose no such a exists. Let $P(n)$ be the predicate $n \notin S$. Then $P(0)$ is true, for $0 \in S$ would imply that 0 is the least element of S by lemma 8.10. Now suppose that $P(m)$ is true for all $m \leq n$, so that if $m \leq n$ then $m \notin S$. If $s \in S$ then $s > n$, so $s \geq n + 1$ by proposition 8.12. We could not have $n + 1 \in S$ since it would then be a least element, so $n + 1 \notin S$ and $P(n + 1)$ is true. By the general principle of induction $P(n)$ is true for all n, that is, S is empty. This is a contradiction. $\square$

Another variation of the induction principle starts not at 0 but at some other $k \in \mathbb{N}_0$. If

$P(k)$ is true, and

the truth of $P(m)$ for $m \geq k$ implies the truth of $P(m + 1)$,

then we may deduce that $P(n)$ is true for all $n \geq k$.

This reduces to the usual induction principle on putting $Q(n) = P(n + k)$. Most often we meet this with $k = 1$. But in the next proposition, which we shall need elsewhere, we require $k = 3$.

Proposition 8.16 (General Associative Law): If $a_1, \ldots, a_n \in \mathbb{N}_0$, then the sum $a_1 + \cdots + a_n$ takes the same value independently of the manner in which brackets are inserted.

Proof: If $n = 3$, there are only two methods of bracketing, namely $(a_1 + a_2) + a_3$ and $a_1 + (a_2 + a_3)$. These are equal by theorem 8.5(a). Suppose the proposition is true for some n. Then without ambiguity we may omit all brackets from a sum of n or fewer numbers. We must therefore consider

$$(a_1 + \cdots + a_k) + (a_{k+1} + \cdots + a_{n+1})$$

and show that the value of this is independent of k. Let

$$a = a_1 + \cdots + a_k$$
$$b = a_{k+1} + \cdots + a_n$$
$$c = a_{n+1}.$$

Then the expression is equal to

$$a + (b + c) = (a + b) + c$$
$$= (a_1 + \cdots + a_n) + a_{n+1}$$

which does not depend on k. This completes the induction step. $\quad\square$

A similar proof works when addition is replaced by multiplication.

Division

Given $m, n \in \mathbb{N}_0$ with $n \neq 0$, it is not always possible to divide n into m and obtain a solution in $\mathbb{N}_0$. For this to happen m must be a multiple of n, that is $m = qn$ for some $q \in \mathbb{N}_0$. If it does not happen, then the division process will yield a remainder.

Theorem 8.17 (Division Algorithm): Given $m, n \in \mathbb{N}_0$ with $n \neq 0$, there exist unique elements $q, r \in \mathbb{N}_0$ such that $m = qn + r$ and $r < n$.

Proof: Use induction on m. Let

$$S = \{m \in \mathbb{N}_0 \mid m = qn + r \text{ for } q, r \in \mathbb{N}_0, r < n\}.$$

Since $0 = 0n + 0$, we have $0 \in S$. Suppose $m \in S$. Then $m = qn + r$ with $r < n$, and

$$m + 1 = qn + r + 1. \tag{8.10}$$

Now $r < n$ implies $r + 1 \leq n$. So either $r + 1 = n$, when (8.10) becomes

$$m + 1 = (q + 1)n + 0,$$

or $r + 1 < n$, when (8.10) becomes

$$m + 1 = qn + (r + 1) \text{ with } r + 1 < n.$$

In either case, $m + 1 \in S$, so by induction $S = \mathbb{N}_0$.

To show that q, r are unique, suppose that

$$m = qn + r = q'n + r'$$

where $r, r' < n$. Then

$$qn \leq m < (q + 1)n$$
$$q'n \leq m < (q' + 1)n.$$

Hence, by transitivity of the order relation, $qn < (q' + 1)n$, so by proposition 8.13, $q < q' + 1$. Then proposition 8.12 implies that $q \leq q'$. Similarly $q' \leq q$, so $q = q'$. Proposition 8.6(a) now implies that $r = r'$. $\qquad \square$

Factorisation

We can now discuss factorisation into primes, and in particular prove uniqueness. Only non-zero numbers are of interest, so for the remainder of this chapter we work in $\mathbb{N} = \mathbb{N}_0 \setminus \{0\}$. First some straightforward definitions are required.

We say that $k \in \mathbb{N}$ is a *factor* or *divisor* of $m \in \mathbb{N}$ if there exists $s \in \mathbb{N}$ such that $m = ks$. We write $k \mid m$. Trivially 1 and m are factors of m; any other factor is called a *proper factor*. We call m prime if $m \neq 1$ and m has no proper factors. (We exclude 1 for convenience, for example in the unique factorisation theorem which follows.) It is easily seen that a factor k of m must lie in the range $1 \leq k \leq m$, for if $k > m$ then since $s \geq 1$ we find that $ks > m$. A proper factor therefore lies in the range $1 < k < m$.

If k is a factor of two numbers $m, n \in \mathbb{N}$, it is called a *common factor*. Now 1 is always a common factor; if it is the only one we say that m and n are *coprime*. Rather than characterising the highest common factor as the largest of the common factors (which indeed it is) we choose to define it in a more useful way.

Definition 8.18: We say that $h \in \mathbb{N}$ is the *highest common factor* of $m, n \in \mathbb{N}$ if h is a common factor with the property that any other common factor k must be a factor of h. We write

$$h = \mathrm{hcf}\,(m, n).$$

The Euclidean Algorithm

The simplest way to prove that any two non-zero natural numbers have a highest common factor is to calculate it explicitly. There is a method for doing this, called the *Euclidean algorithm* for historical reasons, which depends on the following two facts:

(i) If $r_1 = q_1 r_2$ then $r_2 = \mathrm{hcf}\,(r_1, r_2)$.
(ii) If $r_1 = q_1 r_2 + r_3$ with $r_3 \neq 0$, then $\mathrm{hcf}\,(r_1, r_2) = \mathrm{hcf}\,(r_2, r_3)$.

The proofs are easy exercises using the definition of hcf, and in particular (ii) is true since the equation $r_1 = q_1 r_2 + r_3$ shows that any common factor of r_1 and r_2 must also divide r_3, and any common factor of r_2 and r_3 must also divide r_1.

To find the hcf of r_1 and r_2, use the division algorithm repeatedly to find q_i, r_i such that

$$
\begin{aligned}
r_1 &= q_1 r_2 + r_3 && (r_3 < r_2) \\
r_2 &= q_2 r_3 + r_4 && (r_4 < r_3) \\
&\quad \cdots \\
r_i &= q_i r_{i+1} + r_{i+2} && (r_{i+2} < r_{i+1}) \\
&\quad \cdots
\end{aligned}
$$

Since $r_2 > r_3 > r_4 > \ldots$ the process cannot continue indefinitely, for the well-ordering principle tells us that the set of numbers concerned has a least element. Therefore at some stage $r_{i+2} = 0$, $r_{i+1} \neq 0$. This value of r_{i+1} is a highest common factor for r_1 and r_2. This is a consequence of statements (i) and (ii) above, which show that

$$\mathrm{hcf}\,(r_1, r_2) = \mathrm{hcf}\,(r_2, r_3) = \cdots = \mathrm{hcf}\,(r_i, r_{i+1}) = r_{i+1}.$$

As an example, we find the hcf of 612 and 221 (allowing the usual operations of arithmetic as part of our contextual technique, since we have seen that they may be formalised within $\mathbb{N}_0$):

$$612 = 2 \times 221 + 170$$
$$221 = 1 \times 170 + 51$$
$$170 = 3 \times 51 + 17$$
$$51 = 3 \times 17$$

Hence $\mathrm{hcf}(612, 221) = 17$.

Note that this method yields the hcf *without* factorising the numbers into primes, unlike the method often taught in schools.

Proposition 8.19: If h is the hcf of $r_1, r_2 \in \mathbb{N}$, and $n \in \mathbb{N}$, then the hcf of nr_1 and nr_2 is nh.

Proof: If we take the steps in the Euclidean algorithm for hcf (r_1, r_2), as written out above, and multiply through by n, we obtain a system of equations

$$nr_1 = q_1 nr_2 + nr_3 \quad (nr_3 < nr_2)$$
$$nr_2 = q_2 nr_3 + nr_4 \quad (nr_4 < nr_3)$$
$$\ldots$$
$$nr_i = q_i nr_{i+1} \quad \left(\text{recalling that } r_{i+2} = 0\right).$$

Uniqueness of the remainder at each stage implies that this is the Euclidean algorithm for hcf (nr_1, nr_2), so the result is

$$nr_{i+1} = n \times \mathrm{hcf}(r_1, r_2). \qquad \square$$

From this follows a crucial result:

Lemma 8.20: If $m, n \in \mathbb{N}$ and p is a prime dividing mn, then either p divides m or p divides n.

Proof: Suppose that p does not divide m. Since p is prime, its only factors are $1, p$; so the hcf of p and m must be 1. By proposition 8.19 the hcf of nm and np is n. But p divides nm and np, so the definition of hcf implies that p divides n. $\qquad \square$

Corollary 8.21: If $m_1, \ldots, m_r \in \mathbb{N}$ and a prime p divides $m_1, \ldots, m_r$, then p divides at least one of $m_1, \ldots, m_r$.

Proof: Use induction on $r \geq 2$. $\qquad \square$

The final theorem of this chapter states, in formal terms, that the factorisation of a natural number into primes is unique, except for the possibility of writing the factors in a different order.

Theorem 8.22 (Uniqueness of Prime Factorisation): Suppose that $m \in \mathbb{N}, m \geq 2$, and

$$m = p_1^{e_1} \ldots p_r^{e_r} = q_1^{f_1} \ldots q_s^{f_s}$$

for primes p_i, q_j and natural numbers $e_i, f_j \geq 1$. Then $r = s$, and there is a bijection

$$\varphi : \{1, \ldots, r\} \to \{1, \ldots, s\} \text{ such that } p_i = q_{\varphi(i)} \text{ and } e_i = f_{\varphi(i)} \text{ for each } i.$$

Proof: Use induction on $k = e_1 + \cdots + e_r$. If $k = 1$ then $m = p_1, r = 1, e_1 = 1$. Now p_1 divides the product of the q_j's; hence, by corollary 8.21, p_1 divides q_i for some i. Since q_i is prime, $p = q_i$. Using proposition 8.6(b) we may divide through by p_1, obtaining

$$1 = q_1^{f_1} \ldots q_i^{f_i - 1} \ldots q_s^{f_s}$$

which is possible only if $s = 1$, $f_1 = 1$. Hence the two factorisations are given by $m = p_1 = q_1$, and φ may be taken to be the identity.

Now suppose the result true for k, and suppose $e_1 + \cdots + e_r = k + 1$. As for $k = 1$, we have $p_1 = q_i$ for some i. It follows that $e_1 = f_i$, or else, dividing out powers of p_1 using proposition 8.6(b), one side would be divisible by p_1 and the other not. Now we can divide out all powers of p_1 that occur, to get

$$p_2^{e_2} \ldots p_r^{e_r} = q_1^{f_1} \ldots q_{i-1}^{f_{i-1}} q_{i+1}^{f_{i+1}} \ldots q_s^{f_s}.$$

By induction, $r - 1 = s - 1$, and there is a bijection $\varphi : \{2, \ldots, r\} \to \{1, \ldots, i-1, i+1, \ldots, s\}$ such that $p_j = q_{\varphi(j)}$ and $e_j = f_{\varphi(j)}$ for $j = 2, \ldots, r$. It remains only to extend φ to the full set $\{1, \ldots, r\}$ by defining

$$\phi(1) = i$$
$$\phi(j) = \phi(j) \quad \text{for } j = 2, \ldots, r,$$

and the induction step is proved. $\qquad\square$

Reflections

In this chapter we have made significant progress towards a formal approach to mathematics based on set-theoretic definitions and proof. At the beginning of the book you will have had a natural view of the arithmetic of whole numbers based on your experience. You knew all kinds of things, such as the idea that it didn't matter how many terms you were adding together, you could perform the addition in any order and you would always get the same result. Essentially, your experience convinced you of this general property. However, in this chapter we have been able to reduce the whole of arithmetic to a system that satisfies just three axioms (N1), (N2), (N3) and, on the assumption that such a system exists, we have been able to prove all the usual properties of arithmetic from these three axioms. On occasion the journey may have been tortuous, because we needed to base all our arguments on the explicit formal properties that are either axioms, definitions, or theorems proved logically from the axioms and definitions. Now we have established a rich collection of properties of natural numbers that we can use as a foundational context to build new theory.

In future chapters we will use the same techniques to formulate set-theoretic axiomatic structures and further definitions within those structures, safe in the knowledge that properties proved in a given axiomatic structure will continue to hold in new situations which satisfy the given axioms and definitions. We will also use previously established results as part of our technique without explicitly revisiting the detail already proven wherever we are safe in the knowledge that the detail could be filled in as required. This enables us to focus on new ideas and produce increasingly sophisticated theories without obscuring the big picture with established detail.

Exercises

1. Define m^n for $m, n \in \mathbb{N}_0$ by

$$m^0 = 1, m^{n+1} = m^n m.$$

Prove by suitable induction arguments that

$$m^{n+r} = m^n m^r$$
$$m^{nr} = (m^n)^r$$
$$(mn)^r = m^r n^r.$$

2. A sequence of natural numbers is a function $s : \mathbb{N} \to \mathbb{N}_0$. Write s_n instead of $s(n)$ and denote s by (s_n). Given a sequence (s_n), the nth *partial sum* σ_n of (s_n) is defined recursively by

$$\sigma_1 = s_1, \ \sigma_{n+1} = \sigma_n + s_{n+1}.$$

The sum σ_n is also written as $\sigma_n = s_1 + s_2 + \cdots + s_n$.
   Prove by induction that
   (a) $1 + 2 + \cdots + n = \frac{1}{2}n(n+1)$
   (b) $1^2 + 2^2 + \cdots + n^2 = \frac{1}{6}n(n+1)(2n+1)$
   (c) $1^3 + 2^3 + \cdots + n^3 = \frac{1}{4}n^2(n+1)^2$.

3. Define $n!$ for $n \in \mathbb{N}_0$ by

$$0! = 1, (n + 1)! = n! \, (n + 1).$$

Prove by induction on n that $(n - r)!r!$ divides $n!$ for all $0 \le r \le n$.

For all $n, r \in \mathbb{N}_0, 0 \le r \le n$, define $\begin{pmatrix} n \\ r \end{pmatrix} \in \mathbb{N}$ to be $\dfrac{n!}{(n - r)!r!}$.

Show that

$$\begin{pmatrix} n \\ 0 \end{pmatrix} = 1, \; \begin{pmatrix} n \\ 1 \end{pmatrix} = n, \; \begin{pmatrix} n \\ r \end{pmatrix} = \begin{pmatrix} n \\ n - r \end{pmatrix}$$

and

$$\begin{pmatrix} n \\ r \end{pmatrix} + \begin{pmatrix} n \\ r - 1 \end{pmatrix} = \begin{pmatrix} n + 1 \\ r \end{pmatrix}.$$

Use the last equality to prove by induction that for all $a, b, n \in \mathbb{N}_0$:

$$(a + b)^n = a^n + na^{n-1}b + \cdots + \begin{pmatrix} n \\ r \end{pmatrix} a^{n-r}b^r + \cdots + \begin{pmatrix} n \\ n \end{pmatrix} b^n.$$

4. Prove by induction, or otherwise,

(a) $1 \times 1! + 2 \times 2! + \cdots + n \times n! = (n + 1)! - 1$,

(b) $\begin{pmatrix} n \\ 0 \end{pmatrix} + \begin{pmatrix} n \\ 1 \end{pmatrix} + \cdots + \begin{pmatrix} n \\ n \end{pmatrix} = 2^n$,

(c) $\begin{pmatrix} n \\ 1 \end{pmatrix} + 2 \begin{pmatrix} n \\ 2 \end{pmatrix} + \cdots + n \begin{pmatrix} n \\ n \end{pmatrix} = 2^{n-1}n$.

5. Calculate the highest common factor of 2244 and 2145

(a) by the Euclidean algorithm,

(b) by factorising 2244 and 2145 into prime factors.

6. The *Fibonacci numbers* (u_n) are defined recursively by

$$u_1 = 1, u_2 = 2, u_{n+1} = u_n + u_{n-1}.$$

Calculate u_3, u_4, u_5, u_6, and u_7. Prove that every natural number is a sum of Fibonacci numbers. Is this expression unique?

7. If $x_1, \ldots, x_n$ are real numbers, prove that

$$|x_1| + \cdots + |x_n| \ge |x_1 + \cdots + x_n|.$$

8. Let p/q be a fraction in lowest terms such that

$$\frac{1}{n + 1} < \frac{p}{q} < \frac{1}{n}$$

for a natural number n. Show that $\frac{p}{q} - \frac{1}{n+1}$ is a fraction which, in its lowest terms, has numerator less than p. Hence, by induction, prove that every proper fraction p/q, where $p < q$, can be written as a finite sum of distinct reciprocals

$$\frac{p}{q} = \frac{1}{n_1} + \cdots + \frac{1}{n_k}$$

where $n_1, \ldots, n_k$ are natural numbers.

For example, $\frac{14}{15} = \frac{1}{2} + \frac{1}{3} + \frac{1}{10}$.

Use the technique developed in this question to express $\frac{5}{7}$ as a sum of reciprocals.

9. State and prove analogues of the division algorithm and the Euclidean algorithm for polynomials

$$P(x) = a_n x^n + a_{n-1} x^{n-1} + \cdots + a_0$$

with real coefficients. (Hint: If $a_n \neq 0$, then the degree of $P(x)$ is an element of $\mathbb{N}_0$.)

10. *The Tower of Hanoi* is a puzzle consisting of n discs, of different sizes, which can be placed in three heaps A, B, C. A disc may be 'legally' moved from the top of one pile to the top of another provided that it is not placed on top of a smaller disc. Initially all the discs are placed in one pile A, with the largest at the bottom and in decreasing order of size up the pile; the other two piles are empty. Prove that there exists a sequence of legal moves which will transfer all of the discs to pile B.

11. Are the following valid induction proofs?
    (a) Everybody is bald.

    Proof: By induction on the number n of hairs. A man with no hairs is clearly bald. Adding one hair to a bald man is not enough to make him not bald, so if a man with n hairs is bald, so is a man with $n + 1$ hairs. By induction, however many hairs a man has, he is bald. □

    (b) Everybody has the same number of hairs.

    Proof: By induction on the number of people. If this is 0 or 1, the statement is clearly true. Assume it for n. Take $n + 1$ people, remove one, then by the induction hypothesis the remaining n people have the same number of hairs. Remove a different one: the remaining n people again have the same number of hairs, so the first one removed has the same number of hairs as the rest. Hence all $n + 1$ people have the same number of hairs. □

(c) If n straight lines are drawn across a circular disk, such that no three meet in the same point, then they divide the disc into 2^n parts.

Proof: For $n = 1, 2$, the number of parts is 2, 4. Assume the result true for n. Adding another line divides each region it passes through into two, making 2^{n+1} in all. By induction, the statement is proved. $\square$

(d) $n^2 - n + 41$ is a prime number (positive or negative) for every natural number.

Proof:

$$1^2 - 1 + 41 = 41, \ 2^2 - 2 + 41 = 43,$$
$$3^2 - 3 + 41 = 47, \ 4^2 - 4 + 41 = 53,$$
$$5^2 - 5 + 41 = 61, \ 6^2 - 6 + 41 = 71, \ldots \qquad \square$$

(e) $1 + 3 + 5 + \cdots + (2n - 1) = n^2 + 1$.

Proof: If this is true at n, then add $2n + 1$ to each side to get

$$1 + 3 + 5 + \cdots + (2n - 1) + (2n + 1) = n^2 + 1 + (2n + 1)$$
$$= (n + 1)^2 + 1.$$

This is the same formula with n replaced by $n + 1$, so by induction the formula is true for all natural numbers. $\square$

(f) $2 + 4 + \cdots + 2n = n(n + 1)$.

Proof: If $2 + 2 + \cdots + 2n = n(n + 1)$ then

$$2 + 4 + \cdots + 2n + 2(n + 1) = n(n + 1) + 2(n + 1)$$

so

$$2 + 4 + \cdots + 2(n + 1) = (n + 1)(n + 2).$$

By induction the formula is true for all n. $\square$

12. *Induction with a difference.* The arithmetic mean of the n real numbers. $a_1, \ldots, a_n$ is $(a_1 + \ldots + a_n)/n$ and the geometric mean (if they are all non-negative) is $\sqrt[n]{(a_1 a_2 \ldots a_n)}$. Prove that if $a_1, a_2, \ldots, a_n \geq 0$, then

$$(a_1 + a_2 + \cdots + a_n)/n \geq \sqrt[n]{(a_1 a_2 \ldots a_n)}.$$

You may find that a direct induction proof does not work. Try the approach of Cauchy: Let $P(n)$ be the statement '$(a_1 + \cdots + a_n)/n \geq \sqrt[n]{(a_1 \ldots a_n)}$ for all real numbers $a_1, \ldots, a_n \geq 0$'.

First establish by a standard induction that $0 \le a \le b \Rightarrow 0 \le a^n \le b^n$, and deduce that for $a, b \ge 0$, $a^n \le b^n \Rightarrow a \le b$. If $\sqrt[n]{x}$ denotes the positive nth root of $x \ge 0$, deduce that $x, y \ge 0$, $x \ge y \Leftrightarrow \sqrt[n]{x} \ge \sqrt[n]{y}$.

$P(1)$ is trivial, and $P(2)$ may be established by considering the sign of $\frac{1}{4}(a_1 + a_2)^2 - a_1 a_2$. Now prove $P(n) \Rightarrow P(2n)$. (Hint: Use $P(n)$ for $a_1, \ldots, a_n$ and also for $a_{n+1}, \ldots, a_{2n}$ and fit them together using $P(2)$.) Then prove $P(n) \Rightarrow P(n-1)$. (Given $a_1, \ldots, a_{n-1}$, let $a_n = (a_1 + \cdots + a_{n-1})/(n-1)$ and use $P(n)$ to show that

$$\sqrt[n]{(a_1 \ldots a_{n-1} a_n)} \le a_n.$$

Raise to the nth power and simplify to get $P(n-1)$.)

Now deduce that $P(n)$ is true for all $n \in \mathbb{N}$.

13. Proving a statement $P(n)$ true for all $n \in \mathbb{N}$ cannot always be achieved by a simple induction argument. For example, Goldbach's Conjecture that every even integer is the sum of two primes, $2 = 1 + 1$, $4 = 2 + 2$, $6 = 3 + 3$, $8 = 5 + 3$, $10 = 7 + 3$, ... seems plausible (provided that 1 is considered as a prime). Verify Goldbach's Conjecture for every even integer $2n \le 50$. Can you see any pattern that might be amenable for an induction proof? It is not known whether the conjecture is true or false, but the related Odd Goldbach Conjecture—every odd number ≥ 7 is a sum of three primes—was proved by Harald Helfgott in 2013.

CHAPTER 9

Real Numbers

Our intuitive model **R** of the real numbers motivates the properties that are desirable in a rigorous formulation. There should be two binary operations, addition and multiplication, with arithmetical properties that let us define subtraction and division. There should also be an order relation, appropriately related to addition and multiplication, tailored to take account of negative numbers. Finally, we should include the property that distinguishes the real numbers from other number systems like **Z** and **Q**: *completeness*. This property, which is distinctly more technical, was introduced informally in chapter 2. We show that when these three types of property—arithmetic, order, completeness—are formulated precisely, they specify the real numbers uniquely, much like (N1)–(N3) specify $\mathbb{N}_0$.

There are several ways to express the required properties. The experience of the past century's mathematics is that the following system of axioms is one of the best. We define the formal system $\mathbb{R}$ of real numbers as a complete ordered field. We introduce the axioms in order of difficulty: field, then order, then completeness.

Axioms for the Reals: Let $\mathbb{R}$ be a set, equipped with two binary operations + and . (called addition and multiplication). If a, $b \in \mathbb{R}$ we call $a + b$ the *sum* of a and b and $a.b$ the *product*. For traditional reasons we usually omit the dot and write ab for the product.

(a) *Arithmetic*

A set $\mathbb{R}$ with binary operations + and . is said to be a *field* if for all $a, b, c \in \mathbb{R}$

(A1) $a + b = b + a$
(A2) $a + (b + c) = (a + b) + c$
(A3) There exists $0 \in \mathbb{R}$ such that $a + 0 = a$ for all $a \in \mathbb{R}$
(A4) Given $a \in \mathbb{R}$ there exists $-a \in \mathbb{R}$ such that $a + (-a) = 0$
(M1) $ab = ba$

(M2) $a(bc) = (ab)c$

(M3) There exists $1 \in \mathbb{R}$, such that $1 \neq 0$, and $1a = a$ for all $a \in \mathbb{R}$

(M4) Given $a \in \mathbb{R}$, $a \neq 0$, there exists $a^{-1} \in \mathbb{R}$ such that $aa^{-1} = 1$

(D) $a(b + c) = ab + ac$.

The elements 0 and 1 are called the *zero* and *unit* elements of $\mathbb{R}$. By (A1) and (M1) we also have $0 + a = a, (-a) + a = 0, a1 = 1$, and $(a + b)c = ac + bc$.

We define subtraction by

$$a - b = a + (-b)$$

and division by

$$a/b = ab^{-1} \text{ provided that } b \neq 0.$$

(b) *Order*

A field $\mathbb{R}$ is *ordered* if there exists a subset $\mathbb{R}^+ \in \mathbb{R}$ such that

(O1) $a, b \in \mathbb{R}^+ \Rightarrow a + b, ab \in \mathbb{R}^+$

(O2) $a \in \mathbb{R} \Rightarrow a \in \mathbb{R}^+ \text{or} - a \in \mathbb{R}^+$

(O3) $(a \in \mathbb{R}^+) \& (-a \in \mathbb{R}^+) \Rightarrow a = 0$.

These axioms are designed to relate the order to the arithmetic in a sensible way. The set $\mathbb{R}^+$ corresponds to our intuitive idea of the subset of positive elements (recall that 'positive' includes 0). The usual order relation is then defined by

$$a \geq b \Leftrightarrow a - b \in \mathbb{R}^+.$$

We check later that this really is an order relation.

(c) *Completeness*

Recall the following properties, defined for the intuitive concept **R** in chapter 2:

An element $a \in \mathbb{R}$ is an *upper bound* for a subset $S \subseteq \mathbb{R}$ if $a \geq s$ for all $s \in S$.

A set S with an upper bound is said to be *bounded above*. An element λ of $\mathbb{R}$ is a *least upper bound* (lub) for S if

(i) $\lambda \geq s$ for all $s \in S$ (λ is an upper bound)

(ii) $a \geq s$ for all $s \in S \Rightarrow a \geq \lambda$ (λ is the least among the upper bounds).

We can now state the final *completeness axiom*:

(C) If S is a non-empty subset of $\mathbb{R}$ and S is bounded above, then S has a least upper bound in $\mathbb{R}$.

A structure $\mathbb{R}$ satisfying all 13 of the above axioms, (A1)–(A4), (M1)–(M4), (D), (O1)–(O3), and (C), is called a *complete ordered field*. (Later we prove that such a structure is essentially unique.)

We could introduce a new axiom, the existence of a complete ordered field. However, we do not wish to proliferate axioms unnecessarily. It turns out that once we postulate the existence of a system $\mathbb{N}_0$ satisfying Peano's axioms, we can derive from it a related system $\mathbb{R}$ that is a complete ordered field. We first extend $\mathbb{N}_0$ to construct a formal version $\mathbb{Z}$ of the integers $\mathbf{Z}$; then we extend $\mathbb{Z}$ to construct a formal version $\mathbb{Q}$ of the rational numbers $\mathbf{Q}$. Finally we develop $\mathbb{R}$ from $\mathbb{Q}$. This final construction is technically more difficult, mainly because of that vital completeness axiom, but also because we have 13 axioms to check. Each extension is inspired by the intuitive development that we have already encountered at school level.

This sequence of constructions is a part of our mathematical heritage, and all mathematicians should see it at least once in their lives. When first discovered, these constructions resolved frontier questions about the foundations of mathematics, and in particular answered the question 'what is a number?' In retrospect, however, the main importance of these constructions today is to demonstrate that *the existence of $\mathbb{N}_0$, plus set theory, implies the existence of $\mathbb{R}$.*

The main point to appreciate is that this construction is *possible*. Once it has been performed, everything else can be based on the properties (A1)–(A4), (M1)–(M4), (D), (O1)–(O3), and (C). The construction itself is a hangover from the nineteenth century, when the natural numbers were accepted as the basis of mathematics without enquiring about their logical justification, but real numbers were imperfectly understood and therefore seemed mysterious. At that time, it was important to prove that the real numbers are genuine mathematical objects. That demonstration was effected by constructing $\mathbb{R}$ from $\mathbb{N}_0$. Nowadays, having seen that this can be done, the psychological and philosophical problems involved seem less serious: it is logically equivalent to postulate the existence of $\mathbb{R}$, rather than that of $\mathbb{N}_0$. In fact, it is much more convenient to start with $\mathbb{R}$, because it is straightforward to locate within it a chain of subsets

$$\mathbb{R} \supseteq \mathbb{Q} \supseteq \mathbb{Z} \supseteq \mathbb{N}_0$$

that provides the rationals, integers, and natural numbers. On the other hand, the Peano axioms seem very simple and natural, and we find it easy to believe that such a system exists, whereas the 13 axioms for a complete ordered field are harder to swallow.

Chapter 10 explores this alternative approach to the inner constituents of $\mathbb{R}$, and should make its advantages clear. We begin in this chapter with the construction of $\mathbb{R}$ from $\mathbb{N}_0$.

Preliminary Arithmetical Deductions

Before we begin to build an axiomatic structure $\mathbb{Z}$ for the integers, we can obtain some useful clues. Our intuitive model $\mathbf{Z}$ shows that we should not expect all of the properties of a field to hold. Specifically, not all elements of $\mathbb{Z}$ have multiplicative inverses (reciprocals) in $\mathbb{Z}$, axiom (M4). However, we expect all of the other arithmetical axioms to hold.

Some standard algebraic terminology helps us keep track of which properties are under consideration.

Definition 9.1: A set R having two binary operations satisfying (A1)–(A4), (M1)–(M3), and D is a *ring*; more accurately a *commutative* ring. (The word 'ring' is usually applied to any system satisfying a less restrictive set of axioms omitting (M1). Since we rarely deal with non-commutative rings in this text, we omit 'commutative'.)

If, further, there exists a subset R^+ of R satisfying (O1)–(O3), R is an *ordered ring*.

We now make some elementary deductions from these axioms, which, as well as being useful in their own right, are good practice in the axiomatic style.

Proposition 9.2: If R is a ring and for some $x \in R$, $a + x = a$ for all $a \in R$, then $x = 0$. If $xa = a$ for all $a \in R$, then $x = 1$.

Proof: Put $a = 0$, so that $0 + x = 0$. But $0 + x = x$ by (A3) and (A1), so $x = 0$. Similarly $x = x1 = 1$. $\qquad\square$

This proposition shows that the zero and unity elements of R are unique: no other elements have similar properties. In the same way the negative $-a$ of an element a is uniquely determined:

Proposition 9.3: If $x + a = 0$ for elements x, a of a ring R, then $x = -a$.

Proof:

$$
\begin{aligned}
x &= x + 0 & &\text{by (A3)} \\
  &= x + (a + (-a)) & &\text{by (A4)} \\
  &= (x + a) + (-a) & &\text{by (A2)} \\
  &= 0 + (-a) & &\text{since } x + a = 0 \\
  &= -a & &\text{by (A1) and (A3).}
\end{aligned}
$$
$\qquad\square$

If R is a field then multiplicative inverses are uniquely determined (for non-zero elements) and the proof is analogous.

Proposition 9.4: If R is a ring, then for all $a \in R$, $-(-a) = a$.

Proof: By definition, $a + (-a) = 0$. By proposition 9.3, $a = -(-a)$. □

Proposition 9.5: If R is a ring then $a0 = 0a = 0$ for all $a \in R$.

Proof:

$$a0 = a(0 + 0) \quad \text{by(A3)}$$
$$= a0 + a0 \quad \text{by(D)}.$$

Adding $-(a0)$ to each side, we obtain

$$0 = a0 + (-a0) = (a0 + a0) + (-a0) \quad \text{by (A1)}$$
$$= a0 + (a0 + (-a0)) \quad \text{by (A2)}$$
$$= a0 + 0 \quad \text{by (A1)}$$
$$= a0 \quad \text{by (A3)}.$$

Then $0a = 0$ by (M1). □

Proposition 9.6: If R is a ring and $a, b \in R$ then $-(ab) = (-a)b = a(-b)$.

Proof:

$$ab + (-a)b = (a + (-a))\, b \quad \text{by (D) and (M1)}$$
$$= 0b \quad \text{by (A4)}$$
$$= 0 \quad \text{by proposition 9.4}.$$

Hence $(-a)b = -(ab)$ by proposition 9.3. The rest follows by (M1). □

From here it is easy to make further deductions, such as

$$(-a)(-b) = ab, \ (-1)a = -a.$$

If R is a field we may also prove that $(a^{-1})^{-1} = a$ when $a \neq 0$.

Defining subtraction and division as indicated above, we may also verify the expected properties, for example

$$(-a)/b = a/(-b) = -(a/b)$$
$$(a/b) + (c/d) = (ad + bc)/(bd)$$
$$(a/b)(c/d) = (ac)/(bd).$$

The details are left as exercises which will make more sense if you think them through and explain them to yourself.

Next we look at order properties. At this point we take advantage of the discussion of proof in chapter 7, focusing on the new elements of the theory. The arithmetical properties of addition and multiplication will be considered sufficiently well established that we no longer need to quote chapter and verse when using them. We work in a context where the properties of arithmetic can be used without the need for explicit proof, and focus on the new properties of order.

COMMENT. It is common for students to be very successful in manipulating expressions using the operations of arithmetic, but to make unexpected errors when dealing with order relations. For instance, if we know that $ab > c$, we may be tempted to divide by b to get $a > c/b$. That looks plausible, but it is false if b is negative or zero. In the sections that follow, it is important to operate carefully with order relationships using the formal definitions.

Preliminary Deductions about Order

In this section, R is any ordered ring. Its order relation is defined by

$$a \geq b \Leftrightarrow a - b \in R^+. \tag{9.1}$$

It follows that $a \geq 0 \Leftrightarrow a \in R^+$, so

$$R^+ = \{a \in R \mid a \geq 0\}. \tag{9.2}$$

Using (O1)–(O3) we now establish:

Proposition 9.7: The relation $\geq$ is a weak order on R.

Proof: We must verify the three properties

(WO1) $a \geq b$ & $b \geq c \Rightarrow a \geq c$,
(WO2) Either $a \geq b$ or $b \geq a$,
(WO3) $a \geq b$ & $b \geq a \Rightarrow a = b$.

For (WO1), $a \geq b$ & $b \geq c \Rightarrow a - b$, $b - c \in R^+$. By (O1), $(a - b) + (b - c) \in R^+$, so $a - c \in R^+$ so $a \geq c$. (This is our first taste of 'arithmetic without tears': an axiomatic proof that $(a - b) + (b - c) = a - c$ takes several steps, all omitted here.)

For (WO2), (O2) implies that either $a - b \in R^+$ or $b - a = -(a - b) \in R^+$. Therefore $a \geq b$ or $b \geq a$.

For (WO3), if both $a - b$ and $b - a \in R^+$ then $a - b = 0$ by (O3), so $a = b$. $\square$

The order relation behaves appropriately with respect to the arithmetic:

Proposition 9.8: For all $a, b, c, d \in R$,

(a) $a \geq b \,\&\, c \geq d \Rightarrow a + c \geq b + d$,

(b) $a \geq b \geq 0 \,\&\, c \geq d \geq 0 \Rightarrow ac \geq bd$.

Proof: Translate the definition of $\geq$ using (9.1), and do the arithmetic. $\square$

In the definition of an ordered field (or ordered ring) we can replace (O1)–(O3) by the properties stated in propositions 9.6 and 9.7, and use the relation $\geq$ to define the set R^+ by working (9.2) the other way round. Which approach we use is a matter of taste.

The modulus can be defined in an ordered ring by setting

$$|a| = \begin{cases} a \text{ if } a \in R^+ \\ -a \text{ if } -a \in R^+. \end{cases}$$

It can then be proved that $|a| \geq 0$ for all $a \in R$ and, by repeating the argument of chapter 2 in this formal context, that

$$|a + b| \leq |a| + |b|,$$
$$|ab| = |a|\,|b|.$$

Now we have enough technique to carry out the construction of the integers, rationals, and reals.

Construction of the Integers

To get from $\mathbb{N}_0$ to the integers we must introduce negative elements. In fact, we consider differences $m - n$ of natural numbers. These differences are definable *as natural numbers* when $m \geq n$, but not when $m < n$. Our task is to give $m - n$ a meaning no matter which of m or n is larger.

The idea is to relate subtraction to addition, which of course is how we were taught about subtraction in the first place. If $m, n, r, s \in \mathbb{N}_0$ and $m \geq n$, $r \geq s$, then

$$m - n = r - s \Leftrightarrow m + s = r + n.$$

The right-hand side makes sense *without* restrictions on m, n, r, s. This gives a clue. To construct things that behave like differences $m - n$, take the set $\mathbb{N}_0 \times \mathbb{N}_0$ of ordered pairs (m, n), where $m, n \in \mathbb{N}_0$, and define a relation $\sim$ on this set by

$$(m, n) \sim (r, s) \Leftrightarrow m + s = r + n.$$

It turns out that $\sim$ is an *equivalence* relation, and the proof requires only arithmetic in $\mathbb{N}_0$. We can then define the integers $\mathbb{Z}$ to be the set of equivalence classes for $\sim$. The equivalence class of (m, n) corresponds to our intuitive concept of the difference $m - n$, and the formal proof runs as follows.

Let $\langle m, n \rangle$ denote the equivalence class of (m, n). By the definition of $\sim$,

$$\langle m, n \rangle = \langle r, s \rangle \Leftrightarrow m + s = r + n.$$

Define addition and multiplication on $\mathbb{Z}$ by

$$\langle m, n \rangle + \langle p, q \rangle = \langle m + p, n + q \rangle,$$
$$\langle m, n \rangle \langle p, q \rangle = \langle mp + nq, mq + np \rangle.$$

(9.3)

These definitions are motivated by thinking of $\langle m, n \rangle$ as '$m - n$' and translating the sum and product into expressions involving such differences:

$$(m - n) + (p - q) = (m + p) - (n + q),$$
$$(m - n)(p - q) = (mp + nq) - (mq + np).$$

We need to check that the operations (9.3) are well defined in the sense of chapter 4. So suppose that $\langle m, n \rangle = \langle m', n' \rangle$ and $\langle p, q \rangle = \langle p', q' \rangle$. Then $m + n' = m' + n$, $p + q' = p' + q$. Now

$$(m + p) + (n' + q') = (m + n') + (p + q')$$
$$= (m' + n) + (p' + q)$$
$$= (m' + p') + (p + q)$$

Hence $\langle m + p, n + q \rangle = \langle m' + p', n' + q' \rangle$, and addition is well defined. Multiplication is treated in the same way.

It is now a simple but long-winded exercise to show that $\mathbb{Z}$ is an ordered ring, taking

$$\mathbb{Z}^+ = \{\langle m, n \rangle \in \mathbb{Z} \mid m \geq n \text{ in } \mathbb{N}_0\}.$$

Proposition 9.9: With the above operations, $\mathbb{Z}$ is an ordered ring.

Proof: We must check the axioms (A1)–(A4), (M1)–(M3), (D), and (O1)–(O3). In all cases we use the definition of $\mathbb{Z}$ to restate the required property in $\mathbb{N}_0$, and verify it by arithmetic.

(A1) Let $a = \langle m, n \rangle$, $b = \langle p, q \rangle$. Then

$$a + b = \langle m + p, n + q \rangle$$
$$= \langle p + m, q + n \rangle \text{ by arithmetic in } \mathbb{N}_0$$
$$= b + a.$$

(A2)   To prove $ab = ba$, we need to show $\langle m, n \rangle \langle p, q \rangle = \langle p, q \rangle \langle m, n \rangle$. This requires showing that

$$\langle m, n \rangle \langle p, q \rangle = \langle mp + nq, mq + np \rangle$$
$$= \langle p, q \rangle \langle m, n \rangle = \langle pm + qn, qm + pn \rangle,$$

which follows because $mp + nq = pm + qn$ and $mq + np = qm + pn$ in $\mathbb{N}_0$.

(A3)   The simplest way to express 0 as a difference is to form $0 - 0$. So we consider the element $\langle 0, 0 \rangle$ of $\mathbb{Z}$. Now

$$\langle m, n \rangle + \langle 0, 0 \rangle = \langle m + 0, n + 0 \rangle = \langle m, n \rangle.$$

Thus $\langle 0, 0 \rangle$ acts as a 'zero' element.

(A4)   The additive inverse of $m - n$ ought to be $n - m$, so we compute:

$$\langle m,\ n \rangle + \langle n,\ m \rangle = \langle m + n,\ n + m \rangle = \langle m + n,\ m + n \rangle.$$

Are we in trouble? No, because $(m + n, m + n)$ is *equivalent*, under $\sim$, to $(0,\ 0)$. Therefore $\langle m + n, m + n \rangle = \langle 0, 0 \rangle$, and we have proved that

$$\langle m, n \rangle + \langle n, m \rangle = \langle 0, 0 \rangle,$$

which is what we want. Now $\sim$ is starting to make its presence felt.

Proofs of the remaining axioms for arithmetic follow similar lines. We could easily write them out for you. But if we did, you might simply read them through and commit them to memory to pass a test. To make sense of them, it is time to work through them for yourself. The effort of deriving and explaining the links to yourself is more likely to set up a coherent schema of connections in your mind, which you can build on in the future. Mathematics involves active thinking. It is not a spectator sport. $\square$

The next step is to recover the usual notation for integers as positive or negative natural numbers.

Any element of $\mathbb{Z}^+$ is of the form $\langle m, n \rangle$ where $m \geq n$, so can be written as $\langle m - n, 0 \rangle$. Thus every element of $\mathbb{Z}^+$ is of the form $\langle r, 0 \rangle$ for $r \in \mathbb{N}_0$. Now axiom (O2) tells us that for any $a \in \mathbb{Z}$, either $a \in \mathbb{Z}^+$ or $-a \in \mathbb{Z}^+$, hence either $a = \langle r, 0 \rangle$ or $a = -\langle r, 0 \rangle = \langle 0, r \rangle$.

Define a map $f : \mathbb{N}_0 \to \mathbb{Z}^+$ by $f(n) = \langle n, 0 \rangle$. It is easily seen that f is a bijection, and that

$$f(m + n) = f(m) + f(n),$$
$$f(mn) = f(m)f(n),$$
$$m \geq n \Leftrightarrow f(m) \geq f(n).$$

That is, f is an order isomorphism, in the sense of chapter 8.

This leads to a technical problem, which we have met before. We do not have $\mathbb{N}_0 \subseteq \mathbb{Z}$, as we might have hoped: instead, $\mathbb{N}_0$ is *order isomorphic* to the subset $\mathbb{Z}^+ \subseteq \mathbb{Z}$. The elements of $\mathbb{N}_0$ and $\mathbb{Z}^+$ are definitely *different* mathematical objects: the first is a set of numbers, the second is a set of equivalence classes of ordered pairs. However, they behave in exactly the same manner.

There are various ways to get round this problem. One is to replace the elements of $\mathbb{Z}^+$ by the corresponding elements of $\mathbb{N}_0$, creating a hybrid system with the elements 0, 1, 2, … in $\mathbb{N}_0$ as the non-negative integers and the ordered pairs $\langle m, m + n \rangle$ as the negative integers $-n$. The diagram below should make the idea clear.

Fig. 9.1 $\mathbb{N}_0$ as a subset of $\mathbb{Z}$

This hybrid system contains $\mathbb{N}_0$ as a genuine subset, and extends to include elements of the form $\langle m, m + n \rangle$. Such an element is the additive inverse of n, so we can change notation to $-n$ without getting into trouble.

However, this method is inelegant and lacks the aesthetic simplicity desired in mathematics. This kind of complication will escalate as we go on to construct the rational numbers $\mathbb{Q}$ from $\mathbb{Z}$ and then the real numbers $\mathbb{R}$ from $\mathbb{Q}$. At each stage the smaller number system is *isomorphic* to a subsystem of the larger system, but it is not actually a subset as such. We could use a similar trick to replace a subset of $\mathbb{Q}$ by $\mathbb{Z}$, and a subset of $\mathbb{R}$ by $\mathbb{Q}$, but the elegance of the constructions gets lost.

Mathematicians take a more pragmatic route. They 'identify' $\mathbb{N}_0$ and $\mathbb{Z}^+$, that is, they ignore the technical set-theoretic distinction between them for purposes of arithmetic and order. This causes no harm because these two systems have exactly the same mathematical structure as regards arithmetic and order. If we ignore the distinction, we can consider $\mathbb{N}_0$ to be a subsystem of $\mathbb{Z}$. This fits with how the human mind simplifies the situation by thinking of $\mathbb{N}_0$ and $\mathbb{Z}^+$ as different ways to represent the same underlying mathematical concept.

The mathematical justification of this approach will become clear when we construct the rational numbers and the real numbers. We will then prove that the axioms for a complete ordered field define the real numbers *uniquely*, in the sense that any two systems that satisfy all the axioms are order isomorphic. Therefore, up to isomorphism, there is only *one* complete ordered field. Inside it—genuinely inside as subsets—are systems that correspond to the natural numbers, the integers, and the rational numbers. We can then mentally 'throw away' the set-theoretic scaffolding that we are building in this chapter, by replacing the systems we have constructed by these isomorphic subsystems of $\mathbb{R}$.

In case you're wondering why any of this is necessary: actually, you've encountered similar problems before. When you did fractions, at some stage you had to sort out that 2/1 is the same as 2: some fractions can be whole numbers. Similarly, when you did decimals, you got used to replacing $1 \cdot 0000 \ldots$ by 1. Technically the first is an infinite decimal that just happens to have lots of zeros. It *behaves* like the whole number 1, but it's not written that way. But clearly, thinking of it as being *equal* to 1 does no harm.

From a psychological viewpoint, the real numbers are conceived by the human mind as a unique 'crystalline concept' that has specific properties yet can be represented flexibly in different equivalent forms. In this case the real numbers can be defined axiomatically as a list of 13 axioms, represented geometrically as points on a number line, or symbolically as infinite decimals. If the distinction actually matters, you can always sort it out; usually, it doesn't. Once we have this coherent overall structure, we have a perfect platform from which to view 'the' real numbers as a unique, but flexible, mathematical entity.

To reach that stage, however, we must first go through the technicalities of constructing the rationals from the integers and the reals from the rationals. This process shows that all of these number systems are consequences of the Peano axioms, which characterise the natural numbers.

Construction of Rational Numbers

We construct the rational numbers from the integers by following a similar strategy to the one used to construct the integers from the natural numbers. But what matters now is not the difference $m - n$ between two natural numbers, but the quotient m/n of two integers. So, starting from $\mathbb{Z}$, we must introduce a larger set $\mathbb{Q}$ for which quotients m/n are defined.

To do this, let S be the set of all ordered pairs (m, n) where $m, n \in \mathbb{Z}$ and $n \neq 0$. Define a relation $\sim$ by

$$(m, n) \sim (p, q) \Leftrightarrow mq = np.$$

This is inspired by the property that $m/n = p/q$ if and only if $mq = np$. Now define $\mathbb{Q}$ to be the set of equivalence classes for $\sim$. Anticipating the final result, we use the notation m/n for the equivalence class of (m, n). Define operations by

$$m/n + p/q = (mq + np)/nq,$$
$$(m/n)(p/q) = mp/nq.$$

Theorem 9.10: These operations define the structure of a field on $\mathbb{Q}$.

Proof: The details are left to the reader (offering an opportunity to build up the mental links to create a coherent personal schema for these ideas). First, check that the operations are well defined; then go through the whole list of axioms, one by one. Use the proof of proposition 9.8 as a model.

Once more, it really is important to think this proof through for yourself. We could put it in, but it is seldom helpful to read through someone else's long calculations when they are routine. To help, here's a hint: if $n \neq 0$ the multiplicative inverse of m/n is n/m. $\square$

We have now set up the arithmetic of $\mathbb{Q}$, but not its order relation. We define an ordering by specifying the positive elements:

$$\mathbb{Q}^+ = \{m/n \in \mathbb{Q} \mid m, n \in \mathbb{Z}^+, n \neq 0\}.$$

Theorem 9.11: With the above definition, $\mathbb{Q}$ is an ordered field.

Proof: Once more we want you to construct the idea in your own mind, by thinking through the proof for yourself. $\square$

We want the integers to be a subset of the rationals, but once again this is true only up to isomorphism. It's the old problem of $n/1$ being technically different from n, but behaving in exactly the same way. We solve it by proving that the map $g : \mathbb{Z} \to \mathbb{Q}$ defined by $g(n) = n/1$ preserves the arithmetical operations:

$$g(m + n) = g(m) + g(n)$$
$$g(mn) = g(m)g(n)$$
$$m \geq n \Rightarrow g(m) \geq g(n)$$

for all $m, n \in \mathbb{Z}$. All three are straightforward. Therefore g is an order isomorphism from the natural numbers to the elements of $\mathbb{Z}$ of the form $m/1$.

Since every rational m/n can be written as $(m/1)(1/n) = (m/1)(n/1)^{-1}$, identifying n with $n/1$ does not lead to any conflict in notation, and corresponds to the usual intuitive model. This identification lets us think of $\mathbb{Z}$ as a subsystem of $\mathbb{Q}$, just as the natural numbers can be thought of as a subsystem of the integers.

Construction of Real Numbers

The construction of the real numbers is more complicated, and it can be carried out in several different ways. It is possible, though technically awkward, to construct them as infinite decimals, along the lines of chapter 2. However, we saw there that the use of approximating sequences of rationals has technical advantages. Monotonic sequences are especially easy to handle, but we shall use more general 'Cauchy sequences', which we will define in a moment. As in previous sections, many routine details will be omitted, and for the same reason: the broad outline becomes more easily visible when the details merge into the conceptual background. It remains essential for you to think through the relationships for yourself; to understand them in a coherent way that helps to build up a flexible personal insight into the mathematical structure.

Sequences of Rationals

The main idea when constructing the real numbers is to associate each real number with an infinite sequence of rational numbers, which in some sense form better and better approximations to the real number concerned. Truncating an infinite decimal further and further to the right is one way to do this, but the mathematics is simpler if we avoid being that specific.

As in chapter 5, but replacing the informal **N** by the formal version $\mathbb{N}$, a *sequence* of rationals may be formally defined as a function

$$s : \mathbb{N} \to \mathbb{Q}.$$

We write s_n for $s(n)$ and denote the sequence by $(s_n)_{n \in \mathbb{N}}$ or by $(s_1, s_2, s_3, \dots)$, or just by (s_n).

Let S be the set of all sequences of rationals. We define addition and multiplication within S by

$$(a_n) + (b_n) = (a_n + b_n),$$
$$(a_n)(b_n) = (a_n b_n).$$

Lemma 9.12: With these operations, S is a ring.

Proof: The identity is $(1, 1, 1, \dots)$, the zero $(0, 0, 0, \dots)$, and the additive inverse of (a_n) is $(-a_n)$. All verifications are routine. □

We say 'ring' because S is not a field. If all s_n are non-zero, then (s_n) has a multiplicative inverse $(1/s_n)$. But if any term $s_n = 0$, then this does not work. For example, $(0, 1, 1, \ldots)$ cannot have an inverse $(b_1, b_2, b_3, \ldots)$ since

$$(0, 1, 1, \ldots)(b_1, b_2, b_3, \ldots) = (0, b_2, b_3, \ldots) \neq (1, 1, 1, \ldots).$$

As we saw in chapter 2, every real number may be viewed as the 'limit' of a sequence of rationals. In the present context, we can take over the definition of convergence given in that chapter, provided that we insist that the ε in the definition is rational.

Definition 9.13: A sequence of rationals (s_n) *converges* to $l \in \mathbb{Q}$ if, given any $\varepsilon \in \mathbb{Q}$, $\varepsilon > 0$, there exists $N \in \mathbb{N}$ such that

$$n > N \Rightarrow |s_n - l| < \varepsilon.$$

This definition is not yet satisfactory, however: convergence to a *rational* limit is not what really interests us. It fails to deal with real numbers like $\sqrt{2}$, for example. We need a replacement for 'convergent' that does not specify the limit as such.

For the sake of argument, assume that it makes sense to talk of a sequence of rationals converging to a real limit. Certainly this is so in our intuitive models $\mathbf{Q} \subseteq \mathbf{R}$. The catch is that *formally* we do not know what the limit is. Nonetheless, if (s_n) were to converge to a real number l, then there would exist some N such that

$$|s_n - l| < \varepsilon \text{ for all } n > N.$$

Hence also

$$|s_m - l| < \varepsilon \text{ for all } m > N.$$

Combining the two inequalities we obtain

$$|s_m - s_n| < 2\varepsilon \text{ for all } m, n > \mathbf{N}.$$

Now *this* statement does not involve the hypothetical real number l. But it still captures the idea of convergence.

To tidy things up, we start again with $\frac{1}{2}\varepsilon$ instead of ε, and thereby obtain the essential idea:

Definition 9.14: A sequence (s_n) of rational numbers is a *Cauchy sequence* if for any rational $\varepsilon > 0$ there exists N such that

$$m, n > N \Rightarrow |s_m - s_n| < \varepsilon.$$

Intuitively, the terms of such a sequence get closer and closer together.

This concept is named after Augustin-Louis Cauchy, a prolific nineteenth-century French mathematician who made extensive use of such sequences. However, it was Georg Cantor who first realised how to use such sequences to construct the real numbers using the method presented here.

Cauchy sequences may be considered intuitively as sequences of rational approximations to a real number and this provides the raw material for a formal construction of the real numbers starting from the rationals. The proof requires several lemmas. For the first, we say that a sequence (s_n) is *bounded* if there exists a fixed number M such that $|s_n| \leq M$ for all n.

Lemma 9.15: Every Cauchy sequence is bounded in $\mathbb{Q}$.

Proof: Taking $\varepsilon = 1$ in the definition of a Cauchy sequence, there exists N such that $|s_n - s_m| < 1$ for $m, n > N$. Thus for all $n > N$ we have $|s_n - s_{N+1}| < 1$; that is, $|s_n| < |s_{N+1}| + 1$. Hence, for all $n \in \mathbb{N}$,

$$|s_n| \leq \max \left\{ |s_1|, |s_2|, \ldots, |s_N|, |s_{N+1}| + 1 \right\}. \qquad \square$$

Lemma 9.16: If (a_n) and (b_n) are Cauchy sequences, then so are $(a_n + b_n)$, $(a_n b_n)$, and $(-a_n)$.

Proof: If $\varepsilon > 0$ is rational, there exist N_1 and N_2 such that

$$m, n > N_1 \Rightarrow |a_m - a_n| < \tfrac{1}{2}\varepsilon,$$

$$m, n > N_2 \Rightarrow |b_m - b_n| < \tfrac{1}{2}\varepsilon.$$

So for $m, n > N = \max(N_1, N_2)$ we have

$$
\begin{aligned}
\left| (a_m + b_m) - (a_n + b_n) \right| &= \left| (a_m - a_n) + (b_m - b_n) \right| \\
&\leq |a_m - a_n| + |b_m - b_n| \\
&< \tfrac{1}{2}\varepsilon + \tfrac{1}{2}\varepsilon \\
&= \varepsilon,
\end{aligned}
$$

so $(a_n + b_n)$ is Cauchy.

To show that $(a_n b_n)$ is Cauchy, use lemma 9.15 to show that there exist $A, B \in \mathbb{Q}$ such that $|a_n| < A$ and $|b_n| < B$ for all $n \in \mathbb{N}$. Using a little foresight (the authors have seen this proof before!), given $\varepsilon \in \mathbb{Q}, \varepsilon > 0$, observe that $\varepsilon / (A + B) \in \mathbb{Q}, \varepsilon / (A + B) > 0$. Therefore there exist N_1, N_2 such that

$$m, n > N_1 \Rightarrow |a_m - a_n| < \frac{\varepsilon}{A + B},$$

$$m, n > N_2 \Rightarrow |b_m - b_n| < \frac{\varepsilon}{A + B}.$$

If $m, n > N = \max(N_1, N_2)$ then both inequalities hold, so

$$\begin{aligned}
|a_m b_m - a_n b_n| &= |(a_m - a_n)b_m + a_n(b_m - b_n)| \\
&\leq |a_m - a_n||b_m| + |a_n||b_m - b_n| \\
&< \frac{\varepsilon}{A + B}B + A\frac{\varepsilon}{A + B} \\
&= \varepsilon.
\end{aligned}$$

Therefore $(a_n b_n)$ is Cauchy.

Finally, $(-a_n)$ may be proved Cauchy either by a direct calculation, or by putting $b_n = -1$ for all n in the above. $\qquad\square$

Letting **C** denote the set of all Cauchy sequences, we have:

Proposition 9.17: With addition and multiplication of sequences as defined, **C** is a ring.

Proof: If $(a_n), (b_n) \in \mathbf{C}$ then lemma 9.16 says that $(a_n) + (b_n), (a_n)(b_n)$, and $-(a_n) \in \mathbf{C}$. Clearly the zero sequence $(0, 0, \dots)$ and unit sequence $(1, 1, \dots)$ $\in \mathbf{C}$. Looking at the axioms for a ring we see that this takes care of (A3), (A4), and (M3). The remaining axioms hold since, by lemma 9.12, they hold for all sequences of rationals. $\qquad\square$

However, we still do not have a field, for a sequence like $(0, 1, 1, 1, \dots)$ is Cauchy, non-zero, and has no inverse. To overcome this we take note of another problem: intuitively speaking, different Cauchy sequences can converge to the same limit. We have already encountered this in decimal notation:

$$(1, 1, 1, 1, 1, \dots)$$

and

$$(0.9, 0.99, 0.999, 0.9999, 0.99999, \dots)$$

both converge to 1.

Both difficulties evaporate when we introduce one further concept:

Definition 9.18: A sequence (s_n) of rationals is a *null sequence* if it converges to 0. That is, for all rational $\varepsilon > 0$ there exists N such that $|s_n| < \varepsilon$ whenever $n > N$.

If two sequences (a_n) and (b_n) tend to the same limit l, then it is easy to see that the sequence $(a_n - b_n)$ is null. This inspires an equivalence relation on $\mathbf{C}$:

$$a_n \sim b_n \Leftrightarrow (a_n - b_n) \text{ is null.}$$

To check that this *is* an equivalence relation, observe that the properties $(a_n) \sim (a_n)$ and $a_n \sim b_n \Rightarrow b_n \sim a_n$ are trivial. If $a_n \sim b_n$ and $a_n \sim c_n$ then $(a_n - b_n)$ and $(b_n - c_n)$ are null, that is they converge to 0. So $((a_n - b_n) + (b_n - c_n))$ converges to zero, that is, $(a_n - c_n)$ is null, so $(a_n) \sim (c_n)$.

Definition 9.19: $\mathbb{R}$ is the set of equivalence classes of Cauchy sequences, and the equivalence class containing (s_n) is denoted by $[s_n]$ or $[s_1, s_2, \ldots, s_n, \ldots]$. For $q \in \mathbb{Q}$, $[q, q, \ldots, q, \ldots]$ will also be denoted by $\hat{q} \in \mathbb{R}$.

The alternative notation allows us to distinguish clearly between the equivalence class $[s_n]$ corresponding to a given Cauchy sequence (s_n) and the equivalence class $\hat{s}_n$ corresponding to the specific element s_n for a fixed value of n. For instance, for $s_n = 1/n$,

$$[s_n] = [1, \tfrac{1}{2}, \tfrac{1}{3}, \ldots, \tfrac{1}{n}, \ldots]$$

while

$$\hat{s}_n = [\tfrac{1}{n}, \tfrac{1}{n}, \ldots \tfrac{1}{n}, \ldots].$$

Definition 9.20: The operations of addition and multiplication are transferred to $\mathbb{R}$ by defining

$$[a_n] + [b_n] = [a_n + b_n],$$
$$[a_n][b_n] = [a_n b_n].$$

By now, you should, as a reflex, be wondering whether these operations are well defined. Yes, they are. For if $[a_n] = [a'_n]$ and $[b_n] = [b'_n]$ then $(a_n - a'_n)$ and $(b_n - b'_n)$ are null. Hence $((a_n + b_n) - (a'_n + b'_n))$ is null, so $[a_n + b_n] = [a'_n + b'_n]$.

Multiplication is a little less straightforward. By lemma 9.15 there exist rationals A, B such that

$$|a_n| < A, |b'_n| < B, \text{ for all } n \in \mathbb{N}.$$

Given $\varepsilon > 0$ we can find N_1, N_2 such that

$$n > N_1 \Rightarrow |a_n - a'_n| < \varepsilon/(A + B),$$
$$n > N_2 \Rightarrow |b_n - b'_n| < \varepsilon/(A + B).$$

If $n > N = \max(N_1, N_2)$ then

$$
\begin{aligned}
|a_m b_m - a_n' b_n'| &= |a_m(b_n - b_n') + (a_m - a_n')b_n'| \\
&\le |a_m||b_n - b_n'| + |a_m - a_n'||b_n'| \\
&< A\frac{\varepsilon}{A + B} + \frac{\varepsilon}{A + B}B \\
&= \varepsilon.
\end{aligned}
$$

Thus $(a_n b_n - a_n' b_n')$ is null, so $[a_n b_n] = [a_n' b_n']$.

To show that these operations make $\mathbb{R}$ an ordered field, we need to verify all the field axioms (A1)–(A4), (M1)–(M3), (D) and define an order on $\mathbb{R}$ that satisfies the order axioms (O1)–(O3). Most of these are straightforward, but when we attempt to define the subset $\mathbb{R}^+$ of non-negative elements, we need to take care of the possibility that an equivalence class $[a_n]$ may be non-negative even though some of the individual terms a_n are not. (For instance, we may have $a_1 = -1$, $a_n = 1$ for $n > 1$.) We deal with this problem by showing that if a sequence (a_n) is not null, then after a certain stage (say for some $N_0 \in \mathbb{N}$) later terms a_n (for $n > N_0$) are either all strictly positive or all strictly negative. We make this idea precise through the following definition:

Definition 9.21: A Cauchy sequence (a_n) is *strictly positive* if there is a rational number $\varepsilon > 0$ and an $N_0 \in \mathbb{N}$ such that $a_n > \varepsilon$ for all $n > N_0$. It is *strictly negative* if there is a rational number $\varepsilon > 0$ and an $N_0 \in \mathbb{N}$ such that $a_n < -\varepsilon$ for all $n > N_0$.

We can then prove:

Lemma 9.22: If (a_n) is a Cauchy sequence then it is precisely one of the following:

  (i) a null sequence
  (ii) strictly positive
  (iii) strictly negative.

Proof: Because (a_n) is a Cauchy sequence,

$$\forall \varepsilon \in \mathbb{Q}, \ \varepsilon > 0 \ \exists N_0 : m, n > N_0 \Rightarrow |a_m - a_n| < \varepsilon \qquad (9.4)$$

A Cauchy sequence may be null, as in (i).

If not then (a_n) does *not* tend to zero, and (taking a positive rational value, 2ε), the statement

$$\exists N \in \mathbb{N} \ \forall m \in N : m > N \Rightarrow |a_m| < 2\varepsilon$$

is false, so that the following is true:

$$\forall N \in \mathbb{N} \, \exists m \in \mathbb{N}, m > N : |a_m| \geq 2\varepsilon.$$

In particular,

$$\exists m > N_0 : |a_m| \geq 2\varepsilon, \tag{9.5}$$

and combining this with (9.4) gives

$$n > N_0 \Rightarrow |a_m - a_n| < \varepsilon. \tag{9.6}$$

From (9.5), either $a_m \geq 2\varepsilon$, and (9.6) implies

$$n > N_0 \Rightarrow a_n > \varepsilon,$$

which gives (ii), or $a_m \leq -2\varepsilon$ and (9.6) implies

$$n > N_0 \Rightarrow a_n < -\varepsilon,$$

and this gives (iii).

Summing up, if the Cauchy sequence (a_n) does not satisfy (i), it must satisfy precisely one of (ii) or (iii), as required. $\qquad\square$

Proposition 9.23: With the given operations for $[a_n] + [b_n]$, $[a_n][b_n]$, $\mathbb{R}$ is a field.

Proof: Verification of axioms (A1)–(A4), (M1)–(M3), and (D) is straightforward. The zero element is $[0, 0, 0, \dots]$, the unit element $[1, 1, 1, \dots]$, and the negative of $[a_n]$ is $[-a_n]$.

However, the inverse $1/[a_n]$ requires a little ingenuity. By lemma 9.22, if $[a_n] \neq [0]$ then it must be strictly positive or strictly negative, and, in particular, for $n > N_0$, we must have $a_n \neq 0$. We can then define (b_n) by

$$b_n = \begin{cases} 0 & \text{if } n \leq N_0 \\ 1/a_n & \text{if } n > N_0 \end{cases},$$

so that

$$a_n b_n = \begin{cases} 0 & \text{if } n \leq N_0 \\ 1 & \text{if } n > N_0 \end{cases}.$$

Then $(a_n b_n)$ is a sequence whose terms equal 1 for $n > N_0$, so that $[a_n b_n] = [1, 1, 1, \dots]$ and $[b_n]$ is the inverse of $[a_n]$. This completes the proof that $\mathbb{R}$ is a field. $\qquad\square$

The Ordering on $\mathbb{R}$

Lemma 9.22 lets us define an order on $\mathbb{R}$ in which $[a_n] < 0$ if (a_n) is strictly negative, $[a_n]$ is zero if (a_n) is null, and $[a_n] > 0$ if (a_n) is strictly positive. The equivalent weak order can then be defined as follows:

Definition 9.24: $[a_n] \in \mathbb{R}^+$ if and only if either (a_n) is null or strictly positive.

Proposition 9.25: $\mathbb{R}$ is an ordered field.

Proof:

(O1) Suppose that $[a_n], [b_n] \in \mathbb{R}^+$. Then by considering the cases where each of $[a_n], [b_n]$ is null or strictly positive, it is an exercise to prove that $[a_n + b_n] \in \mathbb{R}^+$, $[a_n b_n] \in \mathbb{R}^+$.

(O2) If $[a_n] \in \mathbb{R}$, then by lemma 9.20, the sequence (a_n) is either null, strictly positive, or strictly negative. If it is null or strictly positive then $[a_n] \in \mathbb{R}^+$; otherwise (a_n) is strictly negative, in which case $(-a_n)$ is strictly positive and $-[a_n] \in \mathbb{R}^+$.

(O3) If $[a_n] \in \mathbb{R}^+$ and $-[a_n] \in \mathbb{R}^+$, then by lemma 9.20 the only possibility is $[a_n] = [0]$. $\qquad\square$

Completeness of $\mathbb{R}$

The trickiest property is completeness. Recall that $\mathbb{Q}$ is embedded in $\mathbb{R}$ by defining

$$\hat{q} = [q,\ q, \ldots,\ q, \ldots] \text{ for } q \in \mathbb{Q}.$$

Then $\mathbb{Q}$ is order isomorphic to the subset $\hat{\mathbb{Q}} = \{\hat{q} \in \mathbb{R} \,|\, q \in \mathbb{Q}\}$ of $\mathbb{R}$, an assertion that is readily checked.

Our plan of attack is to show that any non-empty subset $X \subseteq \mathbb{R}$ bounded above by $k \in \mathbb{R}$ can be shown to have a least upper bound. We first show that we can find $l, r \in \mathbb{Q}$ so that $\hat{l} \in \mathbb{R}$ is not an upper bound for X but $\hat{r} \in \mathbb{R}$ is an upper bound. Then we perform a bisection argument to get an increasing sequence (l_n) of rationals which are not upper bounds and a decreasing sequence (r_n) which are upper bounds where

$$0 < r_n - l_n < (r - l)/2^n$$

so that the two Cauchy sequences (l_n) and (r_n) tend to the same limit

$$[l_n] = [r_n],$$

which is the required least upper bound for X.

We get the proof started with a simple lemma:

Lemma 9.26: If $x \in \mathbb{R}$, then $\hat{l} < x < \hat{r}$ for some $\hat{l}, \hat{r} \in \hat{\mathbb{Q}}$.

Proof: Let (a_n) be a Cauchy sequence such that $[a_n] = x$. By lemma 9.15, $|a_n| < A$ for some $A \in \mathbb{Q}$. Choose $l \in \mathbb{Q}$ where $l < -A$ and $r \in \mathbb{Q}$ where $r > A$ then $\hat{l} < [a_n] < \hat{r}$. $\qquad\square$

Theorem 9.27: $\mathbb{R}$ is a complete ordered field.

Proof: By proposition 9.25, $\mathbb{R}$ is an ordered field.

To establish completeness, let X be a non-empty subset of $\mathbb{R}$ bounded above by $k \in \mathbb{R}$.

Because X is nonempty, X must contain an element $x \in \mathbb{R}$. By lemma 9.26, we have $\hat{l} \in \hat{\mathbb{Q}}$ where $\hat{l} < x$ so $\hat{l}$ is not an upper bound of X.

By lemma 9.26 for the upper bound $k \in \mathbb{R}$, we have $k < \hat{r}$ for $\hat{r} \in \hat{\mathbb{Q}}$ and so $\hat{r}$ is also an upper bound of X.

Start with $l_0 = l$ and $r_0 = r$. Suppose that for $n \geq 0$, we have found $l_n \in \mathbb{Q}$ where $\hat{l}_n \in \mathbb{R}$ is not a least upper bound for X and $r_n \in \mathbb{Q}$ where $\hat{r}_n \in \mathbb{R}$ is a least upper bound for X. This is already true for $n = 0$.

Let $m_n = (l_n + r_n)/2 \in \mathbb{Q}$ be the midpoint between l_n and r_n. If $\hat{m}_n$ is not a least upper bound for X, set

$$l_{m+1} = m_n, \ r_{m+1} = r_n,$$

otherwise set

$$l_{m+1} = l_n, \ r_{m+1} = m_n.$$

By induction, this gives:

an increasing sequence (l_n) where $l_n \in \mathbb{Q}$ and $\hat{l}_n \in \mathbb{R}$ is not an upper bound of X,

and

a decreasing sequence (r_n) where $r_n \in \mathbb{Q}$ and $\hat{r}_n \in \mathbb{R}$ is an upper bound of X.

Figure 9.2 shows a particular case where the set X is marked on $\mathbb{R}$ and the sequences of rational numbers $l_0 = l, l_1, \ldots, l_n, \ldots$ and $r_0 = r, r_1, \ldots, r_n, \ldots$ are marked on the rational line $\mathbb{Q}$.

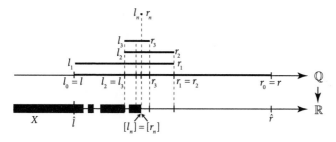

Fig. 9.2 Homing in on the least upper bound of X

Let $d = r - l$. Then the length of the interval from l_n to r_n is $d/2^n$, and for $m, n > N$, we have

$$|l_m - l_n| < d/2^N, \quad |r_m - r_n| < d/2^N, \quad |r_n - l_n| < d/2^n.$$

Both (l_n) and (r_n) are Cauchy sequences and their difference is a null sequence, so they represent the same equivalence class:

$$u = [l_n] = [r_n] \in \mathbb{R}.$$

The element $u \in \mathbb{R}$ is an upper bound of $X \subseteq \mathbb{R}$. (If not, then there would be an element $x \in X$ where $u < x$. Because $(\hat{r}_n)$ tends down to u, we could find $\hat{r}_n$ that is not an upper bound satisfying $u < \hat{r}_n < x$, contradicting the fact that u is an upper bound.) It is also the least upper bound, for if k were an upper bound where $k < u$ then because $(\hat{l}_n)$ tends up to the limit u, we could find an element $k < \hat{l}_n < u$ where $\hat{l}_n$ is not an upper bound, contradicting the fact that u is an upper bound. $\qquad\square$

As before, we have an order isomorphism between the elements $q \in \mathbb{Q}$ and the elements $\hat{q} = [q, q, \ldots, q, \ldots] \in \mathbb{R}$ so that we may identify $\mathbb{Q}$ as a subset of $\mathbb{R}$.

Finally we have a chain of number systems

$$\mathbb{N} \subseteq \mathbb{N}_0 \subseteq \mathbb{Z} \subseteq \mathbb{Q} \subseteq \mathbb{R}$$

as intended.

Exercises

1. First, some bookkeeping.
   (a) Write out a full proof of proposition 9.7.
   (b) Complete the proof of proposition 9.9.

(c) Prove theorem 9.10.

(d) Prove theorem 9.11.

2. If R is a ring with zero element 0_R and unit element 1_R, for $n \in \mathbb{N}_0$ define n_R recursively by

$$0_R = 0_R, (n+1)_R = n_R + 1_R,$$

and for $x \in R, n \in \mathbb{N}_0$ define $x^n \in R$ recursively by

$$x^0 = 1_R, \ x^{n+1} = x^n x.$$

If $\dbinom{n}{r}$ is the binomial coefficient $n! / [r! \, (n-r)!]$, prove that for all x, $y \in R$

$$(x+y)^n = x^n + n_R x^{n-1} y + \cdots + \dbinom{n}{r}_R x^{n-r} y^r + \cdots + y^n.$$

3. If $p \in \mathbb{N}$ is a prime and $p_R = 0_R$ in R, (as is the case, for instance, in $\mathbb{Z}_p$), show that

$$(x+y)^p = x^p + y^p.$$

Give an example of a ring R where $n_R = 0_R$ for $n \neq 0$, but $(x+y)^n \neq x^n + y^n$.

4. If R is an ordered ring, use the definition of order in this chapter to show that

$$x^2 - 5_R x + 6_R \geq 0_R$$

if and only if $x \geq 3_R$ or $x \leq 2_R$.

5. Use the Euclidean algorithm to prove that if $m, n \in \mathbb{N}$ are coprime, that is, have no common factor in $\mathbb{N}$ greater than 1, then there exist $a, b \in \mathbb{Z}$ such that

$$am + bn = 1.$$

Find a, b when $m = 1008, n = 1375$.

6. Prove formally that every positive rational number can be written uniquely in the form m/n for *coprime* $m, n \in \mathbb{N}$. This is called 'expressing a fraction in its lowest terms'. If two rationals $p/q, r/s$ are in lowest terms, is $(ps + qr)/(qs)$? What about $(pr)/(qs)$?

Show that if p/q is in lowest terms, so is p^2/q^2. Use the uniqueness of expression in lowest terms to give a streamlined proof that $\sqrt{2}$ is irrational.

7. Prove the following results in any ordered field.
   (a) $a \leq b \Leftrightarrow -b \leq -a$,
   (b) $a < b \Leftrightarrow -b < -a$,
   (c) $-1 < 0 < 1$,
   (d) if $a \neq 0$, then $a^2 > 0$,
   (e) $0 < a \leq b \Rightarrow 0 < b^{-1} \leq a^{-1}$,
   (f) If $a < 0$ and $b < 0$, then $ab > 0$.

8. Prove that every non-empty *finite* subset X of an ordered field contains a smallest element and a largest element. (A smallest element is an element $x \in X$ such that $x \leq y$ for all $y \in X$; a largest element is defined similarly.) Is the same true if we drop the condition that X be finite?

9. In the definition of the order relation on $\mathbb{R}$, why is it not a good idea to define

$$[a_n] \geq 0 \text{ if } \exists N \in \mathbb{Z}, \text{ such that } a_n \geq 0 \text{ for all } n > N?$$

10. Let $a_n = \frac{1}{2} - \frac{1}{6} + \cdots + (-1)^{n+1}/(n+1)!$. Prove that (a_n) is Cauchy, so tends to some limit l. Prove that each a_n is rational, but l is not.

Real Numbers
as a Complete Ordered Field

In this chapter we show how to reverse the process used in the previous chapter. There we postulated the existence of a set satisfying the basic properties (N1)–(N3) of natural numbers, and eventually constructed a set $\mathbb{R}$ that is a complete ordered field. Here we start by postulating the existence of a complete ordered field, and work down until we reach the natural numbers. This approach is basically simpler from a technical point of view; for example, we really do get $\mathbb{N} \subseteq \mathbb{Z} \subseteq \mathbb{Q} \subseteq \mathbb{R}$ without any fudging with order isomorphisms. However, as remarked in the previous chapter, we have to accept a rather lengthy set of axioms, all interacting with each other, and some are distinctly complicated.

We begin with examples of fields, rings, ordered fields, and ordered rings, to show that a wide variety of such structures exists. Any system that obeys a formal set of axioms is called a *model* for those axioms, and the power of the axiomatic method is that any deduction from the axioms is true in any model for those axioms. So any valid deduction from the axioms for an ordered field will hold in the models $\mathbb{Q}$, $\mathbb{R}$ constructed in the last chapter; indeed in *any* system satisfying the axioms. Therefore we need only perform the deductions *once*, rather than over again for each model.

The axiomatic method has another kind of power: the ability to single out (up to isomorphism) a *unique* model. For example, this happened with axioms (N1)–(N3) for the natural numbers: all systems satisfying them are order isomorphic, so to all intents and purposes the same. The same holds for the axioms for a complete ordered field: they define a *unique* system, up to order isomorphism. It is therefore permissible to call such a system *the* real numbers.

As we think about the system of real numbers, we can now imagine it as a unique crystalline concept whose properties hold together in a coherent manner. It may be represented mentally in many ways: as a system satisfying the 13 axioms for a complete ordered field; symbolically, as infinite decimals; and visually, as points on a number line that satisfy all 13 axioms including completeness. However, there is essentially just one such system that can be represented in different ways.

This is the end of our quest: starting from intuitive ideas about points on a line and decimal expansions we have formulated a set of axioms that defines the required system uniquely. Moreover, within this unique system of real numbers, we can obtain equally simple descriptions of integers and rational numbers at the same time.

Examples of Rings and Fields

Not every system of axioms defines a unique structure, even if isomorphisms are permitted. For example, $\mathbb{Z}$ and $\mathbb{Q}$ are both rings, but they are not isomorphic since $\mathbb{Q}$ is a field and $\mathbb{Z}$ is not. To motivate the narrowing down of possibilities by imposing extra axioms, we describe some further examples.

Example 10.1: $\mathbb{Z}_n$, the ring of integers modulo n. Let n be an integer > 0, and for $r, s \in \mathbb{Z}$ define

$$r \underset{\tilde{n}}{} s \Leftrightarrow r - s = kn \text{ for some } k \in \mathbb{Z}.$$

It is easy to show that $\underset{\tilde{n}}{}$ is an equivalence relation, and we call the set of equivalence classes $\mathbb{Z}_n$. We denote the class containing m by m_n.

The division algorithm implies that if $m, n \in \mathbb{Z}$ and $n > 0$, then there exist $q, r \in \mathbb{Z}$, with $0 \le r < n$, such that $m = qn + r$. Thus $m - r = qn$, so every integer is equivalent under $\underset{\tilde{n}}{}$ to an integer r with $0 \le r < n$. Thus the elements of $\mathbb{Z}_n$ are $0_n, 1_n, \ldots, (n-1)_n$. As in chapter 4 (where we treated the special case $n = 3$), we can define operations on the equivalence classes by

$$r_n + k_n = (r + k)_n,$$
$$r_n k_n = (rk)_n.$$

These operations are well defined and satisfy the axioms for a *ring* with zero element is 0_n and unity 1_n.

If n is not prime, then $\mathbb{Z}_n$ is not a field. For if $n = rk$ with $0 < r < n$, $0 < k < n$, then

$$r_n k_n = n_n = 0_n.$$

Say that an element x of a ring is a *zero-divisor* if $x \ne 0$, but $xy = 0$ for some $y \ne 0$, y in the ring. Then r_n and k_n are zero-divisors. But a field has no zero-divisors, for if $xy = 0$ in a field and $y \neq 0$, then $x = xyy^{-1} = 0y^{-1} = 0$. Thus $\mathbb{Z}_n$ is not a field for composite n.

For instance, in $\mathbb{Z}_6$ we have $2_6 3_6 = 0_6$, where $2_6 \neq 0_6$, $3_6 \neq 0_6$. These elements do not have multiplicative inverses in $\mathbb{Z}_6$, as we can check bare-handed by trying all six possibilities: $2_6 0_6 = 0_6$, $2_6 1_6 = 2_6$, $2_6 2_6 = 4_6$, $2_6 3_6 = 0_6$, $2_6 4_6 = 2_6$, $2_6 5_6 = 4_6$. Nowhere do we get an answer 1_6.

However, if n is *prime* then $\mathbb{Z}_n$ is a field. There are several ways to see this, of which the following is the least sophisticated but the most direct. Given $r_n \neq 0_n$, we look for an inverse by calculating all the products

$$r_n 0_n = 0_n, \ r_n 1_n = r_n, \ \ldots, \ r_n(n-1)_n = ?$$

All of these elements are different, for if

$$r_n k_n = r_n l_n$$

where $0 \leq k < l < n$, then

$$r_n(l-k)_n = 0_n$$

so that n divides $r(l-k)$. But each factor lies between 0 and n, and n is prime—a contradiction.

Now this list of products contains exactly n elements, all different, and since $\mathbb{Z}_n$ only *has* n elements, each must occur precisely once. In particular, 1_n occurs somewhere, say at $r_n k_n = 1_n$; and now k_n is the required inverse. Hence $\mathbb{Z}_n$ is a field if n is prime.

For instance, in $\mathbb{Z}_5$, we look for an inverse for 3_5 by working out $3_5 0_5 = 0_5$, $3_5 1_5 = 3_5$, $3_5 2_5 = 1_5$, $3_5 3_5 = 4_5$, $3_5 4_5 = 2_5$, and the products are precisely the elements of $\mathbb{Z}_5$ in the order 0_5, 3_5, 1_5, 4_5, 2_5. Among them is 1_5, and the inverse of 3_5 is 2_5.

Example 10.2: $\mathbb{Q}(\sqrt{2}) = \{a + b\sqrt{2} \in \mathbb{R} \mid a, b \in \mathbb{Q}\}$. This is a field with zero element $0 + 0\sqrt{2}$ and unity $1 + 0\sqrt{2}$. The additive inverse of $a + b\sqrt{2}$ is $-a - b\sqrt{2}$, and if $a + b\sqrt{2} \neq 0$, its multiplicative inverse is

$$\frac{1}{a + b\sqrt{2}} = \frac{a - b\sqrt{2}}{(a + b\sqrt{2})(a - b\sqrt{2})} = \frac{a}{a^2 - 2b^2} + \frac{-b}{a^2 - 2b^2}\sqrt{2}.$$

(It is an easy exercise to show that if one of a, b is not 0, then $a^2 - 2b^2 \neq 0$; in fact it is the same as proving $\sqrt{2}$ irrational.)

Example 10.3: This will provide a useful counterexample later. It is the field $\mathbb{R}(t)$ of *rational functions* in an indeterminate t. An element of $\mathbb{R}(t)$ is most easily described as a quotient of two polynomials,

$$\frac{a_n t^n + \cdots + a_0}{b_m t^m + \cdots + b_0},$$

where $a_0, \ldots, a_n, b_0, \ldots, b_m \in \mathbb{R}$ and not all bs are zero. We can think of such an expression as giving rise to a function $f : D \to \mathbb{R}$ where

$$D = \{x \in \mathbb{R} \mid b_m x^m + \cdots + b_0 \neq 0\}$$

and

$$f(x) = \frac{a_n x^n + \cdots + a_0}{b_m x^m + \cdots + b_0}.$$

This is a quotient of polynomials in the same way that a rational number is a quotient of integers, hence the name 'rational function'.

A formal definition of $\mathbb{R}(t)$ can be given as follows. First, a polynomial is determined by its coefficients $a_0, \ldots, a_n$, so we can define a *formal polynomial* to be a sequence $s : \mathbb{N}_0 \to \mathbb{R}$ such that for some $n \in \mathbb{N}_0$ we have $s(m) = 0$ for $m > n$. Write $s(m) = s_m$ and denote s by the sequence $(s_0, s_1, \ldots, s_r, \ldots)$ on the understanding that $s_m = 0$ from some point on. Addition and multiplication are defined by

$$(s_0, s_1, \ldots, s_r, \ldots) + (p_0, p_1, \ldots, p_r, \ldots) = (s_0 + p_0, s_1 + p_1, \ldots, s_r + p_r, \ldots),$$
$$(s_0, s_1, \ldots, s_r, \ldots)(p_0, p_1, \ldots, p_r, \ldots) = (s_0 p_0, s_0 p_1 + s_1 p_0, \ldots, q_r, \ldots),$$

where $q_r = s_0 p_r + s_1 p_{r-1} + \cdots + s_r p_0$.

The sequence $(0, 1, 0, 0, \ldots)$ can be denoted by t, and then

$$(s_0, s_1, \ldots, s_r, \ldots) = s_0 + s_1 t + \cdots + s_r t^r + \cdots$$

so we recover the usual notation for a polynomial in t, as long as we identify $s \in \mathbb{R}$ with the sequence $(s, 0, 0, \ldots)$. The formal polynomials constitute a ring.

Using equivalence classes of ordered pairs, in exactly the way that we constructed $\mathbb{Q}$ from $\mathbb{Z}$, we can construct $\mathbb{R}(t)$ from the ring of formal polynomials. The sum and product of rational functions are defined in the customary fashion, and the resulting structure is a field. Finally, we identify $\mathbb{R}$ with the subset of $\mathbb{R}(t)$ consisting of functions $a_0/1$, where $a_0 \in \mathbb{R}$.

It would be possible to exhibit many other interesting rings and fields; however, those listed above are especially pertinent to this chapter.

Examples of Ordered Rings and Fields

Next we try to introduce orderings. We say 'try to' because the attempt doesn't always succeed, and the reasons for the failure are instructive.

(Non-)Example 10.4: $\mathbb{Z}_n$ cannot be ordered in a way that makes it into an ordered ring. (Of course it can be *ordered* in a way that does not fit the arithmetic, for example,

$$0_n < 1_n < 2_n \cdots < (n-1)_n.$$

But this does not lead to an ordered ring, for $1_n > 0_n, (n-1)_n > 0_n$ would then imply $0_n = 1_n + (n-1)_n > 0_n$, which is absurd.)

More generally, suppose we *could* give $\mathbb{Z}_n$ an order relation making it into an ordered ring. Then there is a subset $\mathbb{Z}_n^+$ of positive elements satisfying axioms (O1)–(O3). By (O2) either $1_n \in \mathbb{Z}_n^+$ or $-1_n \in \mathbb{Z}_n^+$. Since $1_n = 1_n \times 1_n = (-1_n)(-1_n)$, either possibility implies that $1_n \in \mathbb{Z}_n^+$ using (O1). Now (O1) and induction lead to $2_n = 1_n + 1_n \in \mathbb{Z}_n^+$, $3_n \in \mathbb{Z}_n^+$, ..., $(n-1)_n \in \mathbb{Z}_n^+$. But this gives the same contradiction as before. Hence $\mathbb{Z}_n$ cannot be given the structure of an ordered ring.

Example 10.5: $\mathbb{Q}(\sqrt{2})$ can be given the structure of an ordered field in two different ways.

The first way is to note that $\mathbb{Q}(\sqrt{2}) \subseteq \mathbb{R}$, and to restrict the usual order relation on $\mathbb{R}$ to $\mathbb{Q}(\sqrt{2})$: clearly this gives $\mathbb{Q}(\sqrt{2})$ the structure of an ordered field.

The second way is more subtle. There is a map $\theta : \mathbb{Q}(\sqrt{2}) \to \mathbb{Q}(\sqrt{2})$ defined by

$$\theta(a + b\sqrt{2}) = a - b\sqrt{2}.$$

Now θ is an isomorphism from $\mathbb{Q}(\sqrt{2})$ to itself (usually called an *automorphism*), that is, θ is a bijection, and for all $x, y \in \mathbb{Q}(\sqrt{2})$

$$\theta(x + y) = \theta(x) + \theta(y),$$
$$\theta(xy) = \theta(x).$$

(Check this!) Denoting the first order relation, defined above, by $\geq$, we define a new relation $\succeq$ by

$$x \succeq y \Leftrightarrow \theta(x) \geq \theta(y).$$

You should check that this, too, gives $\mathbb{Q}(\sqrt{2})$ the structure of an ordered field. For example, if $x, y \succeq 0$ then $\theta(x), \theta(y) \geq \theta(0) = 0$, so $\theta(x)\theta(y) \geq 0$, so $\theta(xy) \geq 0$, so $xy \succeq 0$. The remaining axioms are proved in the same way. Note that in this ordering $\sqrt{2} \prec 0$.

Remark: This example should be offset against the following fact: $\mathbb{Z}$ and $\mathbb{Q}$ can be given the structure of ordered rings (or ordered field in the case of $\mathbb{Q}$) in *only one* way. Here is a quick sketch of the reasoning. As we argued for $\mathbb{Z}_n^+$, we always have $1 > 0$ because $1 = 1^2 = (-1)^2$. Inductively it follows that the ordering on $\mathbb{Z}$ must have all natural numbers positive, hence using (O2) the *usual* negative integers must be negative in the given ordering. So for $\mathbb{Z}$, only the usual ordering works. Since everything in $\mathbb{Q}$ is a quotient of integers, the same goes for $\mathbb{Q}$ (after a little work).

The same holds for $\mathbb{R}$, but the proof requires the fact that every positive real number has a square root, a fact which needs verifying. It is an easy consequence of completeness. If $x \in \mathbb{R}$ and $x > 0$, let

$$L = \{y \in \mathbb{R} \mid y > 0 \,\&\, y^2 < x\}.$$

Then L is easily seen to be bounded above and non-empty. By completeness L has a least upper bound u; a quick contradiction argument shows that $u^2 = x$.

For *any* ordering making $\mathbb{R}$ an ordered field, all elements of the form y^2 ($y \in \mathbb{R}$) must be positive, and all elements $-y^2$ must be negative. By what we have just said, the positive and negative elements of $\mathbb{R}$ (in the usual sense) must also be positive and negative respectively in any other ordering, since they are precisely the elements in the required forms. Thus only one ordering exists making $\mathbb{R}$ into an ordered field.

Example 10.6: We can give the field of rational functions $\mathbb{R}(t)$ an ordering with interesting properties. (This does not give a notion of *size* to a function, but it does not prevent us from imposing an ordering that satisfies axioms (O1)–(O3).) Define

$$\mathbb{R}(t)^+ = \{f(t) \in \mathbb{R}(t) \mid \exists K \in \mathbb{R} : x \in \mathbb{R}, \, x > K \Rightarrow f(x) \geq 0\},$$

which means that $f(t)$ is considered positive if and only if $f(x)$ is positive for all sufficiently large x. (For instance, $(t^2 - 17)/(5t^3 + 4t)$ is positive in this sense, but $(t + 1)/(3t - t^2)$ is negative.) This, it may be verified, makes $\mathbb{R}(t)$ into an ordered held. If we identify $\mathbb{R}$ with the set of constant functions, as above, then the ordering on $\mathbb{R}(t)$ restricts to the usual ordering on $\mathbb{R}$.

Surprisingly, $\mathbb{R} \subseteq \mathbb{R}(t)$ is now *bounded above*. In fact, the function $f(t) = t$ is an upper bound. For if $k \in \mathbb{R}$ then the function $g(t) = f(t) - k = t - k$ has the property that $g(x) > 0$ for all $x > k$, hence $g(t) \in \mathbb{R}(t)^+$. This proves that t is an upper bound for $\mathbb{R}$ in $\mathbb{R}(t)$.

This possibility of having ordered fields that contain the real numbers opens up new avenues in formal mathematics that we will explore later in chapter 15.

Isomorphisms Again

We have already made use of the concepts 'isomorphism' and 'order isomorphism' in special cases, and we now discuss them in general. Recall that if R, S are rings then $\theta : R \to S$ is an *isomorphism* if it is a bijection and if for all $r, s \in R$

$$\begin{aligned} \theta(r + s) &= \theta(r) + \theta(s), \\ \theta(rs) &= \theta(r)\theta(s). \end{aligned} \tag{10.1}$$

Various axiomatic structures have been proved unique *up to isomorphism*. You may wonder why we can't do better and actually make them *unique*. The reason is that this is just too much to ask; in any case, it would present no real advantages. An isomorphism, after all, is just a change of name (from r to $\theta(r)$); so given a ring R, we can find lots of isomorphic rings by finding lots of ways changing the names. In formal terms, let S be *any* set for which there exists a bijection $\theta : R \to S$ (we don't assume S is a ring) and use (10.1) back-to-front, to define ring operations on S by

$$\begin{aligned} \theta(r) + \theta(s) &= \theta(r + s) \\ \theta(r)\,\theta(s) &= \theta(rs). \end{aligned}$$

Then S is isomorphic to R.

How do we know that sets S with suitable bijections exist? Take any element t whatever, and let $S = R \times \{t\}$; define θ by $\theta(r) = (r, t)$. This is always a bijection; and different choices of t lead to different choices of S. This shows how wide a variety of sets S can be found—and this is just *one* very simple way to find them.

Since it is the algebraic operations on a ring that are important, not the elements themselves, an isomorphic ring is just as good as the ring we start from. So it is too restrictive to expect to specify an algebraic structure uniquely. On the other hand, uniqueness up to isomorphism is the most that we ever require.

The same goes for an *order isomorphism* between two ordered rings R and S, which in addition to (10.1) satisfies the condition

$$r \geq s \implies \theta(r) \geq \theta(s).$$

We cease philosophising to point out some useful, simple consequences of (10.1).

Lemma 10.7: If $\theta : R \to S$ is an isomorphism of rings, then for all $r \in R$,

   (a) $\theta(0) = 0$
   (b) $\theta(1) = 1$
   (c) $\theta(-r) = -\theta(r)$
   (d) $\theta(1/r) = 1/\theta(r)$, provided $1/r$ exists.

Proof: For all $r \in R$, we have $r = 0 + r$. Applying θ,

$$\theta(r) = \theta(0 + r) = \theta(0) + \theta(r).$$

Now θ is onto, so every element of S is of the form $\theta(r)$ for some $r \in R$. Hence

$$s = \theta(0) + s$$

for all $s \in S$, so by proposition 9.2 of chapter 9, $\theta(0) = 0$. This proves (a), and (b) is similar. To prove (c),

$$r + (-r) = 0$$

so

$$\theta(r) + \theta(-r) = \theta(0) = 0.$$

By proposition 9.3 of chapter 9, $\theta(-r) = -\theta(r)$. This proves (c), and (d) is similar. $\qquad\qquad\square$

Definition 10.8: If R is a ring, then a *subring* of R is a subset S such that

   (i) $r, s \in S \Rightarrow r + s \in S$
   (ii) $r, s \in S \Rightarrow rs \in S$
   (iii) $s \in S \Rightarrow -s \in S$
   (iv) $1 \in S$.

From (iv), (iii), (i) it also follows that $0 = 1 + (-1) \in S$. For example, $\mathbb{Z}$ is a subring of $\mathbb{Q}$ and $\mathbb{Q}$ is a subring of $\mathbb{R}$.

As with isomorphisms between rings, it is often sufficient to have a subring isomorphic to something, instead of actually being that thing.

Definition 10.9: If R is a field, then a *subfield* S is a subring that satisfies (i)–(iv) and also satisfies

   (v) $s \in S, s \neq 0 \Rightarrow s^{-1} \in S$.

For example, $\mathbb{Q}$ is a subfield of $\mathbb{R}$. These ideas are applied in the next section.

Some Characterisations

Proposition 10.10: Every ring R contains a subring isomorphic either to $\mathbb{Z}$ or to $\mathbb{Z}_n$ for some n.

Remark: Here we must insist that 0 and 1 are different. If we don't, there is also the ring $\{0\}$ in which all operations lead to 0.

Proof: Define $\theta : \mathbb{Z} \to R$ by $\theta(0) = 0$, $\theta(1) = 1$, $\theta(n + 1) = \theta(n) + 1$ (using the recursion theorem) for $n > 0$, then let $\theta(-n) = -(\theta(n))$ for $n > 0$. An induction argument shows that

$$\theta(m + n) = \theta(m) + \theta(n)$$
$$\theta(mn) = \theta(m)\,\theta(n).$$

If θ is an injection, we've finished, for then $\theta(\mathbb{Z})$, the image of $\mathbb{Z}$ under θ, is a subring isomorphic to $\mathbb{Z}$.

However, θ might not be injective. In this case there exist $r > s \in \mathbb{Z}$ such that $\theta(r) = \theta(s)$. Therefore $\theta(r - s) = \theta(r) - \theta(s) = 0$. Using the well ordering property, let n be the smallest natural number such that $n \neq 0$, $\theta(n) = 0$. It follows that $\theta(0)$, $\theta(1)$, $\ldots$, $\theta(n - 1)$ are all different, for if $\theta(r) = \theta(s)$ with $0 < r < s < n$, then $\theta(s - r) = 0$ and this contradicts the definition of n. Also, if

$$u - v = qn \qquad (u, v, q \in \mathbb{Z}),$$

then

$$\theta(u) - \theta(v) = \theta(u - v) = \theta(qn) = \theta(q)\,\theta(n) = \theta(q)0 = 0.$$

Hence, using our notation for $\mathbb{Z}_n$, if $u_n = v_n$ then $\theta(u) = \theta(v)$.

We may therefore define a map $\varphi : \mathbb{Z}_n \to R$ by $\varphi(u_n) = \theta(u)$. The previous remark shows that φ is well defined. Now

$$\varphi(u_n + v_n) = \varphi((u + v)_n) = \theta(u + v) = \theta(u) + \theta(v) = \varphi(u_n) + \varphi(v_n),$$
$$\varphi(u_n v_n) = \varphi((uv)_n) = \theta(uv) = \theta(u)\theta(v) = \varphi(u_n)\varphi(v_n).$$

Since $\theta(0), \ldots, \theta(n - 1)$ are all different, $\varphi(0_n), \ldots, \varphi((n - 1)_n)$ are all different, so φ is an injection. Thus $\varphi(\mathbb{Z}_n)$ is a subring of R isomorphic to $\mathbb{Z}_n$. $\qquad \square$

For fields we get a similar result:

Proposition 10.11: Every field F contains a subfield isomorphic either to $\mathbb{Q}$ or to $\mathbb{Z}_p$ where p is prime.

Proof: Using proposition 10.10, F contains a subring S isomorphic to $\mathbb{Z}$ or to $\mathbb{Z}_n$.

Suppose that S is isomorphic to $\mathbb{Z}$, with $\theta : \mathbb{Z} \to S$ an isomorphism. Define $\varphi : \mathbb{Q} \to F$ by

$$\varphi(m/n) = \theta(m)/\theta(n) \quad (m, n \in \mathbb{Z}, n \neq 0).$$

Notice that $n \neq 0 \Rightarrow \theta(n) \neq 0$, since θ is injective, so the right-hand side makes sense. Now φ is injective, for if $\varphi(m/n) = \varphi(r/s)$ then

$$\theta(m)/\theta(n) = \theta(r)/\theta(s)$$

so

$$\theta(ms) = \theta(m)\theta(s) = \theta(r)\theta(n) = \theta(rn),$$

hence $ms = rn$ and therefore $m/n = r/s$. It is now easy to check that $\varphi(\mathbb{Q})$ is a subfield isomorphic to $\mathbb{Q}$.

Now suppose that S is isomorphic to $\mathbb{Z}_n$. If n is composite, $n = qr$, then $\varphi(q_n)$, $\varphi(r_n)$ are zero-divisors in F. But a field F has no zero-divisors (if $xy = 0$ and $y \neq 0$, then $x = xyy^{-1} = 0y^{-1} = 0$). Therefore n is a prime, say $n = p$; and since $\mathbb{Z}_p$ is a field we have found a subfield of F isomorphic to $\mathbb{Z}_p$. $\qquad\square$

Next we bring in the order relation.

Proposition 10.12: Every ordered ring contains a subring order isomorphic to $\mathbb{Z}$.

Proof: By proposition 10.10 it contains a subring isomorphic to $\mathbb{Z}$ or to $\mathbb{Z}_n$. The proof that $\mathbb{Z}_n$ cannot be made an ordered ring also shows that it cannot be a subring of an ordered ring. The proof that the ordering on $\mathbb{Z}$ is unique shows that the subring isomorphic to $\mathbb{Z}$ is also *order* isomorphic to $\mathbb{Z}$. $\qquad\square$

Similarly:

Proposition 10.13: Every ordered field contains a subfield order isomorphic to $\mathbb{Q}$.

Proof: Eliminate the possibility $\mathbb{Z}_p$ as in proposition 10.11; then use uniqueness of the order on $\mathbb{Q}$. $\qquad\square$

These two propositions give simple axiomatic characterisations of $\mathbb{Z}$ and $\mathbb{Q}$:

$\mathbb{Z}$ is a minimal ordered ring (that is, $\mathbb{Z}$ is an ordered ring with no proper subring);

$\mathbb{Q}$ is a minimal ordered field (that is, $\mathbb{Q}$ is an ordered field with no proper subfield).

These properties define $\mathbb{Z}$ and $\mathbb{Q}$ uniquely up to isomorphism. For by proposition 10.12, any minimal ordered ring must be isomorphic to $\mathbb{Z}$, and by proposition 10.13, any minimal ordered field must be isomorphic to $\mathbb{Q}$.

Finally, we turn to complete ordered fields. To deal with these we must extend to them notions such as 'limit' and 'Cauchy sequence'. Thus let F be an ordered field. By proposition 10.13 it contains a subfield order isomorphic to $\mathbb{Q}$, and by change of notation we may assume without loss of generality that this subfield is $\mathbb{Q}$ itself. We say that a sequence (a_n) of elements of F is *Cauchy* if:

for every $\varepsilon > 0, \varepsilon \in F$, there exists $N \in \mathbb{N}_0$ such that $|a_m - a_n| < \varepsilon$ for m, $n > N$.

The sequence (a_n) tends to a *limit* $\lambda \in F$ if

for every $\varepsilon > 0$, $\varepsilon \in F$, we can find $N \in \mathbb{N}_0$ such that $|a_n - \lambda| < \varepsilon$ for all $n > N$.

These are the previous definitions in a broader context. As before, we write

$$\lim_{n \to \infty} a_n = \lambda \text{ or } \lim a_n = \lambda.$$

The key result is:

Lemma 10.14: In a complete ordered field, every Cauchy sequence has a limit.

Proof: Let (a_n) be a Cauchy sequence in F. By the argument of lemma 9.15 of chapter 9 (carried out in F) the sequence is bounded. Hence so is every subset of elements in the sequence. Define

$$b_N = \text{the least upper bound of } \{a_N, a_{N+1}, a_{N+2}, \ldots\}.$$

This exists by completeness. Clearly

$$b_0 \geq b_1 \geq b_2 \geq \cdots$$

and the sequence (b_n) is bounded below—say, by any lower bound for (a_n). Hence we can define

$$c = \text{the greatest lower bound of } (b_n).$$

We claim that c is the limit of the original sequence (a_n).

To prove this, let $\varepsilon > 0$. Suppose that there exist only finitely many values of n with

$$c - \tfrac{1}{2}\varepsilon < a_n < c + \tfrac{1}{2}\varepsilon.$$

Then we may choose N such that for all $n > N$,

$$a_n \leq c - \tfrac{1}{2}\varepsilon \text{ or } a_n \geq c + \tfrac{1}{2}\varepsilon.$$

But there exists $N_1 > N$ such that if $m, n > N_1$ then $|a_m - a_n| < \tfrac{1}{2}\varepsilon$. Hence

$$\text{for all } n > N_1, \ a_n \leq c - \tfrac{1}{2}\varepsilon,$$

or

$$\text{for all } n > N_1, \ a_n \geq c + \tfrac{1}{2}\varepsilon.$$

The latter condition implies that there exists some m with $a_n > b_m$ for all $n > N_1$, which contradicts the definition of b_m. But the former implies that we may change b_{N_1} to $b_{N_1} - \tfrac{1}{2}\varepsilon$, which again contradicts the definition of b_{N_1}.

It follows that for any M there exists $m > M$ such that

$$c - \tfrac{1}{2}\varepsilon < a_m < c + \tfrac{1}{2}\varepsilon.$$

Since (a_n) is Cauchy, there exists $M_1 > M$ such that $|a_n - a_m| < \tfrac{1}{2}\varepsilon$ for $m, n > M_1$. Hence for $n > M_1$,

$$c - \varepsilon < a_n < c + \varepsilon.$$

But this implies that $\lim a_n = c$ as claimed. $\qquad\square$

The next step is:

Lemma 10.15: Let $F \supseteq \mathbb{Q}$ be a complete ordered field. If $x \in F$ then there exists $p \in \mathbb{Z}$ such that $p - 1 \leq x < p$.

Proof: Suppose $n \leq x$ for all $n \in \mathbb{Z}$. Then $\mathbb{Z}$ is bounded above by x, so by completeness has a least upper bound k. Hence $n + 1 \leq k$ for all $n \in \mathbb{Z}$, because also $n + 1 \in \mathbb{Z}$. This implies that $n \leq k - 1$, so $k - 1$ is a smaller upper bound for $\mathbb{Z}$. This contradicts the definition of k. Therefore $x < n$ for some $n \in \mathbb{Z}$ Similarly $m < x$ for some $m \in \mathbb{Z}$. Since there are only finitely many integers between m and n, we can find an integer p that is the smallest such that $x < p$. Then $p - 1 \leq x < p$. $\qquad\square$

As a final preparatory step:

Lemma 10.16: Let F be a complete ordered field, and let (a_n) and (b_n) be two sequences with limits a and b respectively. Then

(a) $\lim (a_n + b_n) = a + b$
(b) $\lim (a_n b_n) = ab$.

Proof: For (a), copy the proof of theorem 7.1 in chapter 7 and check that formally it still makes sense. For (b), use the argument of lemma 9.15, chapter 9, to show that for all $n \in \mathbb{N}_0$, $|a_n| < A$, $|b_n| < B$ for some $A, B \in F$. Then if $\varepsilon > 0$ we have $\varepsilon/(A + B) > 0$. Hence there exists N_1 such that for $n > N_1$,

$$|a_n - a| < \varepsilon/(A + B)$$

and there exists N_2 such that for $n > N_2$

$$|b_n - b| < \varepsilon/(A + B).$$

Hence for $n > N = \max(N_1, N_2)$,

$$\begin{aligned}
|a_n b_n - ab| &= |(a_n - a)b_n + a(b_n - b)| \\
&< (\varepsilon/(A + B))B + A\varepsilon/(A + B) \\
&= \varepsilon.
\end{aligned}$$

This proves (b). □

For $\mathbb{R}$ we get an even stronger statement than propositions 10.12 and 10.13.

Theorem 10.17: Every complete ordered field is order isomorphic to $\mathbb{R}$.

Proof: Let F be a complete ordered field. By proposition 10.13 it has a subfield order isomorphic to $\mathbb{Q}$. As usual, for notational convenience we identify this subfield with $\mathbb{Q}$, so that without loss of generality $\mathbb{Q} \subseteq F$.

Elements of $\mathbb{R}$ are equivalence classes $[a_n]$ of Cauchy sequences (a_n) of rationals. Define a map $\theta : \mathbb{R} \to F$ by

$$\theta([a_n]) = \lim_{x \to \infty} a_n.$$

First we need to check that this makes sense. The reason for this is that, in the construction of $\mathbb{R}$ from $\mathbb{Q}$, we defined a Cauchy sequence (a_n) to have terms $a_n \in \mathbb{Q}$ and in the definition we used only *rational* values for ε. When we speak of the limit *in F*, we need to allow any $\varepsilon > 0$ that belongs to F. We claim that given $\varepsilon > 0$ where $\varepsilon \in F$, there is a rational ε' with $0 < \varepsilon' < \varepsilon$. To prove this, note that $1/\varepsilon \in F$ and, by lemma 10.15, $1/\varepsilon < p$ for some

$p \in \mathbb{Z}$. Then $p > 0$, and $0 < 1/p \in \mathbb{Q}$. Take $\varepsilon' = 1/p \in \mathbb{Q}$. Now because (a_n) is Cauchy in $\mathbb{Q}$, it follows that there exists $N \in \mathbb{N}_0$ such that for all $m, n > N$

$$|a_m - a_n| < \varepsilon'.$$

Hence for all $m, n > N$,

$$|a_m - a_n| < \varepsilon,$$

and (a_n) is Cauchy in F. By lemma 10.14, $\lim a_n$ exists in F.

It is easy to see, using similar arguments, that θ is well defined and injective; and lemma 10.16 proves that $\theta(\mathbb{R})$ is a subring of F isomorphic to $\mathbb{R}$. It is easy to check that θ preserves the order relation.

It remains to prove that θ is surjective. Let $x \in F$. By lemma 10.15 there exists an integer a_0 with $a_0 \leq x < a_0 + 1$. Inductively (and using lemma 10.15 again) we can find integers a_i between 0 and 9 such that

$$a_0 + \frac{a_1}{10} + \cdots + \frac{a_n}{10^n} \leq x < a_0 + \frac{a_1}{10} + \cdots + \frac{a_n + 1}{10^n}.$$

Then if $b_n = a_0 + \frac{a_1}{10} + \cdots + \frac{a_n}{10^n}$ we have

$$|b_n - x| < 1/10^n$$

and it follows easily (using a similar argument to that in the second paragraph of this proof) that

$$\lim b_n = x.$$

Also (b_n) is Cauchy in $\mathbb{Q}$, hence $[b_n] \in \mathbb{R}$, and

$$\theta(b_n) = \lim b_n = x.$$

Therefore θ is surjective. $\qquad\qquad\qquad\qquad\qquad\qquad\qquad\square$

The Connection with Intuition

We can now tidy up our ideas a little more. We have two types of model of the relevant axiom systems: formal models $\mathbb{N}_0, \mathbb{Z}, \mathbb{Q}, \mathbb{R}$, and informal models **N, Z, Q, R**. Now we explain, plausibly and intuitively, why **R** is a complete ordered field. Then on this intuitive level theorem 10.17 tells us that **R** and $\mathbb{R}$ are isomorphic. That is, the formal construction vindicates intuition, and can be used to justify all of the properties that we expect in **R**. We have therefore reached the stage where it doesn't greatly matter whether we use the informal **R** or the formal $\mathbb{R}$. The work we have done renders both equally safe, and there is now no essential difference between them.

Why, then, did we bother? Because we don't *know* this until we've gone through the constructions.

To summarise: this chapter and the last show between them that we can build up the number systems in two ways. Either we

(a) postulate the existence of $\mathbb{N}_0$ and construct $\mathbb{Z}, \mathbb{Q}, \mathbb{R}$ in turn;

or we

(b) postulate the existence of $\mathbb{R}$ and construct $\mathbb{Q}, \mathbb{Z}, \mathbb{N}_0$ in turn.

By judicious combination of the two methods we can therefore start *anywhere*, such as $\mathbb{Z}$, or $\mathbb{Q}$, and obtain the remaining systems by using chapter 9 to work upwards and this one to work downwards. And the uniqueness theorems proved in this chapter show that it makes no essential difference which method we use: the results are always isomorphic and agree with our intuitive ideas. Precisely where we start has now become a matter of taste rather than a matter of urgency. From any of the different starting points we can provide an equally logical development of all of the usual number systems, and recover all of the standard results of elementary arithmetic, from an axiomatic basis.

Exercises

1. Write out a full proof of proposition 10.13.

2. Prove that in any ordered field F,

$$a^2 + 1 > 0 \qquad \text{for all } a \in F.$$

Deduce that if the equation $x^2 + 1 = 0$ has a solution in a field, that field cannot be ordered. Find all the solutions of $x^2 + 1 = 0$ in the fields $\mathbb{Z}_2$, $\mathbb{Z}_3$, $\mathbb{Z}_5$.

3. Use the Euclidean algorithm to show that given $m, n \in \mathbb{N}$, there is a technique for calculating $a, b \in \mathbb{Z}$ such that $am + bn = h$, where h is the highest common factor of m, n. Deduce that if m, n are coprime, then there exist integers a, b such that $am + bn = 1$. Find a, b when $m = 1008, n = 1375$. Calculate the multiplicative inverse of 1008_{1375} in $\mathbb{Z}_{1375}$.

   Show that m_n has a multiplicative inverse in $\mathbb{Z}_n$ if and only if m and n are coprime.

4. In an ordered ring, prove that for all x, y

$$\|x\| - |y\| \le |x + y| \le |x| + |y|.$$

5. From the axioms of a complete ordered field, prove that every positive element a of $\mathbb{R}$ has a unique positive square root. (*Hint*: consider $\{x \in \mathbb{Q} \mid x^2 \leq a\}$.)

6. Prove by induction that $0 \leq a \leq b \Rightarrow a^n \leq b^n$ in an ordered ring R. Given $a \in R$, $a \geq 0$, show that if there exists an element $r \in R$ such that $r \geq 0$ and $r^n = a$, then it is unique.

7. Show that every positive element in a complete ordered field has a unique nth root. (*Hint*: Consider $\{x \mid x^n \leq a\}$.)

8. Use exercise 7 to define $x^{p/q}$ for a positive element x in a complete ordered field and a rational number p/q.

9. Define a field $\mathbb{Q}(\sqrt{3})$ analogous to $\mathbb{Q}(\sqrt{2})$ and show that there are two different ways of making it into an ordered field.

10. Show that the two orderings mentioned for $\mathbb{Q}\left(\sqrt{2}\right)$ are the only order relations under which it is an ordered field.

11. Find a field with exactly four different orderings which make it an ordered field.

12. Let $\mathbb{R}[t]$ be the ring of polynomials $p(t) = a_n t^n + a_{n-1} t^{n-1} + \cdots + a_0$ with real coefficients. Define the relation $\geq$ by

$$p(t) \geq q(t) \Leftrightarrow p(0) \geq q(0).$$

Does this make $\mathbb{R}[t]$ into an ordered ring?

13. *Cauchy sequences in a general ordered field*
In our construction of $\mathbb{R}$ from $\mathbb{Q}$ we started with Cauchy sequences in $\mathbb{Q}$ and defined a Cauchy sequence using a value of $\varepsilon \in \mathbb{Q}$. In lemma 10.6, we were able to show that in a *complete* ordered field F a Cauchy sequence using any value of $\varepsilon \in F$ will also converge. But what happens in an ordered field that is not complete?

Consider the relationship between the following definitions in any ordered field F:

A sequence (a_n) in F is said to converge to the limit $a \in F$ if given $\varepsilon \in F \; \exists N \in \mathbb{N}$ such that $n > N \Rightarrow |a_n - a| < \varepsilon$.

A sequence (a_n) in an ordered field F is said to be a Cauchy sequence in F if given $\varepsilon \in F \; \exists N \in \mathbb{N}$ such that $m, n > N \Rightarrow |a_m - a_n| < \varepsilon$.

The field F is said to be *Cauchy complete* if all Cauchy sequences in F tend to a limit in F.

(a) Prove using the completeness axiom (C) that a complete ordered field is Cauchy complete.

(b) Prove that a complete ordered field satisfies Archimedes' condition:

$$\text{if } e \in F, \ e > 0, \text{ then } 1/10^n < e \text{ for some } n \in \mathbb{N}.$$

14. Let $F = \mathbb{R}(t)$ be the field of example 10.3, page 214. Let $\varepsilon = 1/t$.
    (a) Show that $0 < \varepsilon < 1/n$ for all $n \in \mathbb{N}$.
    (b) Using the general definition of limit (from question 13), prove that the sequence $(1/n)$ does *not* tend the limit 0 in F.
    (c) Show that in a complete ordered field, the sequence $1/n \to 0$.
    (d) Prove that an ordered field F is complete if and only if it satisfies both Cauchy completeness and Archimedes' condition.

CHAPTER 11

Complex Numbers and Beyond

C omplex numbers are still regarded by some with a mixture of suspicion and awe, but to a modern mathematician they are just a simple set-theoretic extension of the real numbers. In this chapter we show how to construct them from $\mathbb{R}$, completing the standard hierarchy of number systems $\mathbb{N}_0 \subseteq \mathbb{Z} \subseteq \mathbb{Q} \subseteq \mathbb{R} \subseteq \mathbb{C}$.

We could go on to look for an extension of $\mathbb{C}$. The nineteenth century mathematician Sir William Rowan Hamilton found one, which he named the quaternions. We describe this briefly, just to show what's involved. However, the moral of modern mathematics is that we must broaden our horizons and look to axiomatic systems that describe more general mathematical structures. The concept of number is but a part of this study.

Modern algebra concerns itself with axiomatic systems which, broadly speaking, consist of sets with various operations on them. We've already met two, namely rings and fields, but there are many others. This is not an algebra book, so we won't study any of them in detail, but it's worth mentioning the important ones. Looking beyond complex numbers, the more fruitful direction is not towards Hamilton's quaternions, but to the generalised algebraic structures of modern algebra. However, quaternions do have their place, and in some areas of today's mathematics they are important in their own right.

Historical Background

In chapter 1 we mentioned the problems associated with the acceptance of complex numbers as a genuine concept. It's worth pausing briefly to look at a historical outline, because it may help you to become aware of misconceptions that often occur.

At the beginning of the sixteenth century there was much interest in solving algebraic equations, one of which was:

Find two numbers whose sum is 10 and product is 40.

In modern notation this problem leads to the equations

$$x + y = 10,$$
$$xy = 40.$$

Substituting for y from the first equation into the second, we find

$$x(10 - x) = 40,$$

so

$$x^2 - 10x + 40 = 0,$$

with solutions

$$x = \frac{+10 \pm \sqrt{(100 - 160)}}{2} = 5 \pm \sqrt{(-15)}.$$

If $x = 5 + \sqrt{(-15)}$ then $y = 5 - \sqrt{(-15)}$, so the solution is the pair of expressions

$$5 + \sqrt{(-15)}, 5 - \sqrt{(-15)}.$$

Sixteenth-century mathematicians realised that these expressions could not be real numbers. The square of any real number is positive, so –15 is not the square of a real number, and $\sqrt{(-15)}$ cannot be real. Nevertheless, manipulating these expressions, *as if they were numbers*, they found that whatever $\sqrt{(-15)}$ might be, when they added the solutions, the terms $\pm\sqrt{(-15)}$ cancelled, giving

$$(5 + \sqrt{(-15)}) + (5 - \sqrt{(-15)}) = 10,$$

and when they multiplied them, they got

$$(5 + \sqrt{(-15)})(5 - \sqrt{(-15)}) = 5^2 - (\sqrt{(-15)})^2$$
$$= 25 - (-15)$$
$$= 40.$$

In short, by treating $\sqrt{(-15)}$ as an 'imaginary' number and manipulating it algebraically *as if* its square is –15, the expressions $5 + \sqrt{(-15)}$, $5 - \sqrt{(-15)}$ solve the problem.

Any positive real number a has a positive square root $\sqrt{a}$. The square root of a negative real number $-a$ ($a > 0$), if there were such a thing, could be written $\sqrt{(a)} = \sqrt{(-1)}\sqrt{a}$. The eighteenth-century mathematician Leonhard Euler introduced the symbol i for $\sqrt{(-1)}$, so that $\sqrt{(-a)} = i\sqrt{a}$. An expression of the form $x + iy$ where $x, y \in \mathbb{R}$ was called a *complex number*, though it was still not clear what this really was. Using complex numbers, any quadratic equation

$$ax^2 + bx + c = 0 \ (a, b, c \in \mathbb{R})$$

has solutions of the form

$$x = \frac{-b \pm \sqrt{b^2 - 4ac}}{2a} \text{ for } b^2 \geq 4ac,$$

and

$$x = \frac{-b \pm i\sqrt{4ac - b^2}}{2a} \text{ for } b^2 < 4ac.$$

In other words, if $b^2 \geq 4ac$ then the equation has real solutions, but if $b^2 < 4ac$ it does not—but it does have complex ones.

At the time, this discovery set up a dichotomy between real (in the sense of genuine) and imaginary (in the sense of non-existent) solutions to equations. Complex numbers were saddled with the psychological overtones associated with the word 'imaginary'. ('Complex' did not mean 'complicated'; it meant 'composed of several parts', namely x and y. It still means that, but the psychological overtones get worse if you think it means 'complicated'.)

In 1806 the French mathematician Jean-Robert Argand described a complex number $x + iy$ as a point in the plane:

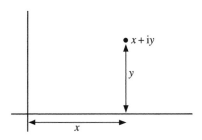

Fig. 11.1 A complex number as a point in the plane

The horizontal axis became the *real axis*, the vertical axis the *imaginary axis*, and the number i was seen as the point one unit up on the imaginary axis.

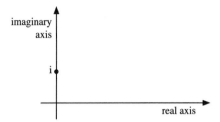

Fig. 11.2 The real and imaginary axes

This description was given the name 'Argand diagram', and it is by this name that the picture of complex numbers as points in the plane is often known today, although the idea was put forward earlier in the doctoral thesis of the great German mathematician Carl Friedrich Gauss (1799), and even this was predated by the little-known work of the Danish surveyor Caspar Wessel (1797)—such are the vagaries of historical acknowledgement. Although the complex numbers were now described concretely as points in the plane, the mystification of earlier eras still shrouded them for most people. Gauss realised that the description could be made even simpler: he clearly regarded a complex number as a pair (x, y) of real numbers. In the 1830s, the Irish mathematician Hamilton canonised complex numbers as 'couples of real numbers' (a couple being his name for an ordered pair). This is the heart of the matter and the key to the modern description: a point in the plane is an ordered pair (x, y), and the symbol $x + iy$ is just another name for that point or that pair. The mysterious expression i is none other than the ordered pair $(0, 1)$.

Construction of the Complex Numbers

We often describe Argand's representation of complex numbers as the 'complex plane', but as a set it is precisely the same as the 'real plane' $\mathbb{R}^2$.[1] However, in this context, it is useful to introduce a special notation $\mathbb{C}$ as an alternative name for $\mathbb{R}^2$, the set of ordered pairs (x, y) for $x, y \in \mathbb{R}$. We then define addition and multiplication on $\mathbb{C}$ by

[1] Modern algebraic geometers call $\mathbb{C}$ the complex *line*. To them the complex plane is $\mathbb{C}^2 = \mathbb{C} \times \mathbb{C}$. You just have to get used to this kind of thing.

$$(x_1, y_1) + (x_2, y_2) = (x_1 + x_2, y_1 + y_2), \qquad (11.1)$$

$$(x_1, y_1)(x_2, y_2) = (x_1 x_2 - y_1 y_2, x_1 y_2 + x_2 y_1). \qquad (11.2)$$

It is a simple matter to check that $\mathbb{C}$ is a field under these operations with zero $(0, 0)$ and unit $(1, 0)$. The negative of (x, y) is $(-x, -y)$ and if $(x, y) \neq (0, 0)$, the multiplicative inverse of (x, y) is

$$\left(\frac{x}{x^2 + y^2}, \frac{-y}{x^2 + y^2} \right).$$

Define $f : \mathbb{R} \to \mathbb{C}$ by $f(x) = (x, 0)$. Then

$$f(x_1 + x_2) = (x_1 + x_2, 0) = (x_1, 0) + (x_2, 0) = f(x_1) + f(x_2)$$

and

$$f(x_1 x_2) = (x_1 x_2, 0) = (x_1, 0)(x_2, 0) = f(x_1) f(x_2).$$

The function f is clearly an injection and so is an isomorphism of fields, from $\mathbb{R}$ onto the subfield $f(\mathbb{R}) \subseteq \mathbb{C}$. This subfield $f(\mathbb{R})$ is none other than the 'real axis' of Argand's description.

As usual, we consider $\mathbb{R}$ to be a subset of $\mathbb{C}$ via this isomorphism, which amounts to regarding the real numbers as the real axis in the complex plane and replacing the symbol $(x, 0)$ by x.

Define i to be the ordered pair $(0, 1)$. Using (11.2),

$$\mathrm{i}^2 = (0, 1)^2 = (-1, 0).$$

Thinking of $(-1, 0)$ as the real number -1, this gives $\mathrm{i}^2 = -1$.

More generally, using (11.1) and (11.2),

$$(x, 0) + (0, 1)(y, 0) = (x, 0) + (0, y) = (x, y).$$

Replacing $(x, 0), (y, 0)$ by $x, y \in \mathbb{R}$ respectively, we get

$$x + \mathrm{i}y = (x, y).$$

The complex number $x + \mathrm{i}y$ is another name for the ordered pair (x, y).

COMMENT There is an occasional misconception that a complex number is $x + \mathrm{i}y$ where x, y are real and $y \neq 0$, reserving the name 'real number' for $x + \mathrm{i}y$ where $y = 0$. Mathematicians regard *all* expressions $x + \mathrm{i}y$ $(x, y \in \mathbb{R})$ as complex numbers, and this *includes* real numbers.

Returning to the definitions of addition and multiplication, (11.1) and (11.2) in this notation, we find

$$(x_1 + iy_1) + (x_2 + iy_2) = (x_1 + x_2) + i(y_1 + y_2),$$
$$(x_1 + iy_1) + (x_2 + iy_2) = (x_1 x_2 - y_1 y_2) + i(x, x_2 + x_2 y_1).$$

We thus recover the usual addition and multiplication rules for complex numbers, which is why definitions (11.1), (11.2) were set up in the first place.

Historically, in the expression $x + iy$, x is referred to as the 'real part' and y as the 'imaginary part'. Both x and y are real numbers, being the first and second coordinates of the ordered pair $(x, y) \in \mathbb{R}^2$. If

$$x_1 + iy_1 = x_2 + iy_2$$

then

$$(x_1, y_1) = (x_2, y_2),$$

and by the usual properties of ordered pairs,

$$x_1 = x_2, \ y_1 = y_2.$$

Historically this deduction was referred to as 'comparing real and imaginary parts'; we now see it as an application of the set-theoretic definition of ordered pairs.

A modern interpretation of the solution of the quadratic $x^2 - 10x + 40 = 0$ is that there are no solutions in $\mathbb{R}$, but if we consider this as an equation in $\mathbb{C}$, there are solutions $5 \pm i\sqrt{15}$. This behaviour is no more 'complex' than what happens with the equation $2x = 1$ in $\mathbb{N}$ and $\mathbb{Q}$. There is no solution in $\mathbb{N}$, but in $\mathbb{Q}$ there is the solution $x = \frac{1}{2}$.

Time and again in mathematics, a problem has no solution in a given context, but it does have one when interpreted in a wider context. Don't be surprised by this phenomenon, or give it unwarranted mystical significance. More gadgets to solve something may lead to more solutions.

Complex Conjugation

A complex number $x + iy$ is also denoted by a single symbol z (or any other suitable letter, for that matter). When we write $z = x + iy$, we always suppose that $x, y \in \mathbb{R}$, unless something is stated to the contrary.

If $z = x + iy$, with $x, y \in \mathbb{R}$, then the *real part* of z is

$$\mathrm{Re}(z) = x$$

and the *imaginary part* of z is

$$\mathrm{Im}(z) = y.$$

We also define the *conjugate* of $z = x + iy$ to be

$$\bar{z} = x - iy.$$

For instance, $\overline{3 + 2i} = 3 - 2i$, $\overline{1 - 2i} = 1 + 2i$, and so on. Conjugation has certain elementary properties, which we collect together as:

Proposition 11.1:

   (a) $\overline{z_1 + z_2} = \bar{z}_1 + \bar{z}_2$
   (b) $\overline{z_1 z_2} = \bar{z}_1 \bar{z}_2$
   (c) $\bar{\bar{z}} = z$
   (d) $z = \bar{z} \Leftrightarrow z \in \mathbb{R}$.

Proof: Elementary checking of definitions. $\square$

If we define $c : \mathbb{C} \to \mathbb{C}$ by $c(z) = \bar{z}$, then proposition 11.1 tells us that c is an automorphism of the field $\mathbb{C}$, and that it is the identity when restricted to $\mathbb{R}$.

The Modulus

If $z = x + iy$ where $x, y \in \mathbb{R}$, then $x^2 + y^2 \geq 0$. Any positive real number has a unique positive square root. The *modulus* or *absolute value* of $z \in \mathbb{C}$ is

$$|z| = \sqrt{x^2 + y^2}.$$

For instance, $|3 + 2i| = \sqrt{3^2 + 2^2} = \sqrt{13}$, $|-5| = \sqrt{25} = 5$. In particular, for any real number x, $|x| = \sqrt{x^2}$ and, since the positive square root is taken, this reduces to the usual definition of modulus in the real case,

$$|x| = \begin{cases} x & \text{for } x \geq 0 \\ -x & \text{for } x < 0 \end{cases} \text{ for } x \in \mathbb{R}.$$

In geometric terms, the modulus is the distance from the origin to the point $x + iy$ in the complex plane.

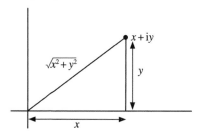

Fig. 11.3 The distance from the origin

If $z_1 = x_1 + iy_1, z_2 = x_2 + iy_2$, then

$$|z_1 - z_2| = |(x_1 - x_2) + i(y_1 - y_2)| = \sqrt{(x_1 - x_2)^2 + (y_1 - y_2)^2}.$$

This is the distance from the point z_1 to the point z_2 in the plane:

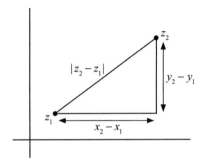

Fig. 11.4 The modulus of a complex number

Proposition 11.2:

(a) $|z| \in \mathbb{R}, |z| \geq 0$ for all $z \in \mathbb{C}$
(b) $|z| = 0 \Leftrightarrow z = 0$
(c) $|z|^2 = z\bar{z}$
(d) $|z_1 z_2| = |z_1||z_2|$
(e) $|z_1 + z_2| \leq |z_1| + |z_2|$.

Proof: Parts (a) and (b) are straightforward. Part (c) follows from the definitions, for if $z = x + iy$ then

$$z\bar{z} = (x + iy)(x - iy) = x^2 - (iy)^2 = x^2 + y^2 = |z|^2.$$

(d) Since $|z| \geq 0$ for all $z \in \mathbb{C}$, it is sufficient to show

$$|z_1 z_2|^2 = |z_1|^2 |z_2|^2.$$

But

$$
\begin{aligned}
|z_1 z_2|^2 &= (z_1 z_2)\overline{(z_1 z_2)} && \text{by 11.2(c)} \\
&= z_1 z_2 \bar{z}_1 \bar{z}_2 && \text{by 11.1(b)} \\
&= z_1 \bar{z}_1 z_2 \bar{z}_2 \\
&= |z_1|^2 |z_2|^2.
\end{aligned}
$$

(e) A frontal attack on this equality leads to some intricate algebra, which can be forced through. However, we can be more refined, but less direct, by writing $z_1 = x_1 + iy_1, z_2 = x_2 + iy_2$ and considering the identity

$$(x_1^2 + y_1^2)(x_2^2 + y_2^2) - (x_1 x_2 + y_1 y_2)^2 = (x_1 y_2 - x_2 y_1)^2,$$

which immediately tells us that

$$(x_1 x_2 + y_1 y_2)^2 \leq (x_1^2 + y_1^2)(x_2^2 + y_2^2) = |z_1|^2 |z_2|^2.$$

Taking square roots yields

$$x_1 x_2 + y_1 y_2 \leq |z_1| |z_2|,$$

which is valid even if $x_1 x_2 + y_1 y_2$ is negative. Hence

$$2(x_1 x_2 + y_1 y_2) \leq 2 |z_1| |z_2|,$$

which gives

$$x_1^2 + 2x_1 x_2 + x_2^2 + y_1^2 + 2y_1 y_2 + y_2^2 \leq x_1^2 + y_1^2 + 2 |z_1| |z_2| + x_2^2 + y_2^2;$$

this simplifies to

$$(x_1 + x_2)^2 + (y_1 + y_2)^2 \leq |z_1|^2 + 2 |z_1| |z_2| + |z_2|^2,$$

which is

$$|z_1 + z_2|^2 \leq (|z_1| + |z_2|)^2.$$

Since the modulus is positive, we can take square roots to give

$$|z_1 + z_2| \leq |z_1| + |z_2|. \qquad \square$$

Part (c) of this proposition gives a nice description of the reciprocal of $z = x + iy$ when $z \neq 0$, for then $|z|^2 = x^2 + y^2 \neq 0$, so the equation $z\bar{z} = |z|^2$ implies

$$z\bar{z}/|z|^2 = 1,$$

so

$$z^{-1} = \bar{z}/|z|^2.$$

It is also worth emphasising that, although part (e) of the proposition involves an inequality, this is between two *real* numbers $|z_1 + z_2|$ and $|z_1| + |z_2|$.

Although the subfield $\mathbb{R}$ is an ordered field, $\mathbb{C}$ is not an ordered field. We can order $\mathbb{C}$ in the sense of chapter 4; for instance, we can define the relation $\geq$ by

$$x_1 + iy_1 \geq x_2 + iy_2 \Leftrightarrow \text{either } x_1 \geq x_2 \quad \text{or} \quad x_1 = x_2 \text{ and } y_1 \geq y_2.$$

This is certainly an order relation. However, it does not blend happily with the arithmetic; for instance

$$z_1 \geq 0, z_2 \geq 0 \nRightarrow z_1 z_2 \geq 0,$$

as is demonstrated by the example

$$i \geq 0, \quad \text{but} \quad i^2 = -1 \ngeq 0.$$

There is no way to define an order on $\mathbb{C}$ which fits with the arithmetic on $\mathbb{C}$ so as to make $\mathbb{C}$ an ordered field in the sense of chapter 9. Doing so would require a subset $\mathbb{C}^+ \subseteq \mathbb{C}$ such that

(i) $z_1, z_2 \in \mathbb{C}^+ \Rightarrow z_1 + z_2 \in \mathbb{C}^+$ and $z_1 z_2 \in \mathbb{C}^+$,
(ii) $z \in \mathbb{C} \Rightarrow z \in \mathbb{C}^+$ or $-z \in \mathbb{C}^+$,
(iii) $z \in \mathbb{C}^+$ and $-z \in \mathbb{C}^+ \Rightarrow z = 0$.

But (ii) gives $i \in \mathbb{C}^+$ or $-i \in \mathbb{C}^+$; in the first case (i) implies $i^2 \in \mathbb{C}^+$, in the second $(-i)^2 \in \mathbb{C}^+$, so in either case $-1 \in \mathbb{C}^+$. Applying (i) again we find $(-i)^2 \in \mathbb{C}^+$, so $1 \in \mathbb{C}^+$. This contradicts (iii) because $1, -1 \in \mathbb{C}^+$ but $1 \neq 0$. Because of this lack of an order on $\mathbb{C}$, inequalities between complex numbers like $z_1 > z_2$ are nonsense unless the numbers involved are real. A formula like $|z_1| > |z_2|$ is perfectly feasible, because $|z_1|, |z_2| \in \mathbb{R}$ and the real numbers are an ordered field.

Euler's Approach to the Exponential Function

In the next section we define the exponential e^z of a complex number z using the real exponential and trigonometric functions. We establish the basic property $e^{z+w} = e^z e^w$. We relate trigonometric functions to the complex exponential, and prove De Moivre's Theorem, an effective way to prove certain basic trigonometric formulas. We use the results to give a geometric interpretation of addition and multiplication of complex numbers. The complex exponential will also be used in chapter 13 to study the symmetries of regular polygons.

First, however, we make a few remarks about the history of the ideas concerned. A lengthy historical development led to the remarkable insight that in the world of complex numbers, trigonometric and exponential functions are intimately related; in fact, two aspects of the same idea. The relationship was first discovered by Euler by manipulating the power series for the complex exponential, sine and cosine functions.

His method was purely algebraic, dealing with infinite series. He wrote the exponential function as

$$e^z = 1 + \frac{z}{1!} + \frac{z^2}{2!} + \frac{z^3}{3!} + \cdots + \frac{z^n}{n!} + \cdots$$

and assumed that the series worked for a complex number z. He then wrote out the power series for sine and cosine (again for a complex number z) as

$$\sin z = z - \frac{z^3}{3!} + \frac{z^5}{5!} \cdots + (-1)^n \frac{z^{2n-1}}{(2n-1)!} + \cdots$$

and

$$\cos z = 1 - \frac{z^2}{2!} + \frac{z^4}{4!} \cdots + (-1)^n \frac{z^{2n}}{(2n)!} + \cdots$$

and substituted $z = i\theta$ to give the remarkable equation

$$e^{i\theta} = \cos \theta + i \sin \theta.$$

He could then take $\theta = \pi$, where $\cos \pi = 0$ and $\sin \pi = -1$, to get the relationship

$$e^{i\pi} = -1.$$

You can bet that Euler was pleased with this!

In fact, if you multiply this equation by minus one, you get the equation

$$-e^{i\pi} = 1$$

in which four of the most problematic aspects of arithmetic—the minus sign, the irrational numbers e and π and the complex number i—all combine together to give the simple number 1.

In this book our goal is to focus on the mathematical foundations, so the study of complex power series is postponed to a later course (see, for example, [34]). In such a course, once the theory of power series has been developed, we can provide very elegant proofs of these results, including the formula for $\cos(A+B)$ and $\sin(A+B)$ for A, B not only real, but also complex. However, at this point, it is instructive to attack the problem in a direct way using only real exponential and trigonometric functions and their properties derived in more elementary mathematics.

Addition Formulas for Cosine and Sine

The most important properties of trigonometric functions, in this connection, are the addition formulas for cosine and sine:

$$\cos(A+B) = \cos A \cos B - \sin A \sin B, \tag{11.3}$$

$$\sin(A+B) = \sin A \cos B + \cos A \sin B. \tag{11.4}$$

You may have seen a geometric proof of these, but this may be in terms of right-angled trigonometry in right-angled triangles where the angles concerned are less than a right angle. In this case, we take an angle A and consider a right triangle with one angle equal to A, as in figure 11.5 (left). Then, with the triangle sides as shown, we define

$$\cos A = x/r, \ \sin A = y/r, \tag{11.5}$$

and remark that by similar triangles these ratios do not depend on r. In particular, we could set $r = 1$ from the start. (The tangent is given by $\tan A = y/x$, but here we focus just on the cosine and sine.)

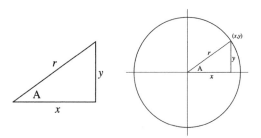

Fig. 11.5 Relationships in a right-angled triangle

This approach initially assumes that A is an acute angle: $0 \leq A \leq \pi/2$. If A is an obtuse angle, so that $\pi/2 < A \leq \pi$, the natural right triangle lies to the left of the y-axis, so x is negative. The internal angle of the triangle is $\pi - A$. Using (11.5) in this new context, we see that

$$\cos A = -\cos(\pi - A) \sin A = \sin(\pi - A) \tag{11.6}$$

for $\pi/2 < A \leq \pi$.

It is possible to continue like this, extending to the ranges $\pi < A \leq 3\pi/2$ and $3\pi/2 < A \leq 2\pi$. Then all other real values for A can be dealt with using periodicity:

$$\cos(A + 2\pi) = \cos A, \ \sin(A + 2\pi) = \sin A.$$

However, this approach involves a lot of cases and gets quite complicated. Here we present an alternative approach, which takes a detour through complex numbers. Along the way, we define cos A and sin A for all real A, and deduce (11.3), (11.4) for all real A, B. Finally, we verify that the resulting extensions of cos and sin to the entire real line agree with those obtained by extending the range of values for A on a case-by-case basis.

Theorem 11.3: If $0 \leq A, B \leq \pi/2$ then (11.3), (11.4) are valid.

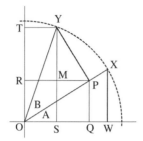

Fig. 11.6 The proof of the formula for sin(A+B)

Proof: We may assume that $r = 1$ for simplicity. Consider figure 11.6, which tacitly assumes not only that $0 \leq A, B \leq \pi/2$, but that $0 \leq A+B \leq \pi/2$. Assume for the moment that this is true.

In figure 11.6, the three important right triangles are

OWX, which tells us about sin A and cos A,
OPY, which tells us about sin B and cos B,
OSY, which tells us about sin(A+B) and cos(A+B).

By the definition of sine and cosine,

$$OW = \cos A,$$
$$WX = \sin A,$$
$$OP = \cos B,$$
$$PY = \sin B.$$

We also need to know that triangles OQP and AWX are similar. The scale factor here is the ratio OP : OX, which is cos B : 1. So triangle OQP has the same shape as AWX, but its size is multiplied by cos B.

Now cos(A+B) = OS = OQ – QS.

By the remark on similar triangles, OQ : OW = cos B, so

$$OQ = OW \cos B = \cos A \cos B.$$

Also, triangle YMP is similar to OWX with scale factor YP : OX = sin B, so

$$QS = PY \sin B = \sin A \sin B.$$

Therefore

$$\cos(A+B) = \cos A \cos B - \sin A \sin B,$$

which is (11.3). The proof for sin(A+B) is similar, based on sin(A+B) = YS = YM + MS.

What if A+B > $\pi/2$? Now the picture is similar to figure 11.6, but Y is to the left of the vertical axis. The same line of argument works, but it is necessary to use (11.6) and be careful about signs. □

We now link the trigonometric functions to the complex exponential function, starting with an angle θ in the range $0 \le \theta \le \pi/2$. For such we define

$$e^{i\theta} = \cos\theta + i\sin\theta. \tag{11.7}$$

We can use (11.3), (11.4) to prove:

Lemma 11.4: If $0 \le \theta, \phi \le \pi/2$, then

$$e^{i(\theta+\phi)} = e^{i\theta} e^{i\phi}.$$

Proof: From (11.7),

$$e^{i(\theta+\phi)} = \cos(\theta + \phi) + i\sin(\theta + \phi).$$

Since $0 \le \theta, \phi \le \pi/2$ we can appeal to (11.3), (11.4) to get

$$\begin{aligned}
e^{i(\theta+\phi)} &= \cos(\theta + \phi) + i\sin(\theta + \phi) \\
&= \cos\theta \cos\phi - \sin\theta \sin\phi + i(\sin\theta \cos\phi + \cos\theta \sin\phi) \\
&= (\cos\theta + i\sin\theta)(\cos\phi + i\sin\phi) \\
&= e^{i\theta} e^{i\phi}.
\end{aligned}$$ □

The next step is to extend the definition of the exponential of $i\theta$ to any real number θ. The main point is that putting $\theta = \pi/2$ in equation (11.7) tells us that

$$e^{i\pi/2} = \cos(\pi/2) + i\sin(\pi/2) = 0 + i.1 = i.$$

We therefore define

$$e^{i(\theta+\pi/2)} = ie^{i\theta}. \tag{11.8}$$

Initially, we know what $e^{i\theta}$ means only when $0 \leq \theta \leq \pi/2$. Use (11.8) with θ in this range to define $e^{i\theta}$ in the range $\pi/2 \leq \theta \leq \pi$. Observe that there is no contradiction at $\pi/2$ since (11.8) applied with $\theta = 0$ gives $e^{i\pi/2} = i$.

Inductively, we can now extend the range to all positive real θ by repeatedly multiplying by i. Moreover, if we replace θ by $-\theta$ in (11.8) and divide through by i we get

$$e^{i(\theta-\pi/2)} = -ie^{i\theta},$$

so we can also extend the definition to negative real numbers θ. Again, the definition is consistent whenever the endpoints of ranges of θ coincide.

A direct consequence of (11.8) is:

Lemma 11.5: For any $\theta \in \mathbb{R}$,

$$e^{i(\theta+2\pi)} = e^{i\theta}.$$

Proof: By (11.8) applied four times,

$$e^{i(\theta+2\pi)} = ie^{i(\theta+3\pi/2)} = i^2 e^{i(\theta+\pi)} = i^3 e^{i(\theta+\pi/2)} = i^4 e^{i\theta} = e^{i\theta},$$

since $i^4 = 1$. $\qquad\qquad\qquad\qquad\qquad\qquad\qquad\qquad\qquad\qquad\qquad\qquad\square$

Having defined the exponential of $i\theta$ for all real θ, we can use (11.7) in the opposite direction to define cos and sin for all real θ:

Definition 11.6:

$$\cos\theta = \text{Re } e^{i\theta}, \quad \sin\theta = \text{Im } e^{i\theta}.$$

This definition makes both sin and cos periodic with period 2π, as we would expect:

Proposition 11.7: For any $\theta \in \mathbb{R}$,

$$\sin(\theta + 2\pi) = \sin\theta,$$
$$\cos(\theta + 2\pi) = \cos\theta.$$

Proof: Use lemma 11.5 and equate real and imaginary parts. $\qquad\qquad\square$

Using the formulae for $\sin(A+B)$ and $\cos(A+B)$, for real values we can then establish:

Proposition 11.8:

$$e^{i(x+y)} = e^{ix}e^{iy} \text{ for } x, y \in \mathbb{R}.$$

Proof:

$$
\begin{aligned}
e^{i(x+y)} &= \cos(x + y) + i\sin(x + y) \\
&= \cos x \cos y - \sin x \sin y + i(\sin x \cos y + \cos x \sin y) \\
&= (\cos x + i\sin y)(\cos y + i\sin y) \\
&= e^{ix}e^{iy}.
\end{aligned}
$$

□

The Complex Exponential Function

The final step in defining e^z for all complex z is to remove the restriction that z should be purely imaginary, that is, $z = i\theta$.

Definition 11.9: Let $z = x + iy \in \mathbb{C}$. Then

$$e^z = e^x \cos y + ie^x \sin y. \tag{11.9}$$

Since x and y are real, this expression makes sense. If $y = 0$, so that $z = x \in \mathbb{R}$, it implies that

$$e^z = e^x \cos 0 + ie^x \sin 0 = e^x,$$

since $\cos 0 = 1$, $\sin 0 = 0$. So the complex exponential reduces to the usual real exponential when z is real, which goes some way towards justifying the notation e^z.

Moreover, (11.9) immediately implies that

$$e^{x+iy} = e^x e^{iy}.$$

We can now establish a basic property of the complex exponential function:

Theorem 11.10: If $z, w \in \mathbb{C}$ then

$$e^{z+w} = e^z e^w.$$

Proof: Let $z = x + iy$, $w = u + iv$, where $x, y, u, v \in \mathbb{R}$. Then

$$e^{z+w} = e^{x+iy+u+iv}$$
$$= e^{(x+u)+i(y+v)}$$
$$= e^{x+u}e^{i(y+v)} \text{ by definition 11.9}$$
$$= e^x e^u e^{iy} e^{iv} \text{ by proposition 11.8}$$
$$= e^x e^{iy} e^u e^{iv}$$
$$= e^{x+iy} e^{u+iv} \text{ by definition 11.9}$$
$$= e^z e^w. \qquad \square$$

We can now prove:

Theorem 11.11 (De Moivre's Theorem): If $n \in \mathbb{N}$ then

$$(\cos\theta + i\sin\theta)^n = \cos n\theta + i\sin n\theta.$$

Proof: By definition 9 this is equivalent to proving that $(e^{i\theta})^n = e^{in\theta}$. Use induction on n. If $n = 1$ both sides are identical. Suppose the result is true for n, and consider:

$$(e^{i\theta})^n e^{i\theta} = e^{in\theta} e^{i\theta}$$
$$= e^{in\theta + i\theta} \text{ by theorem 11.10}$$
$$= e^{i(n+1)\theta}$$

which completes the induction. $\qquad \square$

Examples 11.12: Let $n = 2$. Then $(\cos\theta + i\sin\theta)^2 = \cos 2\theta + i\sin 2\theta$. Expand the first expression as $\cos^2\theta - \sin^2\theta + i(2\cos\theta\sin\theta)$ and equate real and imaginary parts to get

$$\cos 2\theta = \cos^2\theta - \sin^2\theta, \ \sin 2\theta = 2\cos\theta\sin\theta,$$

which are familiar trigonometric formulas.

Let $n = 3$. A similar calculation, expanding the cube of $(\cos\theta + i\sin\theta)$, yields:

$$\cos 3\theta = \cos^3\theta - 3\cos\theta\sin^2\theta, \ \sin 3\theta = 3\cos^2\theta\sin\theta - \sin^3\theta.$$

The method extends to larger multiples of θ.

We can also express the sine and cosine using exponentials:

Theorem 11.13: If $\theta \in \mathbb{R}$ then

$$\cos \theta = \frac{e^{i\theta} + e^{-i\theta}}{2}, \quad \sin \theta = \frac{e^{i\theta} - e^{-i\theta}}{2i}.$$

Proof: Use the equations

$$e^{i\theta} = \cos \theta - i \sin \theta \qquad e^{-i\theta} = \cos \theta - i \sin \theta$$

and solve for $\cos \theta$ and $\sin \theta$. $\qquad\qquad\qquad\qquad\qquad\qquad$ □

The real and imaginary parts x, y of a complex number $z = x + iy$ are the Cartesian coordinates of z in the complex plane. Another useful system, polar coordinates, leads to a different way to represent z:

Theorem 11.14: Every $z \in \mathbb{C}$ has a unique expression in the form $z = re^{i\theta}$, where $r, \theta \in \mathbb{R}$, $r \geq 0$, and $0 \leq \theta < 2\pi$.

Proof: Write $re^{i\theta} = r \cos \theta + ir \sin \theta$ and then solve the equations $r \cos \theta = x$, $r \sin \theta = y$ with the stated conditions (see figure 11.7). $\qquad\qquad\qquad$ □

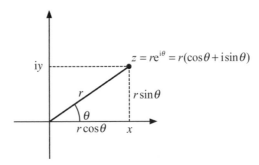

Fig. 11.7 Representing a complex number z in the form $e^{i\theta}$

By Pythagoras' theorem, the number r is equal to the modulus $|z| = \sqrt{x^2 + y^2}$. The angle θ is called the *argument* of z, written arg z.

We can now interpret addition and multiplication of complex numbers geometrically.

For addition, take a fixed but arbitrary complex number $w = u + iv$ and let $z = x + iy$ be any number in $\mathbb{C}$. Consider the map 'add w' from $\mathbb{C}$ to $\mathbb{C}$:

$$\alpha_w(z) = z + w.$$

Then

$$\alpha_w(z) = (x + iy) + (u + iv) = (x + u) + i(y + v).$$

Clearly the effect of this map is to translate z a distance u along the real axis and y along the imaginary axis. So the entire plane slides rigidly so that the origin moves to w.

Multiplication can be described using the Cartesian form, but it makes more sense in the polar coordinate representation. Take a fixed but arbitrary complex number $w = se^{i\theta}$ and let $z = re^{i\phi}$ be any number in $\mathbb{C}$. Consider the map 'multiply by w' from $\mathbb{C}$ to $\mathbb{C}$:

$$\mu_w(z) = zw.$$

Then

$$\mu_w(z) = re^{i\theta}se^{i\phi} = rse^{i(\theta+\phi)}.$$

The effect of this map is to multiply all distances from the origin by a factor s (known as *dilation*) and to rotate the entire complex plane anticlockwise about the origin through and angle ϕ.

Complex numbers manage to combine both Cartesian and polar coordinates in a single mathematical system. Cartesians are best for addition, polars for multiplication. To fill in a final piece of the picture: complex conjugation, mapping $z = x + iy$ to $\bar{z} = x - iy$, reflects the complex plane in the real axis. So the algebra of complex numbers involves the three basic types of rigid motion of the plane (translation, rotation, reflection) and dilations in a natural manner. This makes complex notation a very efficient way to perform calculations involving these transformations of the plane, as we will see in chapter 13.

Quaternions

We might attempt to extend the number system $\mathbb{N}_0 \subseteq \mathbb{Z} \subseteq \mathbb{Q} \subseteq \mathbb{R} \subseteq \mathbb{C}$ further and look for an extension of $\mathbb{C}$. For years in the last century Hamilton followed up his conception of complex numbers as ordered couples (x, y) of real numbers, searching for a system of triples (x_1, x_2, x_3) with similar properties to those of the complex numbers. He never found such a system; we now know that none exists. But in 1843, in a marvellous piece of lateral thinking, he found a system of quadruples (x_1, x_2, x_3, x_4) that is 'almost' a field. It satisfies all the field axioms, *except* for commutativity of multiplication.

Definition 11.15: A *division ring* is an algebraic system consisting of a set D and two binary operations $+, \times$ on D such that for all $a, b, c \in D$, and writing ab for $a \times b$ as usual,

(A₁) $(a + b) + c = a + (b + c)$.
(A₂) There exists $0 \in D$ such that for all $a \in D$, $0 + a = a + 0 = a$.
(A₃) Give $a \in D$, there exists $-a \in D$ such that $a + (-a) = (-a) + a = 0$.
(A₄) $a + b = b + a$.
(M₁) $(ab)c = a(bc)$.
(M₂) There exists $1 \in D$, $1 \neq 0$ such that for all $a \in D$, $a1 = 1a = a$.
(M₃) Given $a \in D$, $a \neq 0$, there exists $a^{-1} \in D$ such that $aa^{-1} = a^{-1}a = 1$.
(D) $a(b + c) = ab + ac$, $(b + c)a = ba + ca$.

Hamilton's quaternions, as he called his system of quadruples, is an example of a division ring. Its multiplication is not commutative: for some elements a, b we have $ab \neq ba$. His discovery can be explained in terms of three symbols i, j, k multiplied according to the rules:

$$i^2 = j^2 = k^2 = -1$$
$$ij = k, jk = i, ki = j$$
$$ji = -k, kj = -i, ik = -j.$$

The last six of these can be described by writing the symbols i, j, k in a clockwise cycle:

Fig. 11.8 Hamilton's quaternions

Then the product of any two in clockwise order is the third, and the product anticlockwise is minus the third.

Hamilton thought of a quadruple of real numbers (x_1, x_2, x_3, x_4) as $x_1 + ix_2 + jx_3 + kx_4$. He added them in the obvious way:

$$(x_1 + ix_2 + jx_3 + kx_4) + (y_1 + iy_2 + jx_3 + ky_4)$$
$$= (x_1 + y_1) + i(x_2 + y_2) + j(x_3 + y_3) + k(x_4 + y_4),$$

and multiplied them using the above rules for multiplying i, j, k. Written out in full this amounts to:

$$(x_1 + \mathrm{i}x_2 + \mathrm{j}x_3 + \mathrm{k}x_4)(y_1 + \mathrm{i}y_2 + \mathrm{j}y_3 + \mathrm{k}y_4)$$
$$= x_1 y_1 - x_2 y_2 - x_3 y_3 - x_4 y_4$$
$$+ \mathrm{i}(x_1 y_2 + x_2 y_1 + x_3 y_4 - x_4 y_3)$$
$$+ \mathrm{j}(x_1 y_3 - x_2 y_4 + x_3 y_1 + x_4 y_2)$$
$$+ \mathrm{k}(x_1 y_4 + x_2 y_3 - x_3 y_2 + x_4 y_1).$$

This can be written in terms of ordered quadruples simply by replacing each $a_1 + \mathrm{i}a_2 + \mathrm{j}a_3 + \mathrm{k}a_4$ by (a_1, a_2, a_3, a_4) in the obvious manner. So formally we can define addition and multiplication of such quadruples by

$$(x_1, x_2, x_3, x_4) + (y_1, y_2, y_3, y_4) = (x_1 + y_1, x_2 + y_2, x_3 + y_3, x_4 + y_4)$$
$$(x_1, x_2, x_3, x_4)(y_1, y_2, y_3, y_4) = (a_1, a_2, a_3, a_4)$$

where

$$a_1 = x_1 y_1 - x_2 y_2 - x_3 y_3 - x_4 y_4,$$
$$a_2 = x_1 y_2 + x_2 y_1 + x_3 y_4 - x_4 y_3,$$
$$a_3 = x_1 y_3 - x_2 y_4 + x_3 y_1 + x_4 y_2,$$
$$a_4 = x_1 y_4 + x_2 y_3 - x_3 y_2 + x_4 y_1.$$

We denote the set of all quadruples, with these operations, by $\mathbb{H}$ (for Hamilton). These quadruples are called *quaternions* or (a more old-fashioned term) *hypercomplex numbers*.

Proposition 11.16: The quaternions $\mathbb{H}$ form a division ring.

Proof: This is simply a matter of checking the axioms (A1)–(A4), (M1)–(M3), (D) for $\mathbb{H}$. They are all straightforward, although we will be the first to admit that the associativity of multiplication (M1) is tedious to say the least. The zero element in (A2) is $(0, 0, 0, 0)$, the negative of (x_1, x_2, x_3, x_4) in (A3) is $(-x_1, -x_2, -x_3, -x_4)$, the unit element in (M2) is $(1, 0, 0, 0)$, and the inverse of $(x_1, x_2, x_3, x_4) \neq (0, 0, 0, 0)$ in (M3) is

$$(x_1, x_2, x_3, x_4)^{-1} = (x_1/a, -x_2/a, -x_3/a, -x_4/a),$$

where $a = x_1^2 + x_2^2 + x_3^2 + x_4^2$. $\qquad\square$

Multiplication in $\mathbb{H}$ need not be commutative; for instance,

$$(0, 1, 0, 0)(0, 0, 1, 0) = (0, 0, 0, 1)$$

but

$$(0, 0, 1, 0)(0, 1, 0, 0) = (0, 0, 0, -1).$$

Writing $i = (0, 1, 0, 0)$, $j = (0, 0, 1, 0)$, $k = (0, 0, 0, 1)$, this amounts to $ij = k$, $ji = -k$ as explained previously. Hamilton's other rules for multiplication of i, j, k also follow, because we set up the rule of multiplication to make it happen that way.

If we look at the subset $C = \{(x, y, 0, 0) \in \mathbb{H} \mid x, y \in \mathbb{R}\}$, we find that multiplication on C reduces to

$$(x_1, y_1, 0, 0)(x_2, y_2, 0, 0) = (x_1 x_2 - y_1 y_2, x_1 y_2 + x_2 y_1, 0, 0),$$

and that this *is* commutative. The map $f : \mathbb{C} \to \mathbb{H}$ given by $f(x + iy) = (x, y, 0, 0)$ is easily seen to be an isomorphism of fields from $\mathbb{C}$ to C. Via this isomorphism we can regard $\mathbb{C}$ as a subset of $\mathbb{H}$. Writing (x_1, x_2, x_3, x_4) as $x_1 + ix_2 + jx_3 + kx_4$ the function $f : \mathbb{C} \to \mathbb{H}$ becomes

$$f(x + iy) = x + iy + j0 + k0.$$

Inclusion $\mathbb{C} \subseteq \mathbb{H}$ regards the complex number $x + iy$ as the quaternion $x + iy + j0 + k0$.

Many properties of $\mathbb{C}$ can be generalised to $\mathbb{H}$, hence the name 'hypercomplex numbers'. For instance the *conjugate* of a quaternion $q = x_1 + ix_2 + jx_3 + kx_4$ is

$$\bar{q} = x_1 - ix_2 - jx_3 - kx_4.$$

This has some of the properties of the complex conjugate, but not all. In particular,

$$\overline{q_1 + q_2} = \bar{q}_1 + \bar{q}_2$$
$$\bar{\bar{q}} = q$$
$$q = \bar{q} \Leftrightarrow q \in \mathbb{R}.$$

However, the rule for the conjugate of a product becomes

$$\overline{q_1 q_2} = \bar{q}_2 \bar{q}_1,$$

as you can check by explicit calculation. Because multiplication is not commutative, we can't straighten this out by reversing the order of $\bar{q}_2$ and $\bar{q}_1$.

We can also define the *modulus* of a quaternion $q = x_1 + ix_2 + jx_3 + kx_4$ to be

$$|q| = \sqrt{x_1^2 + x_2^2 + x_3^2 + x_4^2}.$$

In this case,

$$|q| \in \mathbb{R}, \quad |q| \geq 0 \quad \text{for all } q \in \mathbb{H},$$
$$|q| = 0 \Leftrightarrow q = 0,$$
$$q\bar{q} = |q|^2,$$
$$|q_1 q_2| = |q_1| \, |q_2|,$$
$$|q_1 + q_2| \leq |q_1| + |q_2|.$$

The proofs of these formulas vary in difficulty, though none is truly hard. If you are interested, you should seek to work them out for yourself. They are analogous to the complex case, taking care with the non-commutativity of $\mathbb{H}$.

As with complex numbers, for $q \in \mathbb{H}$, $q \neq 0$, we find $q\bar{q} = |q|^2$ where $|q|^2 \neq 0$, so

$$q\bar{q} / |q|^2 = 1$$

and

$$q^{-1} = \bar{q} / |q|^2.$$

Some properties of the quaternions are startling, to say the least. For instance, we know that $i^2 = j^2 = k^2 = (-i)^2 = (-j)^2 = (-k)^2 = -1$, so the equation $x^2 + 1 = 0$ has at least six solutions in $\mathbb{H}$, namely $\pm i$, $\pm j$, $\pm k$. In fact $(ib + jc + kd)^2 = -b^2 - c^2 - d^2$, so any quaternion $ib + jc + kd$ where $b^2 + c^2 + d^2 = 1$ is a solution of $x^2 + 1 = 0$. There are an *infinite* number of solutions in $\mathbb{H}$.

This is unlike our experience in all previous number systems. In $\mathbb{R}$ the equation $x^2 + 1 = 0$ has no solutions, in $\mathbb{C}$ it has two, and in general, in $\mathbb{R}$ or $\mathbb{C}$, an equation of degree n has at most n solutions. The sudden appearance of more roots in the quaternions completely changes the game as a fondly held belief fails in the new system.

The problem with intuition is that experience in one context need not lead to expected properties in another. As we move through successively larger number systems $\mathbb{N} \subset \mathbb{Z} \subset \mathbb{Q} \subset \mathbb{R} \subset \mathbb{C} \subset \mathbb{H}$, we gain some properties, but we also lose others. In the natural numbers $\mathbb{N}$, subtracting a number leaves a smaller one, but not in the integers $\mathbb{Z}$, where taking away a negative number gives more. In the real numbers $\mathbb{R}$, the square of a non-zero number is always positive, but not in the complex numbers $\mathbb{C}$. Now we find that in the quaternions $\mathbb{H}$, the theorem that every equation of degree n has at most n roots is no longer true.

This continual change of meaning as mathematical systems are generalised can cause serious disorientation for the learner. But it is also the secret of developing more powerful mathematical systems.

Prior to the introduction of the quaternions, the idea that multiplication of numbers is independent of their order was always considered to be a self-evident, preordained law. The discovery of the quaternions revealed an algebraic system worthy of study in its own right, but it also revealed that it is possible to have algebraic systems in which 'a times b' need not equal 'b times a'. This led to the many new algebraic structures studied in modern mathematics. For example, matrix multiplication is not commutative, and the theory of vectors and matrices is an essential feature of advanced algebra.

The Change in Approach to Formal Mathematics

At this point we free ourselves from the mental block that mathematics must work in a preordained natural way, by specifying new axiomatic systems in terms of some set with various prescribed operations that have specific properties. We have already exemplified this process with rings, fields, ordered rings, ordered fields, and so on. The rationals, reals, and complex numbers are all examples of fields, so *any* theorem that we prove true for all fields will be true in all of these specific systems. There may also be theorems that are true in one field but not in another; for instance, that a Cauchy sequence in $\mathbb{R}$ or $\mathbb{C}$ will converge to a limit, but not in $\mathbb{Q}$, or the square of a non-zero number is always positive in $\mathbb{Q}$ or $\mathbb{R}$ but not in $\mathbb{C}$.

The formal approach, starting from a list of axioms for a system, may seem complicated and abstract. However, once we have shifted to a formal approach, the reverse is often true. Any theorems we prove in a given axiomatic system will remain true in any new system that satisfies the given axioms. This enables us to build more sophisticated structures based on established formal theories.

It may also happen that some of the theorems we prove offer new ways of visualising and symbolising the ideas, giving us new ways of imagining the structure and operating with the elements of a formal system. As we shall see in the next part of the book, not only do intuitive ideas build into formal concepts, the resulting formal concepts may take us back to natural ways of visualising and operating symbolically with the axiomatic systems, now supported by the power of formal proof.

Exercises

1. If $z_1, \ldots, z_n$ are complex numbers, prove that

$$|z_1 + \cdots + z_n| \leq |z_1| + \cdots + |z_n|.$$

2. Let ω be the complex number defined by $\omega = (1 + \sqrt{-3})/2$. Prove that $\omega^3 = 1$ and $1 + \omega + \omega^2 = 0$.

3. Let $\omega = e^{i\theta}$ where $\theta = 2\pi/n$ for $n \in \mathbb{N}$. Show that $z = \omega^r$ satisfies $z^n = 1$ and draw a picture showing the positions of $\omega, \omega^2, \ldots, \omega^n$ round the unit circle. (These numbers are called the nth roots of 1.)

   Show that $1 + \omega + \cdots + \omega^{n-1} = 0$.

   Factorise $z^n - 1$ into linear factors over $\mathbb{C}$. By showing that

   $$(z - \omega^r)(z - \omega^{n-r}) = z^2 - 2\cos\theta + 1,$$

   factorise $z^n - 1$ into linear and quadratic factors.

   In particular, factorise the real polynomial $x^5 - 1$ into real linear and quadratic factors:

   $$x^5 - 1 = (x - 1)(x^2 - 2\cos(2\pi/5) + 1)(x^2 - 2\cos(4\pi/5) + 1).$$

4. Use De Moivre's theorem to find formulae for $\cos 4\theta$ and $\sin 4\theta$ in terms of $\sin\theta$ and $\cos\theta$.

5. For quaternions p, q verify
   (a) $\overline{p + q} = \bar{p} + \bar{q}$
   (b) $\overline{pq} = \bar{q}\bar{p}$
   (c) $\bar{\bar{q}} = q$
   (d) $q = \bar{q} \Leftrightarrow q \in \mathbb{R}$.

6. For $a, b \in \mathbb{H}$, show $(a + b)^2 = a^2 + ab + ba + b^2$. Give an example to show that we cannot replace this by $(a + b)^2 = a^2 + 2ab + b^2$ in general. If $a \in \mathbb{H}, b \in \mathbb{R}$, prove that $(a + b)^2 = a^2 + 2ab + b^2$. Solve the equation $x^2 + 2x + 1 = 0$ in $\mathbb{R}, \mathbb{C}$, and $\mathbb{H}$. (Let $x = y - 1$ and solve for y.) By the substitution $x = y + 1$, solve the equation $x^2 - 2x + 2 = 0$ in $\mathbb{R}, \mathbb{C}$, and $\mathbb{H}$.

7. Solve the equation $x\left(1 + j\right) + k = 2 + i$ for the quaternion x.

8. Solve the equation $ixj + k = 3 + 2j$ for the quaternion x.

9. Find $x, y \in \mathbb{H}$ such that $3ix - 2jy = -1, xk + y = 0$.

10. Define *complex quaternions* $\mathbb{H}_\mathbb{C}$ to be quadruples (a_1, a_2, a_3, a_4) of complex numbers, with the same addition and product rules as given for $\mathbb{H}$ on page 249. Which of the axioms for a field does $\mathbb{H}_\mathbb{C}$ satisfy?

11. Prove that the complex numbers are *Cauchy complete*, in the following sense: if (a_n) is a sequence of complex numbers such that for all $\varepsilon \in \mathbb{R}$, $\varepsilon > 0$, there exists $N \in \mathbb{N}$ such that $|a_m - a_n| < \varepsilon$ for all $m, n > N$, then (a_n) tends to a limit in $\mathbb{C}$. (Hint: Show that $x_n + iy_n \to x + iy \Leftrightarrow x_n \to x \ \& \ y_n \to y$.)

12. Define a binary operation $\wedge$ on $\mathbb{R}^3$ (known as the *vector product*) as follows:

$$(a, b, c) \wedge (d, e, f) = (bf - ce, cd - af, ae - bd).$$

Prove for all $x,\ y \in \mathbb{R}^3$,

$$x \wedge y + y \wedge x = 0,$$
$$(x \wedge y) \wedge z + (y \wedge z) \wedge x + (z \wedge x) \wedge y = 0.$$

13. Consider the ordered pairs of complex numbers (z_1, z_2) with addition and multiplication defined by

$$(z_1, z_2) + (w_1, w_2) = (z_1 + w_1,\ z_2 + w_2),$$
$$(z_1, z_2) \times (w_1, w_2) = (z_1 w_1 - z_2 w_2,\ z_1 w_2 + z_2 w_1).$$

Show that this is isomorphic to the quaternions. (Hint: Consider how the complex numbers were constructed by imagining $x + iy$ as the ordered pair (x, y) with an appropriate addition and multiplication, and generalise this to ordered pairs of complex numbers.)

14. Look up the octonions on the internet and see how the above construction might be extended to ordered pairs of quaternions. The quaternions lack the axiom of commutative multiplication. What is lost in extending to the octonions? Which is a more productive way to develop: generalising from real to complex to quaternions to octonions and so on, or to develop the general theory of vector spaces of dimension n over a field F?

PART IV
Using Axiomatic Systems

In part I we began by building on earlier experiences in mathematics. In part II we used our experiences to build ideas of set theory, logic, and proof. Then we used these principles in part III to give formal constructions of natural numbers and successively larger number systems.

Our earlier metaphors of building houses and growing plants depend on the foundations on which we base our activities. Building on intuition may include implicit beliefs that cause our house to have weak foundations or our plants to grow in unsuitable soil.

The way ahead is to specify clearly the foundations of set-theoretic axioms and definitions being used in a particular theory and to focus *only* on properties that can be deduced from them by formal proof. These properties remain true, not only in systems that are already familiar, but also in any future examples that satisfy the given axioms. This needs to be done with care to make sure that we do not use any implicit ideas that have not been proved from the axioms and definitions.

Having built up a framework of theorems proved from the foundational axioms and definitions, some theorems, called *structure theorems*, may prove that the system also has visual and symbolic properties, allowing us to imagine formal structures in more natural ways. This enables us to imagine new possibilities that we may then seek to prove formally.

In the case of a complete ordered field, we were led to the visual structure of a number line, and the symbolic structure of decimal arithmetic. Structure theorems let us complement formal theory with natural visual and symbolic models that we can use to imagine new possibilities.

The first chapter of this part examines these general ideas. The next three chapters consider examples of formal systems and their natural interpretations. The first deals with the notion of a group, a central idea in formal mathematics. The other two describe extensions of the natural number system $\mathbb{N}$ to infinite cardinal numbers, and of the real numbers $\mathbb{R}$ to larger ordered fields. Both have structural properties that extend natural ideas to give new intuitions now based on formal axioms and proof. This reveals the great power of formal mathematics, with structures that may be imagined visually and symbolically, now supported by formal proof.

CHAPTER 12

Axiomatic Systems, Structure Theorems, and Flexible Thinking

As we construct successively larger number systems $\mathbb{N} \subseteq \mathbb{Z} \subseteq \mathbb{Q} \subseteq \mathbb{R} \subseteq \mathbb{C} \subseteq \mathbb{H}$, each stage gains by generalising some properties, but other meanings change. We can talk of prime numbers and factorisation in the natural numbers, but this has no relevance for real numbers in general. Fondly held beliefs may fail in more general structures as we found when we shifted from the natural numbers to introduce negative or complex numbers.

While the generalisation of ideas can give great power and pleasure, changes in meaning can cause serious disorientation, not only for learners, but also for research mathematicians. Even a change in a single property, such as the loss of the commutative law in the quaternions, has unforeseen consequences. For instance, we saw that a quaternionic polynomial can have an infinite number of roots, and it is not immediately clear how the loss of commutativity leads to that effect.

These long-term changes in meaning are not only problematic for you the reader, they have also occurred in the beliefs of communities of mathematicians as ideas evolve over the generations. Not only did they occur in the past; they continue to occur in the present, and will undoubtedly continue into the future as the boundaries of mathematics keep expanding.

When the Greeks began to formulate geometry, they imagined that points, lines, and planes had subtle meanings more perfect than a particular picture drawn on paper or scratched in sand. A point for the Greeks was not just a mark on paper, it represented a unique location in the plane or in space. A straight line was not just the practical result of drawing a pencil line by hand guided by a ruler, it was a representation of a perfect straight line with a Platonic existence beyond any human capacity to represent it physically. A circle was more perfect than a curve drawn physically using a pair of compasses: it was the locus of a sizeless point moving on a plane at a fixed distance from its centre.

Likewise, whole numbers can be represented practically by counting pebbles and placing them in patterns that reveal theoretical structures. For example, given a number of pebbles, we can sometimes place them physically as a rectangular array and sometimes we cannot, leading to the conceptions of composite and prime numbers, and eventually to the formal proof that there is an infinite number of primes, and that every whole number can be uniquely represented as a product of primes.

The ancient Greek conception of mathematics was based on phenomena that occur in nature, yet were imagined to have perfect Platonic properties that were physically unattainable. In this sense, their mathematics is *natural*, in that it is based on observed natural phenomena. Yet they sought a perfect theoretical foundation, arising in their imagination, which takes them beyond what is physically attainable in nature.

As they contemplated more general numbers—which to their way of thinking had to be done using geometry—they first imagined numbers as magnitudes that measured lengths, areas, and volumes. They related these magnitudes to ratios of whole numbers, based on experience in other areas, such as music where causing a string to vibrate at a half, a third, or two thirds of its length produces harmonics that provide the basis of musical theory. But then they discovered that the hypotenuse of a right-angled triangle with unit sides is not rational, so they had to take this into account in developing their mathematical theories.

Subsequent communities of mathematicians broadened these ideas by introducing new number systems, each of which was accompanied by language that expressed concerns about the new meanings: positive and *negative* numbers, rationals and *irrationals*, real and *complex* numbers that have real and *imaginary* parts. The italicised words all have negative connotations. At every stage, these new number systems were initially imagined to be more abstract than the old ones, and did not appear to relate to naturally occurring phenomena. But at a later stage, as mathematicians became more sophisticated, they found new ways of imagining negative numbers as points on an extended number line and complex numbers as points in the plane. Moreover, the familiar older concepts started to look just as puzzling as the new ones. By the time mathematicians finally understood what a complex number was, they had started to wonder about real numbers.

Geometric ideas continued to be based on imagining points as entities that can be marked *on* lines and lines that go *through* points. Even when

Descartes envisaged points in the plane using a pair of numbers (x, y), the Greek view of points and lines continued as the natural basis of geometric thinking.

Newton explained naturally occurring phenomena, such as gravity and the movement of the planets, using a combination of Greek geometry and symbolic algebra to build his ideas in the calculus. Leibniz imagined quantities that could be infinitesimally small and produced a powerful symbolism for the calculus that has stood the test of time, despite widespread concerns about its logical foundation. Later giants in mathematical development focused on different aspects. Euler manipulated symbols algebraically using power series and complex numbers, and Cauchy imagined infinitesimals geometrically as variable quantities on the line or in the plane that become arbitrarily small. His approach led to major advances, using a blend of visual and symbolic methods in real and complex analysis, but it also generated significant criticism about the precise meaning. The critics had a point: the meaning had not then been fully worked out. What prevailed was more an act of faith, that everything would work out much as it always had done. Many of Euler's published papers would have caused him to fail today's examinations and Cauchy's ideas of infinitesimal quantities were later heavily criticised.

In the latter part of the nineteenth century and the early twentieth, a shift occurred from natural mathematics to more formal methods. Mathematical entities were introduced through set-theoretic definitions and their properties were deduced solely through mathematical proof. A seminal moment occurred when David Hilbert, taking refreshment with colleagues in the Berlin railway station after a lecture on the foundations of geometry, is reputed to have said, 'One must be able to say at all times—instead of points, straight lines, and planes—tables, chairs, and beer-mugs' [7]. The significance of his insight was that mathematics did not need to refer only to naturally occurring phenomena. The focus of attention changed from what the objects *are* to focus on their formally defined properties.

Instead of thinking of points being marked *on* lines, the real number line was seen to be *a set* that consisted *of* points. While natural mathematics sensed points moving about smoothly on the line, formal mathematics re-interpreted numbers as fixed entities that make up the set of real numbers.

At this period in the history of mathematics, new ways of thinking were introduced that would apply not only to naturally occurring situations but also to systems described only in terms of their formally stated properties. A range of possible developments occurred, with emphases on different aspects of mathematics, including:

Intuitionism: natural mathematics based on human perception and construction; in particular, a construction must be performed explicitly by a finite sequence of operations and proof by contradiction is not allowed.

Logicism: mathematics is based on formal logic without any reliance on natural intuition.

Formalism: mathematics has a formal set-theoretic basis, which Hilbert acknowledged could be inspired by natural intuitive experiences, but which must be fully formulated in terms of set-theoretic definitions and formal proof.

Subsequent mathematics has expanded into a diverse range of specialities as mathematicians focus on particular areas of interest. Applied mathematicians look at problems and formulate mathematical models that they use to find solutions. Physicists consider natural phenomena such as gravity or magnetism, and formulate mathematical models in terms of Newtonian mechanics or the four-dimensional space-time of Einstein's theory of relativity. They contemplate the origins of the universe in terms of the Big Bang, a mathematical model of an expanding universe. They imagine the structures of atoms, formulate models involving subatomic particles, perform sophisticated experiments to see whether their model matches the physical world. Climate scientists develop mathematical models of natural changes in long-term patterns of weather. Economists model and predict the change and growth of economies. Sometimes the models are good predictors, sometimes not. If they prove to be inadequate, better models that make more accurate predictions are sought.

Meanwhile, pure mathematicians seek to formulate precise theories that work consistently in well-specified contexts. They allow their imaginations to range over any phenomena that may intrigue them, seeking patterns and relationships to solve problems. At various times, some may use existing theories to solve problems, some may build naturally on their previous experience to suggest new possibilities, some may reflect on established theories to seek new theorems, to make new formal definitions and establish new formal theories. Many will use a combination of approaches depending on the context, as individuals develop their own preferences for ways of operating mathematically.

Students taking courses in different topics are likely to encounter significant differences in approach. You should take these differing approaches in your stride. Diversity is an advantage. Mathematics is difficult: we need as many different ways to think about it as we can find. The more tools and methods you have at your fingertips, the more you can create.

To help you to develop from a 'natural' attitude to mathematics at school to the wide variety of more sophisticated mathematics encountered at university, this book builds on natural experiences that are familiar to you at the outset, and works towards a formalist approach while building links with the underlying logic.

Once the formalist approach is mastered, two *complementary* modes of operation become available. You don't have to choose one: you can use whichever seems most helpful or fruitful in a given context. A natural approach builds formal structures inspired by intuition; a formal approach builds them by proving their properties from set-theoretic definitions. You should be flexible enough to use whichever approach is appropriate in a given context.

Agreed, a natural approach based on familiar mental images and symbolic operations may be easier for the human mind to grasp as a whole, but it still requires formal proof to show that the properties concerned do actually follow from the formal definition. You may also find new possibilities that you never dreamt of before. For instance, the complex numbers extend familiar decimals to a new system that has a square root of minus one, and the extension from the complex numbers to the quaternions produces a system where multiplication is no longer commutative and quadratic equations can have an infinite number of roots. The formal approach will provide the structures needed to place these new ideas on a sound foundation.

A formal approach pays attention to the precision of logical deduction from specific assumptions and can be used to build subtle schemas of mental relationships. It is helpful to give these schemas visual and symbolic meanings that make natural sense. This may occur through proving certain *structure theorems* that prove that a given formal structure has formally deducible properties that may be used to picture the ideas or to represent them as symbols that can be manipulated to solve problems.

This allows mathematics to expand in various ways, based on logical deduction or thinking about formal systems naturally, using visual or operational ideas, now supported by the power of formal proof.

Structure Theorems

We have already established the axiomatic properties of the familiar number systems $\mathbb{N}$, $\mathbb{Z}$, $\mathbb{Q}$, $\mathbb{R}$, and their extensions to $\mathbb{C}$ and $\mathbb{H}$. In chapter 8 we proved a structure theorem for the natural numbers:

> Any system that satisfies the Peano axioms is order isomorphic to the natural numbers $\mathbb{N}$.

This theorem tells us that the natural numbers are unique up to isomorphism, and lets us use the word 'the'. Indeed, for any system where we start with a single element, then move on to another and another, always different from any that came before, we have a potentially infinite set that is isomorphic to the natural numbers.

In chapter 10 we introduced a number of axiomatic systems as sets with various prescribed operations satisfying specified properties, including rings, fields, ordered rings, ordered fields. A number of theorems were proved characterising these structures as follows:

> Every *ring* contains a subfield isomorphic either to $\mathbb{Z}$ or to $\mathbb{Z}_n$ for some natural number n (proposition 10.10).

> Every *field* contains a subfield isomorphic either to $\mathbb{Q}$ or to $\mathbb{Z}_p$ for some prime number p (proposition 10.11).

> Every *ordered ring* contains a subring isomorphic to $\mathbb{Z}$ (proposition 10.12).

> Every *ordered field* contains a subfield isomorphic to $\mathbb{Q}$ (proposition 10.13).

> Every *complete ordered field* is isomorphic to the real numbers $\mathbb{R}$ (theorem 10.17) and so can be represented visually as points on a number line and symbolically as infinite decimals.

All of these results are *structure theorems*. That is, they prove that each of these structures contains a *specific* system up to isomorphism—here one of $\mathbb{Z}, \mathbb{Z}_n, \mathbb{Q}, \mathbb{R}$ as appropriate. In such cases, this subsystem has a visual representation as points on a number line (or round a circle in the case of $\mathbb{Z}_n$) and corresponding symbolic representations as whole numbers, integers modulo n, rationals, or infinite decimals.

Of course, when using visual representations, we need to be aware that the visual picture alone, as seen with our finite human vision, does not provide the full structure. For example, the rational numbers and the real numbers can both be represented as a number line, yet structurally they have very different symbolic and set-theoretic properties. Visual and symbolic thinking offer us modes of operation that can inspire formal structures that have well-defined consequences. On the other hand, using structure theorems, we can think about systems in visual and or symbolic ways now supported by formal deduction.

Structure theorems also tell us that we should relax. The human brain naturally makes links between mental concepts. Structure theorems let us think about formal systems in more brain-friendly visual and operational ways. When we started with the natural numbers $\mathbb{N}_0$ and constructed $\mathbb{Z}, \mathbb{Q}$, and $\mathbb{R}$, we did so by setting up equivalence relations and showing that there is an isomorphism between each system and a subsystem of the next. Subsequently we saw that we could start from the top with $\mathbb{R}$, and then find $\mathbb{Q}, \mathbb{Z}$, and $\mathbb{N}_0$ as subsystems of $\mathbb{R}$, without any need to talk about isomorphisms.

How we 'identify' one system with an isomorphic copy is sometimes called 'abuse of notation'. Far from this being an abuse of the process of mathematical thinking, it uses notation in a flexible way, helping the human brain to work more simply. Isomorphic systems represent the same underlying crystalline concept, which satisfies the required properties. In this way we obtain more concise natural ideas, such as:

Every *ring* contains either $\mathbb{Z}$ or $\mathbb{Z}_n$ for some natural number n.

Every *field* contains either $\mathbb{Q}$ or $\mathbb{Z}_p$ for some prime number p.

Every *ordered ring* contains $\mathbb{Z}$.

Every *ordered field* contains $\mathbb{Q}$.

In terms of crystalline concepts, the natural numbers are the *unique* system satisfying the Peano Axioms, and the real number system is the *unique* complete ordered field.

Psychological Aspects of Different Approaches to Mathematical Thinking

Just as mathematicians develop their own personal ways of operating mathematically, students also vary in their approach. On occasions they may build *naturally* on their previous experiences, *formally* through deduction from set-theoretic definitions and formal proof, *procedurally*, based on committing proofs to memory to pass the examinations, or use a combination of these and other techniques [6]. You may find it helpful to reflect on how you make sense of formal mathematics to become aware of why you may have certain difficulties and how you might work to improve your understanding.

You may prefer a *natural* approach based on your previous experiences. This can work well, but it is wise to reflect on the changes of meaning as new structures are encountered. It is helpful to develop a flexible understanding how old ideas may need to be rationalised to work in a new context where new ideas clash with previous experience and cause confusion. Some lecturers tell students to 'forget all you know and start afresh from the formal definitions', but this is difficult for anyone whose mind is full of earlier ideas with deeply embedded mental connections that behave subtly differently. It is important to be resilient and think carefully about strange new ideas, to make them your own by explaining to yourself how and why they work in the new context.

You may prefer a *formal* approach, building a schema of ideas based only on definitions and proofs. These may be motivated by intuition, but the proofs have to lead to a formal schema of theorems that builds up the entire knowledge structure. Some students manage to do this successfully, but many have trouble with new ideas that may either clash with previous experience or involve complicated quantifiers that prove too difficult to cope with.

You may prefer a *procedural* approach, learning specific procedures for solving routine problems in the course, and remembering theorems by heart to reproduce in examinations without being over-concerned about the meaning.

You may use a combination of methods depending on the context.

In every case you can enhance your understanding by reflecting on the proofs and explaining to yourself how and why the deductions work.

Each approach has subtle aspects that affect your understanding of the mathematics. For example, when seeking to make sense of the convergence of a real number sequence (a_n) to a limit a, a natural approach might imagine plotting the points a_n for $n = 1, 2, \ldots$, with a horizontal line $y = a$ representing the limit value, and lines $y = x - \varepsilon$ and $y = a + \varepsilon$ above and below representing the allowable range of values for a given $\varepsilon > 0$. The definition then says that the sequence converges if for any ε the terms a_n lie within the allowable range of values from some $n = N$ onwards.

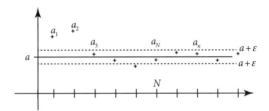

Fig. 12.1 A natural limit

This diagram needs to be imagined *dynamically*. The values of the sequence are plotted first; then the horizontal line is placed in its appropriate place with the range $\pm \varepsilon$ above and below, then the value of N is sought so that the values of a_n for $n > N$ lie in the desired range. Then imagine ε having a smaller value, and repeat. The phenomenon must occur for a fixed value of a while ε is taken as small as is desired. So as ε shrinks towards zero, N gets bigger, and the relevant terms are sandwiched between the two horizontal lines. It's as though the terms of the sequence are being sucked into an ever-narrowing funnel.

While this natural approach may work well for some, it is problematic in a number of ways. For example, many students have difficulty with nested quantifiers. Instead of writing the definition of the limit as

Given $\varepsilon > 0$ there exists N such that $n > N$ implies $|a_n - a| < \varepsilon$,

one student wrote:

A sequence (a_n) tends to a limit a for $\varepsilon > 0$ if there exists $N \in \mathbb{N}$ such that $|a_n - a| < \varepsilon$ provided $n \geq N$,

while another wrote:

If $a_n \to a$, then there exists $\varepsilon > 0$, such that $|a_n - a| < \varepsilon$ for all $n \geq N$, where N is a large positive integer.

At the very least, it is essential to be able to reproduce the limit definition correctly, and even then, there are subtleties. Many examples of limits involve a formula, and this can give the impression that a sequence *approaches* a limit, but is never equal to it. Some students taking a natural approach believe that a constant sequence cannot *tend* to a limit 'because it is already there'. Others may cope by separating convergence into two distinct ideas where some sequences *approach* the limit while others are *at* the limit.

An alternative interpretation, showing real mathematical insight, occurred when one student concentrated on the formal definition and realised that in computing a value for N, some sequences get within a given value of ε for smaller values of N than others, and so 'converge at a faster rate'. This led to the insight that a constant sequence is 'the fastest converger of them all', because it is already there. Unlike his colleagues who saw a constant sequence as an exceptional case, this student saw the constant sequence as the simplest central example of all convergent sequences. It is a mark of true mathematical insight to include exceptional cases as part of the general theory.

Some professors introduce convergence by interpreting it as a numerical calculation: given a numerical value of $\varepsilon > 0$, calculate a numerical value of N. For instance, given the sequence $(1/n)$ and $\varepsilon = 1/1000$, work out that $N = 1000$ will do the job, then generalise this idea for general ε, by taking N bigger than $1/\varepsilon$.

A numerical approach can be a helpful first step, but only learning how to perform a procedural solution can fail in a more general situation such as:

Given a sequence (a_n) that tends to 1, show that there exists a value of N such that $a_n > \frac{3}{4}$.

In this case the numerical calculation is not appropriate since, as no formula for a_n is specified, it is not possible to calculate a numerical value for N.

A subtler problem arises when proving that a sequence is *not* convergent, using here the idea of negating quantifiers by replacing $\neg\forall$ (not for all) by $\exists\neg$ (there exists not), and $\neg\exists$ by $\forall\neg$, as in chapter 6.

The statement that a sequence is not convergent takes the form

$$\neg\,(\forall \varepsilon > 0 \exists N \in N : \forall n \geq N \,|a_n - a| < \varepsilon)\,.$$

Successively moving the 'not' symbol $\neg$ past the quantifiers gives

$$\exists \varepsilon > 0 \,\neg\exists\, N \in N : \forall n \geq N \,|a_n - a| < \varepsilon$$
$$\exists \varepsilon > 0 \,\forall N \in N : \neg\forall n \geq N \,|a_n - a| < \varepsilon$$
$$\exists \varepsilon > 0 \,\forall N \in N : \exists n \geq N \,\neg\,|a_n - a| < \varepsilon$$

and finally

$$\exists \varepsilon > 0 \,\forall N \in N : \exists n \geq N \,|a_n - a| \geq \varepsilon.$$

This can now be expressed in words, as finding an $\varepsilon > 0$ such that, for all N, we can find $n \geq N$ such that $|a_n - a| \geq \varepsilon$.

Such a technique can be developed by natural thinking about the definition, formal manipulation of the quantifiers, or procedural learning of the rules manipulating quantifiers. However, the technique is enhanced by making sense of how it operates, to overcome possible limitations of intuitive meanings and to develop flexible forms of reasoning that will support coherent mathematical thinking.

Building Formal Theories

In the remainder of this chapter we offer an overview of how formal mathematical theories can be organised efficiently by focusing first on a small list of related axioms to prove properties that can be deduced from them. These proven properties can then be used in new contexts that satisfy the given axioms, to build increasingly sophisticated theories.

Several axiomatic systems start from a set with a single operation: here are a few.

Semigroups and Groups

Examples like the integers under addition or the non-zero integers under multiplication suggest thinking about a set X and a binary operation $*$ on X. Possible properties might include:

(1) $(a * b) * c = a *(b*c)$ for all $a, b, c \in X$. In this case $*$ is *associative*.
(2) There exists an element $e \in X$ satisfying $a * e = e * a = a$ for all $a \in X$. Such an element is an *identity*.
(3) If an identity e exists, then for all $a \in X$ there exists $b \in X$ such that $a * b = b * a = e$. Such an element b is called an *inverse* for a.
(4) $a * b = b * a$ for all $a, b \in X$. In this case $*$ is *commutative*.

A set X with a binary operation $*$ satisfying (1) and (2) is a *semigroup*. If (3) is also satisfied, it is a *group*. If (1)–(4) all hold then it is a *commutative* (or *abelian*) *group*.

Several of these properties are already familiar in various contexts.

Examples 12.1:

(i) $\mathbb{N}_0$ is a semigroup with identity 0 under the binary operation +.
(ii) $\mathbb{N}_0$ is a semigroup under multiplication with identity 1.
(iii) $\mathbb{Z}$ is a semigroup under multiplication.
(iv) $\mathbb{Z}$ is a group under addition. The identity is zero and the inverse of $n \in \mathbb{Z}$ is $-n$ because $n + 0 = 0 + n = n$ and $n + (-n) = (-n) + n = 0$.
(v) The non-zero elements of $\mathbb{Z}$ form a semigroup under multiplication with identity element 1.
(vi) The non-zero elements of $\mathbb{Q}$ (or of $\mathbb{R}$ or $\mathbb{C}$) form a group with identity 1, and the inverse of r is $1/r$.
(vii) The non-zero elements of $\mathbb{H}$ form a group under multiplication. The identity is 1, and the inverse of $q \in \mathbb{H}\setminus\{0\}$ is $\bar{q} / |q|$.

Examples (i)–(vi) are commutative; example (vii) is non-commutative. We consider groups in greater detail in chapter 13 to show how they arise naturally in number systems and in many other contexts. We include formal deductions that reveal structural features of groups.

Rings and Fields

Rings and fields have already been introduced in chapter 9. They can be described more succinctly using the notions of group and semigroup.

In these terms, a ring consists of a set R and two binary operations $+$ and $\times$, such that R is a commutative group under $+$, a semigroup under $\times$ (where $a \times b$ is written as ab), and the two operations are related by the distributive laws $a(b + c) = ab + ac$, $(b + c)a = ba + ca$ for all $a, b, c \in R$. If multiplication is commutative then R is called a commutative ring. Thus $\mathbb{Z}, \mathbb{Q}, \mathbb{R}$, and $\mathbb{C}$ are commutative rings and $\mathbb{H}$ is a (non-commutative) ring.

A field is a set F with two commutative operations $+$ and $\times$ such that F is a group under $+$ with identity 0, $F\backslash\{0\}$ is a group (with identity 1) and the two operations are related by the distributive law $a(b + c) = ab + ac$ for all $a, b, c \in F$.

Examples include $\mathbb{Q}$, $\mathbb{R}$, and $\mathbb{C}$, but not $\mathbb{Z}$ (because there are non-zero elements without multiplicative inverses) or $\mathbb{H}$ (because multiplication is non-commutative).

However, all of these systems are also division rings, consisting of a set D with operations $+$, $\times$ such that D is a commutative group under $+$ with identity 0, $D\backslash\{0\}$ is a group under $\times$ (not necessarily commutative) and the distributive laws hold: $a(b + c) = ab + ac$, $(b + c)a = ba + ca$ for all $a, b, c \in D$. Examples include $\mathbb{Q}$, $\mathbb{R}$, $\mathbb{C}$, and $\mathbb{H}$.

Other formal systems can be designed based on our intuitive experiences, inspiring set-theoretic definitions that may lead to interesting structures. Such activities used to be quite an extensive industry when mathematicians were coming to grips with axiomatic structures, but nowadays new axiom systems have to prove their worth by helping to advance other areas of the subject.

Vector Spaces

As an example of an axiomatic system that arises in a wide range of situations, yet has a clear natural structure, we consider how points in three-dimensional space can be described symbolically by selecting axes and using coordinates x, y, z.

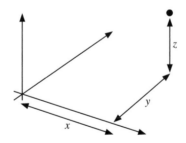

Fig. 12.2 Points in three-dimensional space

A point in three-dimensional space corresponds to an ordered triple of real numbers (x, y, z). So we can regard space as *being* the set $\mathbb{R}^3$ of ordered triples of real numbers. We add such triples using the obvious rule

$$(x_1, y_1, z_1) + (x_2, y_2, z_2) = (x_1 + x_2, y_1 + y_2, z_1 + z_2).$$

Addition is associative and commutative, the triple $(0, 0, 0)$ is an identity, and the additive inverse of (x, y, z) is $(-x, -y, -z)$. Therefore $\mathbb{R}^3$ is a commutative group under addition. We can also multiply a triple (x, y, z) by an element $a \in \mathbb{R}$ to get $a(x, y, z) = (ax, ay, az)$. This operation relates to addition and multiplication on $\mathbb{R}$ according to the rules:

$$(a + b)(x, y, z) = a(x, y, z) + b(x, y, z),$$
$$(ab)(x, y, z) = a(b(x, y, z)),$$
$$1(x, y, z) = (x, y, z),$$

and to addition of vectors by:

$$a((x_1, y_1, z_1) + (x_2, y_2, z_2)) = a(x_1, y_1, z_1) + a(x_2, y_2, z_2).$$

Since we live in three-dimensional space—at least, that's the natural image on a human scale—it may seem strange to talk about higher dimensions. However, Einstein's theory of relativity uses time as a fourth variable, so that a point (x, y, z) at time t is given by the ordered quadruple (x, y, z, t). What is the fifth dimension? The answer is that this approach is a diversion from the mainstream of mathematics. Time is *a* fourth dimension, not the only one. 'The' fourth dimension is a misnomer, so the question makes even less sense for 'the' fifth. Newtonian and relativistic physics both constrain us to live in three-dimensional space with time as some sort of fourth dimension, but higher dimensions have genuine mathematical significance. (If string theory—one of the most popular proposals to unify relativity with quantum mechanics—is correct, space might really have 10 or perhaps 11 dimensions. For various reasons, the extra ones don't show up in daily life.) There are sound mathematical reasons for defining spaces with any number of dimensions, even infinity. Such spaces arise naturally from mainstream mathematics.

For instance, describing the positions of *two* independent points (x_1, y_1, z_1) and (x_2, y_2, z_2) in three-dimensional space requires six real numbers. These can be put in order as a single sextuple, $(x_1, y_1, z_1, x_2, y_2, z_2)$ which now describes the position of them both. So the 'configuration space' for the two particles—the set of possible arrangements—has six dimensions, and it makes sense to denote it by $\mathbb{R}^6$.

Now consider a rigid body in space—for example, an asteroid in the asteroid belt. To describe its position uniquely, we have to specify the positions of *three* non-collinear points P, Q, R in the body.

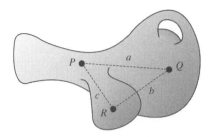

Fig. 12.3 A rigid body in space

Suppose that the distances PQ, QR, RP are a, b, c. Then we can place P at the point (x_1, y_1, z_1), move Q to any point (x_2, y_2, z_2) subject only to the restriction that the distance from (x_1, y_1, z_1) to (x_2, y_2, z_2) is a. By Pythagoras' theorem (in three dimensions, or applied twice in two planes), the distance between (x_1, y_1, z_1) and (x_2, y_2, z_2) in $\mathbb{R}^3$ is $\sqrt{(x_1 - x_2)^2 + (y_1 - y_2)^2 + (z_1 - z_2)^2}$. So the distance condition can be stated as

$$(x_1 - x_2)^2 + (y_1 - y_2)^2 + (z_1 - z_2)^2 = a^2. \tag{12.1}$$

Finally we can rotate the body around the axis PQ to put R at a point (x_3, y_3, z_3) subject only to the restrictions $QR = b$, $RP = c$:

$$(x_2 - x_3)^2 + (y_2 - y_3)^2 + (z_2 - z_3)^2 = b^2, \tag{12.2}$$

$$(x_3 - x_1)^2 + (y_3 - y_1)^2 + (z_3 - z_1)^2 = c^2. \tag{12.3}$$

Thus the position of the rigid body is determined by the nine coordinates x_1, y_1, z_1, x_2, y_2, z_2, x_3, y_3, z_3, subject to the equations (12.1)–(12.3). It is possible, and by no means bizarre, to consider this as an ordered 9-tuple

$$(x_1, y_1, z_1, x_2, y_2, z_2, x_3, y_3, z_3) \in \mathbb{R}^9,$$

so that the rigid body's position is a point in $\mathbb{R}^9$ subject only to equations (12.1)–(12.3).

Examples like this in mathematics are legion. Far from restricting 'spaces' to $\mathbb{R}^3$, it is a positive advantage to consider the set $\mathbb{R}^n$ of all n-tuples of real numbers, for any $n \in \mathbb{N}$. It should now be obvious how to do this. Define addition and multiplication by real numbers in $\mathbb{R}^n$ by:

$$(x_1, x_2, \ldots, x_n) + (y_1, y_2, \ldots, y_n) = (x_1 + y_1, x_2 + y_2, \ldots, x_n + y_n)$$
$$a(x_1, x_2, \ldots, x_n) = (ax_1, ax_2, \ldots, ax_n).$$

These operations satisfy the same properties that we listed for $\mathbb{R}^3$. For convenience write $v = (x_1, x_2, \ldots, x_n)$, $w = (y_1, y_2, \ldots, y_n)$. Then these properties can be stated as:

$$(a + b)v = av + bv$$
$$(ab)v = a(bv)$$
$$1v = v$$
$$a(v + w) = av + aw \quad \text{for all } a, b \in \mathbb{R}, v, w \in \mathbb{R}^n.$$

This is the genesis of the idea of a vector space. Consider a set V with a binary operation $+$. Then require a map $m : \mathbb{R} \times V \to V$, where for convenience we write $m(a, v)$ as av. V is said to be a *vector space* over $\mathbb{R}$ if:

(VS1) V is a commutative group under $+$.
(VS2) For all $a, b \in \mathbb{R}, v, w \in \mathbb{R}^n$ $(a + b)v = av + bv$
$\qquad (ab)v = a(bv)$
$\qquad 1v = v$
$\qquad a(v + w) = av + aw$.

These axioms hold for $\mathbb{R}^n$, but there are many other interesting examples of vector spaces.

For instance, let V be the set of all functions from $\mathbb{R}$ to $\mathbb{R}$. Then $f \in V$ means that $f : \mathbb{R} \to \mathbb{R}$. We can add two functions $f, g \in V$ to get $f + g : \mathbb{R} \to \mathbb{R}$ by defining $(f + g)(x) = f(x) + g(x)$ for all $x \in \mathbb{R}$. For example, if $f(x) = x^3 + x^2$, $g(x) = 3x + 2$, then $f(x) + g(x) = x^3 + x^2 + 3x + 2$. Multiplication by $a \in \mathbb{R}$ is given by $(af)(x) = a(f(x))$ for all $x \in \mathbb{R}$. An example is $f(x) = x^3 + x^2$, $a = -3$, in which case $(af)(x) = -3(x^3 + x^2)$. The set V is a vector space over $\mathbb{R}$ according to the given definition. In this case the elements of V are functions.

Vector spaces occur in unexpected places, too. Suppose, for example, we try to find the solution $y = f(x)$ of the differential equation

$$\frac{d^2 y}{dx^2} + 95\frac{dy}{dx} + 1066y = 0.$$

(Here we assume familiarity with calculus.) Then for differentiable functions $f : \mathbb{R} \to \mathbb{R}, g : \mathbb{R} \to \mathbb{R}$ and real numbers $a, b \in \mathbb{R}$, we find

$$\frac{d}{dx}\big(af(x) + bg(x)\big) = a\frac{df(x)}{dx} + b\frac{dg(x)}{dx},$$

$$\frac{d^2}{dx^2}\big(af(x) + bg(x)\big) = a\frac{d^2 f(x)}{dx^2} + b\frac{d^2 g(x)}{dx^2}.$$

Therefore

$$\frac{d^2}{dx^2}\big(af(x) + bg(x)\big) + 95\frac{d}{dx}\big(af(x) + bg(x)\big) + 1066(af(x) + bg(x))$$

$$= a\left\{\frac{d^2f(x)}{dx^2} + 95\frac{df(x)}{dx} + 1066f(x)\right\} + b\left\{\frac{d^2g(x)}{dx^2} + 95\frac{dg(x)}{dx} + 1066g(x)\right\}.$$

This implies that if $y = f(x)$ and $y = g(x)$ are solutions of the differential equation, then each of the expressions in curly brackets is zero, so $y = af(x) + bg(x)$ is also a solution.

Let S be the set of differentiable functions that are solutions of the differential equation. Then (putting $a = b = 1$)

$$f, g \in S \Rightarrow f + g \in S$$

and it is easily seen that S is a commutative group under +. Similarly (putting $b = 0$),

$$a \in \mathbb{R}, f \in S \Rightarrow af \in S.$$

Checking axioms (VS1) and (VS2), we see that the set of solutions S of this differential equation is a vector space over $\mathbb{R}$. (There is a solution corresponding to each initial condition $x(0) = p$, $x'(0) = q$, for any $p, q \in \mathbb{R}$. So there are plenty of solutions, and this statement is not vacuous.)

Our brief description of mathematical structures might give the impression that modern algebra is just an arid catalogue of axioms. To counteract that impression, we mention some of the striking deductions that have been made using this approach.

For two thousand years, since the time of the ancient Greeks, mathematicians wondered whether it is possible to trisect any angle using ruler and compass alone. It took an intriguing blend of vector space theory and field theory to show that the angle $60°$ (and many others) cannot be trisected in this way (for details, see [32]).

The method for solving a quadratic equation

$$ax^2 + bx + c = 0$$

by a process that we now encapsulate in the formula

$$x = \frac{-b \pm \sqrt{b^2 - 4ac}}{2a}$$

was, in effect, known to the ancient Babylonians more than 3000 years ago. In the sixteenth century, Italian mathematicians developed more complicated algebraic formulas for any cubic

$$ax^3 + bx^2 + cx + d = 0$$

and any quartic

$$ax^4 + bx^3 + cx^2 + dx + e = 0.$$

For well over two centuries the search continued for an algebraic formula for the solution of a quintic

$$ax^5 + bx^4 + cx^3 + dx^2 + ex + f = 0.$$

In the nineteenth century an intricate chain of deduction using field theory and group theory showed that no algebraic formula for the quintic exists (see [32]).

Various generalisations of the notion of vector space over $\mathbb{R}$ are possible. For instance if in the definition of a vector space we replace $\mathbb{R}$ by a field F, we get a *vector space over F*. If we replace it by a ring R, then we get the notion of a *module* over R. The study of these systems and their applications is central to modern algebra.

However, not only can we deduce properties formally: we can seek to prove a structure theorem. Here is an example for a vector space V over a field F.

Say that $v = a_1 v_1 + \cdots + a_n v_n$ where $a_1, \ldots, a_n \in F$ is a *linear combination* of the vectors $v_1, \ldots, v_n \in V$. For example, (a, a, b) is a linear combination of $(1, 1, 0)$ and $(0, 0, 1)$ for all $a, b \in F$, because $(a, a, b) = a(1, 1, 0) + b(0, 0, 1)$.

More generally, any vector $(x, y, z) \in \mathbb{R}^3$ is a linear combination of the three vectors $\mathbf{i} = (1, 0, 0)$, $\mathbf{j} = (0, 1, 0)$, $\mathbf{k} = (0, 0, 1)$, because

$$(x, y, z) = x\mathbf{i} + y\mathbf{j} + z\mathbf{k}.$$

If a vector space V has a set of vectors $v_1, \ldots, v_n$ so that every vector $v \in V$ can be written as a linear combination,

$$v = a_1 v_1 + \cdots + a_n v_n \text{ where } a_1, \ldots, a_n \in F,$$

then this set of vectors is called a *spanning set* for V.

For instance, the vectors $\mathbf{i}$, $\mathbf{j}$, $\mathbf{k}$ form a spanning set for $\mathbb{R}^3$. They have another special property. A set of vectors $v_1, \ldots, v_n \in V$ is *linearly independent* if

$$a_1 v_1 + \cdots + a_n v_n = 0 \text{ implies } a_1 = \cdots = a_n = 0.$$

If a spanning set is also linearly independent, then the linear representation is unique, for if a vector $v \in V$ has two possible representations,

$$v = a_1 v_1 + \cdots + a_n v_n = b_1 v_1 + \cdots + b_n v_n,$$

then

$$(a_1 - b_1) v_1 + \cdots + (a_n - b_n) v_n = 0$$

and, by linear independence,

$$a_1 - b_1 = 0, \ldots, a_n - b_n = 0,$$

so

$$a_1 = b_1, \ldots, a_n = b_n.$$

A set of vectors that is both a spanning set and linearly independent is called a *basis* for the vector space V and the vector space is said to be *finite dimensional*.

A first course in vector space theory ('linear algebra') usually concentrates on finite-dimensional vector spaces over the fields $\mathbb{R}$, $\mathbb{C}$, or a general field F, and proves that any two bases of a given finite-dimensional vector space have the same number of elements. This number is called the *dimension* of the vector space. Now, if $v_1, \ldots, v_n$ is a basis then any element $v \in V$ can be written uniquely as

$$v = a_1 v_1 + \cdots + a_n v_n.$$

So the map $f : V \to \mathbb{R}^n$ for which $f(a_1 v_1 + \cdots + a_n v_n) = (a_1, \ldots, a_n)$ is a structure-preserving isomorphism. This leads to a *structure theorem* for finite-dimensional vector spaces:

Theorem 12.2: Every finite-dimensional vector space V over a field F is isomorphic to F^n.

We omit the proof since the theorem is for illustrative purposes only, but the above discussion includes the key ideas.

This theorem provides a natural symbolic interpretation of a finite-dimensional space, in which the vectors are given by coordinates. It then turns out that linear maps between vector spaces are given by matrices. If $F = \mathbb{R}$ and $n = 2$ or 3, the vectors can be represented visually in two- or three-dimensional space.

The details can be found in any first course on vector spaces. They lay out a template for later courses that study other axiomatic structures.

The Way Ahead

In more sophisticated developments of algebra, the systems studied invariably consist of sets with various operations defined on those sets, with functions from one system to another preserving the structure. These structures have many applications; indeed, the applications often dictate the most profitable structures. Applications come from physics, engineering, biology, chemistry, economics, statistics, computing, social science, psychology, and many other areas. We now stand on a springboard, ready to leap into the higher realms of mathematical thought. As examples of how formal systems operate in mathematics, the next three chapters study three typical formal structures: *groups*, *infinite cardinal numbers*, and *infinitesimal quantities*.

The notion of a group occurs naturally in many areas, including symmetries of geometric objects and permutations of sets. We prove a structure theorem proving that any group can be viewed as a group of permutations of a set.

The second example is a generalisation of finite counting to infinite sets, using infinite cardinal numbers. These have an arithmetic of their own, deriving from the arithmetic of finite counting but with some significant differences.

The third concept generalises the idea of finite measurement by placing the real numbers in an even larger ordered field K that contains $\mathbb{R}$ as an ordered subfield. There are many possible candidates for K, but all of them share a unique structure theorem. An element $x \in K$ is said to be *finite* if $a < x < b$ for $a, b \in \mathbb{R}$. It is said to be an *infinitesimal* if $0 < |x| < a$ for all positive $a \in \mathbb{R}$. The structure theorem says that if x is finite, then x is uniquely of the form $x = c + e$ where $c \in \mathbb{R}$ and e is zero or infinitesimal.

This has a profound consequence in the longer-term development of mathematics. While formal mathematics tells us that there are no infinitesimals in the real numbers, it also tells us that any larger ordered field *must* contain infinitesimals. It is possible to develop a theoretical framework (called non-standard analysis) that allows the logical use of infinitesimals, but this requires a strengthening of the logical foundations. (We said in chapter 1 that mathematics grows like a tree, not only do its branches grow up, its roots must also become stronger to support the larger structure.)

The existence of a structure with infinitesimals alongside a theory of infinite cardinals which excludes infinitesimals is not contradictory because they occur in two different contexts. Cardinal numbers generalise counting in $\mathbb{N}$ and (apart from 1) the elements do not have inverses in $\mathbb{N}$. Non-standard analysis generalises measuring in $\mathbb{R}$, where multiplicative inverses do exist. This is typical of what happens when we generalise familiar systems in different ways. In mathematical analysis within the real numbers there are no infinitesimals, so the set-theoretic epsilon–delta method is appropriate. In fields larger than the reals, infinitesimals occur and they can be used to develop the theory of non-standard analysis. Meanwhile, applied mathematicians can work well with 'arbitrarily small numerical quantities' in a natural way. What matters is that the approach used is appropriate for the speciality concerned.

The approach followed in this book is to offer a foundation for a full range of mathematical thinking, combining natural visual and symbolic methods with formal definitions and formal proof. It offers a preparation for various future developments, be they in pure mathematics with a focus on mathematical analysis, an alternative logical approach using infinitesimal methods, or a more pragmatic approach that justifies the intuition of engineers and physicists, clearly based on sound formal foundations.

Exercises

This chapter offers a broader picture of a formal approach to mathematics. Given the broad sweep of ideas, we do not set a list of specific examples to practise at this time. Far more important is to reflect and consider how your ideas are progressing. A useful exercise is to re-read the opening chapter and look at the notes you may have kept on the exercises in that chapter. How are your views changing?

Look through this chapter again and write some notes to help you think through the advantages and disadvantages of a formal approach to mathematics. Don't just take our word for it, explain to yourself the reasons in favour of a formal approach, so that you grasp how the formal approach works for you and make explicit any problems that may concern you. Share your insights and concerns with others so that you can make better sense of the reasons for a formal approach. Then you can use these insights to help make sense of the ideas in the next part of the book.

CHAPTER 13

Permutations and Groups

In this chapter we develop some basic aspects of group theory to illustrate how axiomatic systems can be used to generalise features found in concrete mathematical systems. Groups are absolutely fundamental to modern mathematics. They originated from advanced areas of algebra and geometry, but the underlying concept turned out to be very simple, though sophisticated.

We begin with a practical example: the notion of a permutation. This is a way to rearrange the elements of a set X. For example, if $X = \{1, 2, 3\}$ then we might rearrange the order 1, 2, 3 to 1, 3, 2. We formulate this concept as a bijection $\sigma : X \to X$. The set of all permutations of a fixed set X has several pleasant algebraic properties, and from these and other examples we derive a short list of axioms to define the formal notion of a group. We then prove some basic theorems about groups, including a structure theorem showing that every group can be considered as a group of permutations. This theorem tells us that a group is not merely an abstract concept: there are ways to imagine groups visually and to manipulate their elements symbolically. We can then build up new insights in the theory that involve both formal proof of theorems and also natural ways of making sense of their structure. These ideas will arise in a range of different situations in later mathematical courses, so it is worth gaining experience of their general properties.

Permutations

In everyday life, we often find ourselves arranging some set of objects in different ways, or choosing one arrangement out of many possibilities. For example, several guests are coming to dinner and we have to decide who sits where. Or we're playing a game of cards and start by shuffling the pack. Initially, mathematicians thought of a permutation as one such arrangement. For example, if the objects were the symbols x, y, z, then the permutations were ordered triples like (z, y, x) or (y, x, z). There are six such triples:

$$(x, y, z) \quad (x, z, y) \quad (y, x, z) \quad (y, z, x) \quad (z, x, y) \quad (z, y, x).$$

However, mathematicians now focus not on the order of the numbers, but on the way one arrangement is changed to get another one. For example, when the order is reversed, so that the triple (x, y, z) is changed into (z, y, x), the symbol in position 1 is initially x, but ends up as z. Similarly the symbol in position 2 is initially y, and ends up as y, while the symbol in position 3 is initially z, but ends up as x. This change can be described by a function

$$\sigma : \{1, 2, 3\} \rightarrow \{1, 2, 3\}$$

defined by

$$\sigma(1) = 3, \quad \sigma(2) = 2, \quad \sigma(3) = 1.$$

What matters here is the numbers that give the *positions* of the symbols in the list. The symbols themselves tell us how to change these numbers. The modern approach is more elegant, leading to a simple and precise definition:

Definition 13.1: A *permutation* of a set X is a bijection $\sigma : X \rightarrow X$.

When X is finite (and in practice not too large) there is a useful notation for permutations, which rewrites a list of rules like $\sigma(3) = 1$ in a compact form:

$$\begin{pmatrix} 1 & 2 & 3 \\ 3 & 2 & 1 \end{pmatrix} \tag{13.1}$$

The top row lists the elements of X. Beneath each element x is its image $\sigma(x)$ under σ.

In this notation, the elements of X can be listed in any order in the top row. Provided we stick to the rule that $\sigma(x)$ is written underneath x, changing the order of the elements makes no difference to σ. For example, the permutation in (13.1) could be written as

$$\begin{pmatrix} 2 & 1 & 3 \\ 2 & 3 & 1 \end{pmatrix},$$

which conveys the same information. The possibility of changing the order in this manner, which we will shortly see is very useful, is one of the reasons why focusing on some specific ordering of the elements of X is not a good idea.

First we note that the composition[1] of functions $\sigma \circ \tau$ is defined to be $\sigma \circ \tau(x) = \sigma(\tau(x))$, which means, first do τ and then σ.

Permutations of a given set X have three basic properties:

Theorem 13.2:

    (1) The identity i_X is a permutation of X.

    (2) Every permutation σ of X has an inverse σ^{-1}, which is also a permutation of X.

    (3) If σ and τ are permutations of X, then so is their composition $\sigma \circ \tau$.

Proof:

    (1) The identity is obviously a bijection, as remarked in chapter 5 when it was defined.

    (2) Since σ is a bijection, it has an inverse by theorem 5.17(c) of chapter 5. Clearly σ^{-1} is also a bijection.

    (3) This follows from proposition 5.20 of chapter 5.   □

When X is finite, we can calculate the inverse of a permutation and the composition of two permutations using the notation introduced above. For example, suppose that $X = \{1, 2, 3, 4, 5, 6, 7\}$ and

$$\sigma = \begin{pmatrix} 1 & 2 & 3 & 4 & 5 & 6 & 7 \\ 7 & 6 & 5 & 4 & 3 & 2 & 1 \end{pmatrix},$$

$$\tau = \begin{pmatrix} 1 & 2 & 3 & 4 & 5 & 6 & 7 \\ 5 & 7 & 3 & 1 & 4 & 6 & 2 \end{pmatrix}.$$

To find τ^{-1}, we just swap the two rows:

$$\tau^{-1} = \begin{pmatrix} 5 & 3 & 6 & 1 & 4 & 7 & 2 \\ 1 & 2 & 3 & 4 & 5 & 6 & 7 \end{pmatrix}.$$

If necessary, we can rearrange the columns so that the numbers in the first row are in numerical order 1–7. Or we can observe that the number lying above 1 is 4, the number lying above 2 is 7, the number lying above 3 is 2, and so on. Either way, we get the equivalent expression

$$\tau^{-1} = \begin{pmatrix} 1 & 2 & 3 & 4 & 5 & 6 & 7 \\ 4 & 7 & 2 & 5 & 1 & 3 & 6 \end{pmatrix}.$$

[1] Algebraists sometimes write $\sigma(x)$ the other way round as $(x)\sigma$ so that σ followed by τ is written as $(x)\sigma\tau = ((x)\sigma)\tau$. This makes the rule for composition read more naturally, so that $\sigma\tau$ means 'first do σ, then τ'. However, the notation $\sigma(x)$ is much more widely accepted, so we accept the minor irritation of composing permutations from right to left.

To calculate the composition of functions $\sigma \circ \tau$ we first do τ and then σ. This involves running through the numbers $x = 1, x = 2, \ldots, x = 7$, working out what $\tau(x)$ is, and then what happens to that number when we apply σ. For example,

$$\tau(1) = 5 \text{ then } \sigma(5) = 3$$
$$\tau(2) = 3 \ldots \quad \sigma(3) = 5$$
$$\tau(3) = 6 \ldots \quad \sigma(6) = 2$$
$$\tau(4) = 1 \ldots \quad \sigma(1) = 7$$
$$\tau(5) = 4 \ldots \quad \sigma(4) = 4$$
$$\tau(6) = 7 \ldots \quad \sigma(7) = 1$$
$$\tau(7) = 2 \ldots \quad \sigma(2) = 6$$

Therefore

$$\sigma \circ \tau = \begin{pmatrix} 1 & 2 & 3 & 4 & 5 & 6 & 7 \\ 3 & 5 & 2 & 7 & 4 & 1 & 6 \end{pmatrix}.$$

Another way to see how to get this result is to rewrite σ so that the top row is listed in the same order as the bottom row of τ, like this:

$$\tau = \begin{pmatrix} 1 & 2 & 3 & 4 & 5 & 6 & 7 \\ 5 & 3 & 6 & 1 & 4 & 7 & 2 \end{pmatrix},$$

$$\sigma = \begin{pmatrix} 5 & 3 & 6 & 1 & 4 & 7 & 2 \\ 3 & 5 & 2 & 7 & 4 & 1 & 6 \end{pmatrix},$$

and then combine the top row of τ with the bottom row of σ, which gives the same result. After a little practice you can do this by writing down the top row in order 1, 2, 3, ... and trace your finger along σ to find what is beneath each successive $\tau(x)$, to write this down on the bottom line.

Permutations as Cycles

A permutation can be written in a more compact form by tracing where each element goes. For example, in the case of the permutation

$$\tau = \begin{pmatrix} 1 & 2 & 3 & 4 & 5 & 6 & 7 \\ 5 & 7 & 3 & 1 & 4 & 6 & 2 \end{pmatrix}$$

we have 1 goes to 5, 5 goes to 4, 4 goes to 1. At the same time 2 goes to 7 and 7 goes to 2, while both 3 and 6 remain unchanged. We can represent the transformation as

$$1 \to 5 \to 4 \to 1, 2 \to 7 \to 2, 3 \to 3, \text{ and } 6 \to 6.$$

Each of these short lists determines a permutation in its own right, and is called a *cycle*. The first is written as (1 5 4) on the understanding that each number in the cycle goes to the next and the last number returns to the first. The product of these cycles is then written as

$$(1\ 5\ 4)\ (2\ 7)\ (3)\ (6).$$

Because cycles with only one element effectively do nothing, they can be omitted to write the product as

$$(1\ 5\ 4)\ (2\ 7).$$

Remembering that composition reads from right to left, this notation operating on an element x means first see what happens in the cycle (2 7) then in the cycle (1 5 4). For instance, operating on 4, the cycle (2 7) doesn't change it, but the cycle (1 5 4) takes 4 to 1.

In this case the cycles are *disjoint*; that is, they have no elements in common. In this case, the order does not matter. However, if two cycles have an element in common, then the order *does* matter. The product

$$(1\ 2)(2\ 3)$$

operating on the element 2 first takes 2 to 3 in the cycle (2 3) and then 3 is unchanged by the cycle (1 2), so 2 goes to 3 overall. But the product

$$(2\ 3)(1\ 2)$$

operating on the element 2 takes 2 to 1 in the cycle (1 2), then 1 is unchanged in the cycle (2 3), so 2 goes 1 overall.

This means that the product of two permutations σ, τ need not be commutative, so we may have $\sigma \circ \tau \neq \tau \circ \sigma$.

On the other hand, it is easy to write down the inverse of a cycle. It operates a cycle in reverse order. For example the inverse of (1 5 4) is (4 5 1). Check this by working it through for yourself.

Group Properties for Permutations

Theorem 13.2 established three basic properties of the set of all permutations of any set X, which we write as follows:

Definition 13.3: A set G of permutations of a set X is a *permutation group* or has the *group property* if:

(PG1) The identity $i_X \in G$.
(PG2) If $\sigma \in G$ then $\sigma^{-1} \in G$.
(PG3) If $\sigma, \tau \in G$ then $\sigma \circ \tau \in G$.

Classically, X was always taken to be a finite set, and in this case the first two properties are consequences of the third (exercise 12). This is why the phrase '*the* group property' came to be used.

This definition applies to the set of all permutations on X, which we denote by $\mathbb{S}_X$. When $X = \{1, 2, 3, \dots, n\}$ we use the simpler notation $\mathbb{S}_n$. We can then restate theorem 13.2 as:

Theorem 13.4: For any X, the set $\mathbb{S}_X$ is a permutation group.

We have worded the definition of a permutation group on X carefully, so that it does not have to be the whole set of permutations on X. For instance, if we take the subset $\{i_X, s\}$ of the set $\mathbb{S}_7$ of all permutations of $\{1, 2, 3, 4, 5, 6, 7\}$, where $s = (2\ 3)$, we find that $(2\ 3)(2\ 3)$ is the identity, so $s^{-1} = s$. In fact, the set consisting of i_X and s satisfies *all* of the properties in the definition 13.3, so it is itself a permutation group.

The following subsets of $\mathbb{S}_3$ are also permutation groups where i denotes the identity:

$$\{i\}, \{i, (2\ 3)\}, \{i, (3\ 1)\}, \{i, (1\ 2)\}, \{i, (1\ 2\ 3), (1\ 3\ 2)\}, \mathbb{S}_3.$$

You can check that these satisfy definition 13.3 by hand.

To be able to operate fluently with permutation groups, it may be useful to imagine them visually in a way that enables you to grasp the structure. For example, we might visualise $\mathbb{S}_3$ by permuting the three corners of an equilateral triangle ABC. Initially we mark the three vertices as lying at positions 1, 2, 3.

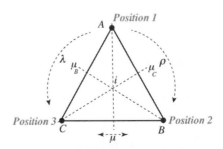

Fig. 13.1 Permuting the corners of an equilateral triangle

Then we pick up the triangle and rotate it or turn it over to place the corners in different positions to give six possible symmetries:

the identity i, which leaves the triangle unchanged.

Two are rotational symmetries:

clockwise rotation ρ by a third of a full turn is the permutation (1 2 3),
anticlockwise rotation λ by a third of a turn is the permutation (1 3 2).

Three are mirror symmetries:

flip μ over the line of symmetry through A, given by (2 3),
flip μ_B over the line of symmetry through B, given by (3 1),
flip μ_C over the line of symmetry through C, given by (1 2).

However, by performing various combinations of a rotation and a mirror image, we can obtain all of these symmetries. For instance, using combinations of the rotation ρ and the mirror image μ, we can obtain:

the identity i
the rotation ρ as (1 2 3)
the rotation λ as (1 3 2) or ρ^2
the flip μ as (2 3)
the flip μ_B as (1 3), which can also be written as $\mu\rho$ or $\rho^2\mu$
the flip μ_C as (1 2) or $\rho\mu = \mu\rho^2$.

You should all check these statements for yourself. Either write the permutations as cycles and carry out the composition symbolically as explained above, or cut out an equilateral triangle from paper or card and physically carry out the sequences of rotations and flips.

This calculation shows that all six elements of the group $\mathbb{S}_3$ may be written in the form $\rho^p\mu^q$ where $0 \le p \le 2$ and $0 \le q \le 1$. In fact, all we need are the expressions

$$i, \rho, \rho^2, \mu, \rho\mu, \rho^2\mu.$$

This observation simplifies computations with the elements of $\mathbb{S}_3$. We can think of them as being 'generated' by the two elements ρ, μ subject to the 'relations'

$$\rho^3 = i, \mu^2 = i, \mu\rho = \rho^2\mu.$$

In general we can simplify any product of powers of ρ, μ to these six distinct elements by using these relations. For instance, if we have a product such as

$$\rho\mu\rho$$

then we can write it as

$$\rho(\mu\rho) = \rho(\rho^2\mu) = \rho^3\mu = \mu.$$

Not only do we have $\mu\rho = \rho^2\mu$, we also have $\mu^2\rho = \rho\mu$. So manipulating the symbols in this case is easy. The commutative law does not hold in general, so we cannot change the order of the terms in a product, but we can pass a term ρ over a term μ provided we replace ρ by ρ^2 when we do it. In this way we can reduce any product of powers of ρ and μ to the form $\rho^p\mu^q$ where $0 \le p \le 2$ and $0 \le q \le 1$.

This rule of thumb only works in this particular group, but it is fruitful to think of various other groups in terms of generators and relations. For instance, the group of symmetries of a regular polygon with n sides is generated by two symmetries, a rotational symmetry ρ shifting one corner round to the next position and a mirror symmetry μ flipping the polygon over an axis of symmetry through one of the corners.

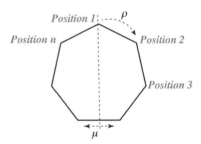

Fig. 13.2 Symmetries of a regular polygon

This group is generated by ρ and μ subject to the relations

$$\rho^n = i, \mu^2 = i, \rho\mu = \mu\rho^{n-1}.$$

In other groups, with many generators and relations, calculations like these can become very complicated. It then becomes essential to give a formal definition of the general concept of group, and to build up theorems about its structure.

Axioms for a Group

The three conditions that define a permutation group prove to be good candidates for a more general mathematical structure. However, this generality requires us to get rid of the condition that the elements being operated on are permutations. Indeed, we do not even require them to be functions. Moreover, the operation concerned need not be composition.

Historically, many examples arose in various areas of algebra, complex analysis, geometry, and topology that essentially satisfied properties like those in the definition of a group of permutations.

For example, the set $\mathbb{Z}$ of integers has the following properties:

If $n \in \mathbb{Z}$ then $0 + n = n$.

If $n \in \mathbb{Z}$ then $n + (-n) = 0$.

If $m, n \in \mathbb{Z}$ then $m + n \in \mathbb{Z}$.

The first states that the analogue of the identity element is the number 0, because *adding* 0 maps n to itself. The second similarly states that the 'inverse' of n is $-n$. And the third is like the condition on composing two permutations or rigid motions, except that now we add the numbers instead of composing them.

The set of rational numbers has similar properties with respect to addition. Moreover, the set $\mathbb{Q}\backslash\{0\}$ of *non-zero* rational numbers has similar properties with respect to multiplication:

If $r \in \mathbb{Q}\backslash\{0\}$ then $1r = r$.

If $r \in \mathbb{Q}\backslash\{0\}$ then $r(1/r) = 1$.

If $r, s \in \mathbb{Q}\backslash\{0\}$ then $rs \in \mathbb{Q}\backslash\{0\}$.

Now the 'identity' is 1, the inverse of r is its reciprocal $1/r$ (hence the restriction to non-zero numbers), and we use multiplication instead of composition. We encountered these features of $\mathbb{Q}$ in chapter 9.

These examples are the tip of a gigantic iceberg. During the early part of the twentieth century it became clear that it was pointless to keep proving the same theorems over and over again in many different contexts, especially since the proofs were often identical. The whole topic was crying out for an axiomatic approach, which would bring all of these different systems under one heading and define concepts and prove theorems with as much generality as possible. The history is complicated, but the final outcome is amazingly simple.

Less obvious—or perhaps *too* obvious, because it happens by default in all of the above examples—is the associative law. This law holds for composition of functions (chapter 5, proposition 5.14), and addition, and multiplication (see chapter 9).

One part of the group property—the one that got the whole subject started—is so basic that it is now incorporated directly into the definition of a group. This is 'closure' under the operation: the idea that when you compose permutations, or add integers, or multiply non-zero rationals, you get another object of the same kind. Recall from chapter 5 that a binary operation on a set A is a function $f : A \times A \to A$. As we stated when defining a binary operation, examples include composition, addition, and multiplication.

Definition 13.5: A *group* is a set G together with a binary operation $*$ on G, satisfying the following conditions:

(G1) There exists an *identity*: an element $1_G \in G$ such that for all $g \in G$

$$1_G * g = g, \ g * 1_G = g$$

(G2) For all $g \in G$ there exists an *inverse* element $g^{-1} \in G$ such that

$$g * g^{-1} = 1_G, \ g^{-1} * g = 1_G$$

(G3) The operation $*$ is *associative*: for all $g, h, k \in G$

$$(g * h) * k = g * (h * k).$$

If we want to be really formal, we can write the group as a pair $(G, *)$ to make it clear which binary operation is being considered.

Examples 13.6: All the following are groups:

- the set of all permutations $\mathbb{S}_X$ on a set X with the binary operation $\circ$
- the set $\mathbb{Z}$ with the binary operation $+$
- the set $\mathbb{Q}$ with the binary operation $+$
- the set $\mathbb{Q} \backslash \{0\}$ with the binary operation $\times$
- the real numbers $\mathbb{R}$ under (that is, with respect to the operation of) addition
- the complex numbers $\mathbb{C}$ under addition
- the integers $\mathbb{Z}_n$ modulo n under addition
- the non-zero real numbers $\mathbb{R} \backslash \{0\}$ under multiplication
- the non-zero complex numbers $\mathbb{C} \backslash \{0\}$ under multiplication
- the non-zero integers $\mathbb{Z}_p \backslash \{0\}$ modulo p under multiplication when p is prime.

Only the last of these needs any effort to check and we leave this as an exercise.

Many of these examples satisfy the commutative law $g * h = h * g$ and are given a special name:

Definition 13.7: If a group G satisfying (G1)–(G3) of definition 13.5 also satisfies the *commutative* law:

$$g * h = h * g \text{ for all } g, h, k \in G$$

then G is an *abelian group* (named after the mathematician Niels Hendrik Abel).

The definition of a group is clearer if we introduce an unambiguous and general notation, such as $*$, for the binary operation. However, its continued use is clumsy, and we normally replace $f * g$ by fg unless there is a serious danger of this being confused with some other meaning of the product. We also replace 1_G by 1, which usually causes no problem when the operation is thought of as a 'product', although we realise that in some of the above examples 1 is an identity map, in others it may be 1 and in others it is 0. In contexts where the main groups that arise are all abelian, it is common to use + for the binary operation and 0 for the identity element. We reserve the right to use whatever notation is appropriate in any given context.

We do not require the set G to be finite. It is when $G = \mathbb{S}_n$ for finite n, but $\mathbb{Z}$ and $\mathbb{Q}$ are infinite.

In the early stages of an axiomatic theory, quite a lot of effort has to be expended to sort out basic book-keeping issues: making sure that results that seem obvious are actually true. The commutative law is an instructive example, because it is often false. So any deduction that tacitly makes use of the commutative law must be viewed with suspicion unless there is another way to get the same result, or the group is known to be abelian. For example, our algebraic instincts could easily lead us to write

$$(fg)^2 = f^2 g^2$$

for two elements f, g of some group G. But this equation is not always correct—as can be seen by working out the operations $(\rho\mu)^2$ and $\rho^2\mu^2$ in the group $G = \mathbb{S}_3$.

Our next task is to sort out a number of useful properties that are true. We collect them in one jumbo package:

Theorem 13.8: Let G be a group. Then:

(1) The identity element is unique. That is, if $fg = g$ for all $g \in G$, or just for one such g, then $f = 1$. The same goes if $gf = g$.
(2) The inverse of any element of G is unique.
(3) If $f, g \in G$ then $(fg)^{-1} = g^{-1}f^{-1}$.

(4) *General associative law*: If brackets are inserted into any product $g_1 g_2 \ldots g_n$ so that it makes sense, the result is always the same. We can therefore omit the brackets and write this unique value as $g_1 g_2 \ldots g_n$.

(5) *General commutative law*: If the elements $g_1, g_2, \ldots, g_n \in G$ *commute*, that is, $g_i g_j = g_j g_i$ for all $1 \leq i, j \leq n$, then the product $g_1 g_2 \ldots g_n$ has the same value if the elements are permuted in any way.

Proof: We prove parts (1)–(3) and outline inductive proofs for (4)–(5), with discussion and examples.

(1) If $fg = g$ then

$$f = f1 = f(gg^{-1}) = (fg)g^{-1} = gg^{-1} = 1.$$

Similarly if $gf = g$.

(2) Suppose that $gh = 1$. Then

$$g^{-1} = g^{-1}1 = g^{-1}(gh) = (g^{-1}g)h = 1h = h,$$

and similarly if $hg = 1$.

(3) For all f and g,

$$(g^{-1}f^{-1})(fg) = g^{-1}(f^{-1}(fg)) = g^{-1}((f^{-1}f)g) = g^{-1}(1g) = g^{-1}g = 1.$$

Now use the uniqueness of inverses to conclude that $g^{-1}f^{-1} = (fg)^{-1}$.

(4) We already know that we can write the product of three terms f, g, h as fgh (without brackets) by the associative law. Suppose that we can write the product of n terms without brackets for some $n \geq 3$. Then, if we have a product of $n + 1$ terms $g_1, g_2, \ldots, g_n, g_{n+1}$, it may either be of the form $(g_1 \ldots g_n)g_{n+1}$, where the product of the n terms $g_1, \ldots, g_n$ is the same whatever the position of the brackets, or it is of the form $(g_1 \ldots g_r)(g_{r+1} \ldots g_{n+1})$ where $r < n$. Since each bracket has fewer than n terms, it is independent of the order of bracketing and we may write

$$g = g_1 \ldots g_r$$
$$h = g_{r+1} \ldots g_n$$
$$k = g_{n+1}$$

and use the associative law $g * (h * k) = (g * h) * k$ to write

$$(g_1 \ldots g_r)(g_{r+1} \ldots g_{n+1}) = (g_1 \ldots g_n)g_{n+1},$$

so the associative law holds for $n + 1$ terms and, by induction, it holds for all $n \geq 3$.

(5) Again, the general case can be formulated for elements $g_1, g_2, \ldots, g_n$ in any order, and proved using induction on $n \geq 2$. As the elements

commute in pairs, this is true for $n = 2$. For the induction step, use a series of swaps to move g_1 to the front; then the induction hypothesis shows that the rest can be arranged in the order $g_2 g_3 \ldots g_n$ and the proof is complete. □

From now on we use all of the above facts without further comment as part of the context of building the theory of groups. They are basic reflexes for every group-theorist. If you're wondering why we've bothered with, say, the general associative law, find out what calculations look like when the associative law is false. Look up 'non-associative algebra' on the internet, or borrow a suitable book. We can tell you the main point now: it gets very complicated. With enough motivation, you can learn to love it, and some special kinds of non-associative operation are actually very useful, although they usually satisfy some weaker version of associativity. Non-associative algebra is an acquired taste.

Subgroups

Recall that we discovered that several subsets of $\mathbb{S}_3$ also form permutation groups under the same operation, which in this case is composition. This phenomenon is very common, so we give it a name:

Definition 13.9: Let G be a group. A subset $H \subseteq G$ is a *subgroup* of G if:

(1) $1_G \in H$.
(2) If $h \in H$ then $h^{-1} \in H$.
(3) If $h, k \in H$ then $hk \in H$.

That is, H contains the identity and is *closed* under inverses and products.

There is a more efficient way to verify that a subset is a subgroup:

Theorem 13.10: A subset $H \subseteq G$ is a subgroup if and only if H is non-empty and $hk^{-1} \in H$ whenever $h, k \in H$.

Proof: Suppose that H is a subgroup. Then $1_G \in H$ so H is non-empty. Moreover, $k^{-1} \in H$ so $hk^{-1} \in H$.

Conversely, suppose H is non-empty and $hk^{-1} \in H$ whenever $h, k \in H$. Since H is non-empty there exists $h \in H$. Set $k = h$: then $1_G = hh^{-1} \in H$. Then set $h = 1_G$ to get $k^{-1} \in H$. Finally, observe that $hk = h(k^{-1})^{-1}$. □

Proposition 13.11: Suppose that H is a subgroup of G. Then H is a group under the operation $*$ of G, restricted to $H \times H$. It has the same identity element and inverses as G.

Proof: Check the axioms systematically and verify the properties required for identity and inverses. All are straightforward. □

One important way to obtain subgroups of G is to pick an element g and see what else the subgroup must contain. There's 1, of course, but also

$$g^2 = gg$$
$$g^3 = ggg$$
$$\cdots$$
$$g^{-1}$$
$$g^{-2} = g^{-1}g^{-1}$$
$$g^{-3} = g^{-1}g^{-1}g^{-1},$$

and so on. This motivates the definition of powers g^n of g, for any integer n (positive, negative, or zero):

Definition 13.12: Let G be a group and $g \in G$. For any $n \in \mathbb{Z}$ define g^n inductively by:

(1) $g^0 = 1$
(2) $g^{n+1} = gg^n$ $(n > 0)$
(3) $g^{-n} = (g^n)^{-1}$ $(n < 0)$.

We would be astonished if the following theorem were not true. Fortunately it is.

Theorem 13.13: Let G be a group, $g \in G$, and $m, n \in \mathbb{Z}$. Then $g^m g^n = g^{m+n}$.

Proof: Use induction on n. □

We introduce the notation

$$\langle g \rangle = \{g^n \mid n \in \mathbb{Z}\}$$

because the set of all powers of g is always a subgroup:

Theorem 13.14: Let G be a group and let $g \in G$. Then $\langle g \rangle$ is a subgroup.

Proof: Take any two elements $g^m, g^n \in \langle g \rangle$. Then $g^m(g^n)^{-1} = g^{m-n} \in \langle g \rangle$. Now appeal to theorem 13.10. □

Definition 13.15: Let G be a group and $g \in G$. We call $\langle g \rangle$ the *subgroup generated by* g.

Clearly any subgroup that contains g must contain $\langle g \rangle$, so $\langle g \rangle$ is the unique *smallest* subgroup that contains g. Moreover, it is commutative by theorem 13.13.

If we throw in an extra element, not a power of g, nothing as simple can be proved. The possibilities are very complicated, except for commutative groups.

What does $\langle g \rangle$ look like? Let's try a few examples. Suppose $G = \mathbb{S}_3$ and $g = \rho$. The powers of ρ are $\rho^0 = i$, $\rho^1 = \rho$, $\rho^2 = \rho\rho$. But $\rho^3 = i$, and from here on, the powers of ρ just cycle repeatedly through i, ρ, ρ^2. Moreover, $\rho^{-1} = \rho^2$ so negative powers provide nothing new. In short, in this case

$$\langle \rho \rangle = \{i, \rho, \rho^2\}.$$

This should not be a great surprise since we already know that $\{i, \rho, \rho^2\}$ is a subgroup, so it must contain all powers of ρ. What drives this phenomenon is the fact that $\rho^3 = i$.

In contrast, suppose that $G = \mathbb{Z}$ under addition and $g = 1$. Then $g^n = n$ because the group operation is addition. Now all the 'powers' g^n are distinct, and the subgroup generated by 1 is $\langle 1 \rangle = \mathbb{Z}$. This is an infinite group.

A group $\langle g \rangle$ generated by a single element g is called a *cyclic group*. It consists of all the powers of g. Its structure is easy to classify:

Proposition 13.16: A cyclic group $\langle g \rangle$ generated by a single element g is either finite with n distinct elements $\{1, g, g^2, \ldots, g^{n-1}\}$ where $g^n = 1$, or it is infinite of the form $\{g^n \mid n \in \mathbb{Z}\}$ where $g^m \neq g^n$ for $m \neq n$.

Proof: Either there are two distinct values of m and n for which $g^m = g^n$, or all powers of g are distinct. In the first case we may take $n \leq m$. If $k = m - n$ then $g^k = (g^m)(g^n)^{-1} = 1$. Now let n be the *smallest* power such that $g^n = 1$. All powers g^r for $0 \leq r < n$ must then be different, for if $g^r = g^s$ for $0 \leq r < s < n$, then $g^{s-r} = 1$, where $s - r < n$, contrary to n being the smallest power of g with this property. In this case we therefore have the cyclic group $\langle g \rangle$ with n distinct elements $\{1, g, g^2, \ldots, g^{n-1}\}$ where $g^n = 1$.

On the other hand, if all the powers are distinct, then $\langle g \rangle$ is $\{g^n \mid n \in \mathbb{Z}\}$ where $g^m \neq g^n$ for $m \neq n$. $\qquad\square$

Isomorphisms and Homomorphisms

Sometimes two technically different groups have essentially the same structure. For example, the subgroups $\{i, (2\ 3)\}$, $\{i, (3\ 1)\}$, $\{i, (3\ 2)\}$ of $\mathbb{S}_3$ all consist of two elements, the identity and a second element whose square is the identity.

In these examples, the groups concerned work in the same way, except for a change of notation. This preserves the operations in the sense that changing notation for two elements and then multiplying them has the same effect as multiplying them and then changing notation. This motivates the following definition:

Definition 13.17: An *isomorphism* between two groups G, H is a bijection

$$\phi : G \to H$$

such that

$$\phi(g_1 g_2) = \phi(g_1)\phi(g_2) \quad \forall g_1, g_2 \in G.$$

If such a bijection ϕ exists, we say that G is *isomorphic* to H. Symbolically, we write $G \cong H$.

If two groups are isomorphic, all of their abstract properties—those that do not depend on the notation—are the same. Moreover, corresponding elements have the same abstract properties. The next theorem lists some examples.

Theorem 13.18: Let G, H be groups and suppose that there is an isomorphism $\phi : G \to H$. Then:

(1) $\phi(1_G) = 1_H$.
(2) If $g \in G$ then $\phi(g^m) = (\phi(g))^m$.
(3) If $g \in G$ then $\phi(g^{-1}) = (\phi(g))^{-1}$.
(4) If K is a subgroup of G then $\phi(K)$ is a subgroup of H.

Proof: We leave the proofs, which are straightforward, as exercises. □

More generally, we can consider a map $\phi : G \to H$ between groups which preserves the operation but need not be bijective.

Definition 13.19: A *homomorphism* ϕ between two groups G, H is a map $\phi : G \to H$ such that

$$\phi(g_1 g_2) = \phi(g_1)\phi(g_2) \quad \forall g_1, g_2 \in G.$$

If ϕ is injective, then it is a *monomorphism*. If ϕ is surjective then it is an *epimorphism*.

For example, inclusion $i : \mathbb{Z} \to \mathbb{Q}$ from the integers under addition to the rationals under addition is a monomorphism. The map $\phi : \mathbb{Z} \to \mathbb{Z}_n$ from the integers under addition to the integers modulo n under addition, mapping an integer to its remainder modulo n, is an epimorphism.

Monomorphisms and epimorphisms are important concepts in group theory. For example, a monomorphism $\phi : G \to H$ is an isomorphism between G and the image $\phi(G) = \{\phi(g) \in H \mid g \in G\}$. This lets us identify G with the subgroup $\phi(G)$ of H, because there is a bijection between the elements and a precise correspondence between the operations.

Now our careful definition of a permutation group pays off. Recall that a permutation group on a set X was defined to be a set G of permutations of X satisfying:

(1) the identity $i_X \in G$
(2) if $\sigma \in G$ then $\sigma^{-1} \in G$
(3) if $\sigma, \tau \in G$ then $\sigma \circ \tau \in G$.

This can now be seen to be a *subgroup* of the permutation group S_X. More generally, we can prove:

Theorem 13.20 (Structure Theorem for a Group as a Group of Permutations): Every group is isomorphic to a permutation group.

Proof: Let G be a group. For fixed but arbitrary $g \in G$ define

$$\pi_g : G \to G$$

by

$$\pi_g(x) = gx \, (x \in G).$$

Informally, this map is 'left multiplication by g'.

This map is clearly injective: if $\pi_g(x) = \pi_g(y)$ then $gx = gy$, so $g^{-1}gx = g^{-1}gy$ and $x = y$. It is also surjective: if $y \in G$ then, for $x = g^{-1}y$, $\pi_g(x) = g(g^{-1}y) = y$. So π_g is a bijection and therefore a permutation of the set G.

Define the map $\phi : G \to \mathbb{S}_G$ by $\phi(g) = \pi_g$. We claim that ϕ is a monomorphism.

The map ϕ is injective: if $\phi(g) = \phi(h)$, then $gx = hx$, so $gxx^{-1} = hxx^{-1}$ and $g = h$. To show that it is a homomorphism, observe that $\phi(hg)$ maps x to $(hg)x$, and

$$(hg)x = h(gx) = \pi_h(\pi_g(x)) \text{ for all } x \in G,$$

so

$$\phi(hg) = \phi(h)\phi(g)$$

and ϕ is a homomorphism.

An injective homomorphism is a monomorphism from G to $\mathbb{S}_G$. It is therefore an isomorphism from G to its image, the subgroup $\phi(G)$ of $\mathbb{S}_G$. By definition, this is an isomorphism between G and a permutation group. $\qquad\square$

This theorem shows us that any abstract group can be viewed as a permutation group. In particular, for finite groups, every group is isomorphic to a subgroup of $\mathbb{S}_n$. In principle, this means that we can derive properties of groups from permutation groups, particularly for small values of n. For example, we can gain insight by exploring the properties of permutations as cycles. But in practice, for larger values of n, the possibilities become far more complicated, and for infinite groups even more possibilities occur.

Partitioning a Group to Obtain a Quotient Group

So far we have introduced the notion of *sub*group: a subset of a group that is itself a group. It is also possible to define a group structure by clumping group elements together and defining an operation on the set of clumps. We have already seen this idea in action when defining $\mathbb{Z}_n$, the integers modulo n. The clumps are equivalence classes of integers, where two integers are equivalent if they are congruent modulo n. To add two clumps, we choose an element from each, add those, and see which clump the result belongs to. This construction provides a second way to analyse a group in terms of simpler groups, and it turns out to be intimately related to homomorphisms.

We formalise the clumps by working with a *partition* of a group. By analogy with integers modulo n, we take a partition P of a group G and try to use the group operation to define an operation on the equivalence classes of the partition P. However, this procedure can run into trouble, because different choices of elements from the clumps may lead to inconsistent results. The construction of $\mathbb{Z}_n$ works because the clumps have a very regular structure: elements in the same clump differ by a multiple of n. If we try the same trick with a less regular partition, it may not work.

For example, suppose we partition $\mathbb{Z}$ into $\{0, \ 1\}$, $\{2, \ 5, \ 6\}$, $\{3, \ 8\}$, and various other disjoint pieces. What should $\{0, 1\} + \{2, 5, 6\}$ be? If we choose representative elements 0 and 2, then $0 + 2 = 2$ so the sum ought to be the clump $\{2, \ 5, \ 6\}$. But if we choose elements 1 and 2, then $1 + 2 = 3$ so the sum ought to be the clump $\{3, \ 8\}$. Therefore 'sum' is not well defined in this case, and the attempt fails.

Working with a general group G, we want to divide the set G into subsets that themselves operate as a group. When this is possible, the result is called a *quotient group*. The reason for this name will become clearer later.

A partition P of G is a set of disjoint (non-empty) subsets of G, so that every element of G lies in precisely one of the subsets in the partition. These subsets are called equivalence classes, and theorem 4.9 of chapter 4 provides a structure theorem for partitions: every partition corresponds to an equivalence relation $\sim$ in which $a \sim b$ if and only if a, b belong to the same equivalence class. We denote this equivalence class by E_a, where $E_a = E_b$ if and only if $a \sim b$.

For this partition P to inherit a product from the group, we obtain the product ST of two equivalence classes S, T in P by taking elements $x \in S$ and $y \in T$ and defining ST to be the subset that contains xy. This may also be written by defining $E_x E_y$ to be the subset E_{xy}.

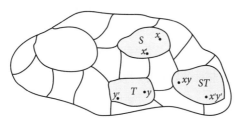

Fig. 13.3 Defining a product on a partition of a group

To be able to do this independently of the choice of the individual elements, we need to know that if we take other elements $x' \in S$ and $y' \in T$ then the product $x'y'$ needs to be in the same subset ST as xy. Another way of saying the same thing is to assert that if $x' \in E_x$ and $y' \in E_y$ then $x'y' \in E_{xy}$, or if $x' \sim x$ and $y' \sim y$ then $x'y' \sim xy$. Only then can the product of elements in the group be used to define a product on the equivalence classes as elements of the group P.

If we can define a group structure on the partition, this must have the properties we have proved about groups in general. For instance, the identity of any group is unique and the inverse of any element is unique. This immediately restricts how a group structure can be defined on a partition.

For example, there is only one candidate for the identity. The equivalence class I contains the identity 1_G, so I^2 must contain $(1_G)^2 = 1_G$ which implies $I^2 = I$. Therefore I must be the identity element.

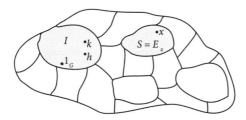

Fig. 13.4 The identity element for a partition

We can prove more:

Theorem 13.21: If a partition P of a group G has a group structure in which the product of equivalence classes E_x and E_y is defined to be E_{xy}, then the identity element of the partition I is the equivalence class containing 1_G and I is a subgroup of G.

Proof: If $h, k \in I$, then

$$E_{hk} = E_h E_k = I,$$

so $hk \in I$, so I is closed under the group operation.

Also, for any $g \in G$,

$$E_g E_{g^{-1}} = E_{gg^{-1}} = I.$$

If $g \in I$ then $E_g = I$, so this reduces to

$$I E_{g^{-1}} = I$$

which implies that

$$E_{g^{-1}} = I.$$

Therefore $g^{-1} \in I$. So I must be a *subgroup* of the whole group G. $\qquad\square$

These conditions are therefore necessary to set up a group structure on P in the stated manner. However, they are still not sufficient. The other equivalence classes must also have a special structure. To state what it is we require a new construct: the notion of a *coset* of a subgroup.

Definition 13.22: Let H be a subgroup of G and let $x \in G$. Then

$$\text{the } \textit{left coset} \text{ of } x \text{ is } xH = \{xh \in G \mid h \in H\}$$

and

$$\text{the } \textit{right coset} \text{ of } x \text{ is } Hx = \{hx \in G \mid h \in H\}.$$

Proposition 13.23: Let G be a group, and let H be a subgroup of G. The left cosets of H partition the set G. The right cosets of H also partition the set G.

Proof: First consider the left cosets $\{xH \mid x \in G\}$. Each coset is non-empty, since it contains $x1_G = x$. Every element $g \in G$ lies in at least one coset, namely gH. If two cosets xH, yH contain a common element $g = xh_1 = yh_2$, then $x = yh_2 h_1^{-1} = yh$ where $h = h_2 h_1^{-1} \in H$, because H is a subgroup. Any element $g \in xH$ is therefore of the form $g = xk$ where $k \in H$, so $g = xk = yhk$. Because H is a subgroup, $hk \in H$, giving $g = yhk \in yH$. Thus $xH \subseteq yH$.

A corresponding argument shows that $yH \subseteq hH$, hence the left cosets xH and yH are the same.

The proof for right cosets follows the same pattern. ☐

The Number of Elements in a Group and a Subgroup

When we partition a group G into left and right cosets of a subgroup H, it is clear that the map between two (left) cosets xH and yH which maps xh to yh for all $h \in H$ is a bijection. If G is finite, then all the cosets will be the same size.[2] This means that G is subdivided into a number of equal-sized subsets, which leads to a direct relationship between the number of elements in G and the number of elements in H.

Definition 13.24: The *order* of a finite group G is the number of elements in the group and is denoted by the symbol $|G|$.

The use of the term 'order' should not be confused with its use in other contexts, such as an order relation or the order of elements in a permutation. It is part of the traditional theory of groups and you should just get used to it.

Proposition 13.25: If H is a subgroup of a finite group G then the order of H divides the order of G.

Proof: Let $n = |H|$ be the order of H. Then every left coset has n elements, and the cosets are disjoint subsets of the same size that include all the elements of G. If there are m distinct cosets, the number of elements in G is therefore mn. ☐

This result is very helpful when seeking subgroups of a given group. For a subset to be a subgroup, its order must divide the order of the group. For example, when considering the possible subgroups of the permutation group $\mathbb{S}_3$, which has six elements, the subgroups must have order 1, 2, 3, or 6 *and there are no others*. So the subgroups must either be the identity (order 1), the whole group (order 6), or subgroups of order 2 or 3, which have all been identified earlier.

This proposition has an important consequence for elements of the group. Let $H = \langle g \rangle$, the cyclic subgroup generated by g. This is of the form $\{1, g, g^2, \ldots, g^{n-1}\}$, where $g^n = 1$ and all listed elements are distinct. Therefore the order of this cyclic subgroup is n. This leads to:

---

[2] This property also holds for infinite groups, but to explain this requires us to define what we mean by 'the number of elements in an infinite set'. We consider this in chapter 14.

Definition 13.26: An element $g \in G$ is said to be of *finite order* if there exists $n \in \mathbb{N}$ such that $g^n = 1$. The smallest such n is said to be the *order* of g.

An immediate consequence is:

Theorem 13.27: If g is an element of a finite group G then the order of g divides the order of G. □

Partitions that Define a Group Structure

Now that we can partition a group G using left or right cosets, we ask when it is possible to define a group operation on the partition, using the operation

$$xH * yH = xyH \tag{13.2}$$

for left cosets, or

$$Hx * Hy = Hxy \tag{13.3}$$

for right cosets. We show that this is possible if and only if the left and right cosets are the same. In such a case we would have

$$xH = Hx \text{ for all } x \in G,$$

and the two rules (13.2) and (13.3) will give the same result.

We make the following definition:

Definition 13.28: A subgroup H of a group G is a *normal subgroup* if the left and right cosets gH and Hg are equal for all $g \in G$.

The condition $gH = Hg$ does *not* mean that $gh = hg$ for every $h \in G$. It simply requires that $gh = kg$ for some $k \in H$. This means that the element $k = ghg^{-1}$ lies in H, which gives rise to the following:

Alternative Definition 13.29: A subgroup H of a group G is a *normal subgroup* if for every $h \in H$ and $g \in G$, the element $ghg^{-1} \in H$.

Symbolically, if H is normal, we write $H \triangleleft G$.

Example 13.30: Consider our old friend $\mathbb{S}_3$ and the two subgroups

$$H = \{i, \mu\}, K = \{i, \rho, \rho^2\},$$

where the mirror symmetry μ satisfies $\mu^2 = i$ and the rotation ρ satisfies $\rho^3 = i$.

Case 1: the subgroup $H = \{i, \mu\}$.

Here the left coset ρH is $\{\rho i, \rho \mu\} = \{\rho, \mu \rho^2\}$. But the right coset $H\rho$ is $\{i\rho, \mu\rho\} = \{\rho, \mu\rho\}$. These cosets are *not* equal. If we take the six elements of $\mathbb{S}_3$ and partition them into left and right cosets of H, then we get the following:

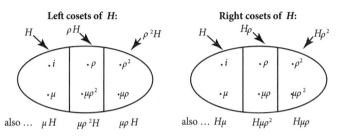

Fig. 13.5 Different left and right cosets

Using the relations $\mu\rho = \rho^2\mu$ and $\rho\mu = \mu\rho^2$, we see that $\rho H \neq H\rho$ and $\rho^2 H \neq H\rho^2$ so the subgroup H is not normal. If we try to define the product of two left cosets, say H and ρH, by selecting an element in each and multiplying them together, the results may lie in different cosets. For instance, if we select $i \in H$ and $\rho \in \rho H$, then their product $i\rho = \rho \in \rho H$, but if we choose $\mu \in H$, and $\rho \in \rho H$, then their product $\mu\rho \in \rho^2 H$.

On the other hand, subgroup K is different:

Case 2: the subgroup $K = \{i, \rho, \rho^2\}$.

The left coset ρK is

$$\rho K = \{\rho i, \rho\rho, \rho\rho^2\} = \{\rho, \rho^2, i\},$$

and the right coset $K\rho$ is

$$K\rho = \{i\rho, \rho\rho, \rho^2\rho\} = \{\rho, \rho^2, i\}.$$

Using the element μ instead, the left and right cosets are still the same. The left coset μK is

$$\mu K = \{\mu i, \mu\rho, \mu\rho^2\},$$

and the right coset $K\mu$

$$K\mu = \{i\mu, \rho\mu, \rho^2\mu\} = \{\mu, \mu\rho^2, \mu\rho\}.$$

In this case, the partition of $\mathbb{S}_3$ has two equivalence classes, K and μK, which can be written in different ways as follows:

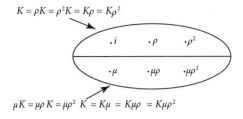

$$K = \rho K = \rho^2 K = K\rho = K\rho^2$$

$$\mu K = \mu\rho\, K = \mu\rho^2\ K = K\mu = K\mu\rho = K\mu\rho^2$$

Fig. 13.6 Identical left and right cosets

Now H is a *normal* subgroup, and the set of cosets forms a group with elements K and μK where K is the identity element and $(\mu K)^2 = K$.

This is true in general for normal subgroups:

Theorem 13.31: If G is a group and N is a normal subgroup, then the partition P consisting of the subset N and the cosets gN for all $g \in G$ forms a group under the product

$$gN\, hN = ghN.$$

Proof: It is essential in this proof to build on the formal definition of a group and to ascertain that the operations are all well defined.

First, suppose that $xN = x'N$ and $yN = y'N$, then $x' = xh$, $y' = yk$ for h, $k \in N$. So

$$x'y' = xhyk = x(yy^{-1})hyk = xy(y^{-1}hy)k = xyn \text{ where } n = (y^{-1}hy)k.$$

Because N is normal, $y^{-1}hy \in N$, and because $k \in N$ and N is a subgroup,

$$n = (y^{-1}hy)k \in N.$$

Therefore

$$x'y' = xyn \in xyN$$

and the cosets $x'y'N$ and xyN are the same.

The remainder of the proof is simple. The identity element is N and the inverse of xN is $x^{-1}N$. Associativity for multiplication of equivalence classes follows from associativity in G. □

Now we can see why this is called a *quotient* group. This theorem shows that for any normal subgroup N of G we can partition the group G into its cosets and define a group structure on them. The partition is denoted by G/N. These cosets are all the same size, in the sense that there is a one-to-one correspondence between any two of them. In particular, if G is a finite group of order $|G|$, then each coset has the same number of elements as the order $|N|$ of the normal subgroup. The order of G/N is therefore

$$|G/N| = |G|/|N|.$$

Theorem 13.31 tells us that if N is a normal subgroup of a group G, then the group operation on G naturally leads to a group structure on the quotient group G/N. We can say something stronger: that this is the *only* way that a partition of G can be given a group structure inherited from that of G.

To understand why, we introduce a more general notation for multiplying any two subsets X, Y of a group G. (We do not assume any further properties; they need not be subgroups, for example.) The product is

$$XY = \{xy \in G \mid x \in X, y \in Y\}.$$

For example, in $\mathbb{S}_3$

$$\{\rho, \mu\}\{i, \rho^2\} = \{\rho i, \mu i, \rho\rho^2, \mu\rho^2\}.$$

Similarly, if $X \subseteq G$ and $g \in G$, we define

$$X^{-1} = \{x^{-1} \in G \mid x \in X\}$$
$$gX = \{g\}X = \{gx \in G \mid x \in X\}$$
$$Xg = X\{g\} = \{xg \in G \mid x \in X\}.$$

Multiplication of subsets is obviously associative, and the general associative law applies. Therefore, if $g, h \in G$ then gNh is defined unambiguously (as either $g(Nh)$ or $(gN)h$, which are equal). If N is a normal subgroup, we can now write:

$$(gN)(hN) = g(Nh)N = g(hN)N = ghN^2 = ghN.$$

Now it is evident why multiplication of cosets works for a normal subgroup N. Multiplication of elements in the group may not be commutative, but multiplication of any element $g \in G$ by N *does* commute. So the operation on cosets given by

$$(gN)(hN) = ghN$$

is well defined.

This leads to the major structure theorem for quotient groups and normal subgroups:

Theorem 13.32 (Structure Theorem for a Partition of a Group):
If G is a group and P is a partition of the underlying set G, then P is a group with the operation inherited from G *if and only if* N is a normal subgroup and P is the quotient group G/N.

Proof: Theorem 13.31 proves that if N is a normal subgroup, then the partition G/N is a group under the operation $(gN)(hN) = ghN$.

Conversely, we have shown above that whenever a partition P inherits a group structure, the identity element in P must be a normal subgroup, and the other equivalence classes must be its left (and right) cosets. $\quad\square$

The Structure of Group Homomorphisms

We now prove a structure theorem for group homomorphisms, relating them to normal subgroups. Suppose that $\phi : G \to H$ is a homomorphism. A homomorphism need not be injective, and it need not be surjective. Not being injective makes homomorphisms worth studying, because this can partition a complicated group G into simpler pieces. But not being surjective makes very little difference, as the image

$$\mathrm{im}(\phi) = \{\phi(g) \mid g \in G\}$$

(which we previously denoted by $\phi(G)$) is simply a subgroup of H:

Proposition 13.33: If G, H are groups and $\phi : G \to H$ is a homomorphism, then $\mathrm{im}(\phi)$ is a subgroup of H.

Proof: If $g, h \in G$ then, by theorem 13.18(3), $\phi(h^{-1}) = (\phi(h))^{-1}$, so

$$\phi(g)(\phi(h))^{-1} = \phi(g)(\phi(h^{-1})) = \phi(gh^{-1}) \in \mathrm{im}(\phi).$$

By theorem 13.10, $\mathrm{im}(\phi)$ is a subgroup. $\quad\square$

The homomorphism ϕ also gives rise to a special subgroup in G:

Definition 13.34: Let $\phi : G \to H$ be a homomorphism. The *kernel* of ϕ is

$$\ker(\phi) = \{g \in G \mid \phi(g) = 1_H\}.$$

We can then prove:

Theorem 13.35: Let $\phi : G \to H$ be a homomorphism. Then the kernel of ϕ is a normal subgroup of G.

Proof: If $h \in \ker(\phi)$, then $\phi(h) = 1_H$, so for any $g \in G$,

$$\phi(ghg^{-1}) = \phi(g)\phi(h)\phi(g^{-1}) = \phi(g)1_G\phi(g^{-1}) = \phi(g)\phi(g^{-1}) = 1_G.$$

Therefore $ghg^{-1} \in \ker(\phi)$, and $\ker(\phi)$ is a normal subgroup. $\quad\square$

This leads immediately to:

Theorem 13.36 (Structure Theorem for Group Homomorphisms): Let G and H be groups and $\phi : G \to H$ be a homomorphism. Then

$$G/\ker(\phi) \cong \mathrm{im}(\phi).$$

Proof: Let $N = \ker(\phi)$, which is a normal subgroup of G. Then G/N consists of the left cosets gN for $g \in G$, and the group operation is setwise multiplication. Define the map $\mu : G/N \to \mathrm{im}(\phi)$ by

$$\mu(gN) = \phi(g).$$

This is certainly well defined, for if $gN = hN$ then $g = hn$ for $n \in I$, so $\phi(n) = 1_H$ and

$$\phi(g) = \phi(hn) = \phi(h)\phi(n) = \phi(h)1_H = \phi(h).$$

μ is a homomorphism because

$$\mu(xN\,yN) = \phi(xy) = \phi(x)\phi(y) = \mu(xN)\mu(yN).$$

It is injective because given $\mu(gN) = \mu(hN)$, then $\phi(g) = \phi(h)$, so

$$\phi(gh^{-1}) = \phi(g)\phi(h^{-1}) = \phi(g)\phi(h)^{-1} = 1_H.$$

So $g^{-1}h \in N$, implying $g^{-1}hN = N$, so $gN = hN$ and μ is injective.

It is also surjective, because, given any $k \in \mathrm{im}(\phi)$, then $k = \phi(g)$ for some $g \in G$ and so

$$k = \phi(g) = \mu(gN).$$

Hence μ is an isomorphism. $\qquad\square$

Example 13.37: For the additive group of integers $\mathbb{Z}$, the set $n\mathbb{Z} = \{nm \in \mathbb{Z} \mid m \in \mathbb{Z}\}$ of all multiples of n is a subgroup of $\mathbb{Z}$ under addition. Here the operation is addition and the cosets should be written as $k + n\mathbb{Z}$. For instance, if $n = 3$, then the cosets are

$$3\mathbb{Z} = \{\dots, -6, -3, 0, 3, 6, \dots\}$$
$$1 + 3\mathbb{Z} = \{\dots, -5, -2, 1, 4, 7, \dots\}$$
$$2 + 3\mathbb{Z} = \{\dots, -4, -1, 2, 5, 8, \dots\}.$$

For $n \geq 1$, we have

$$\mathbb{Z}/n\mathbb{Z} \cong \mathbb{Z}_n.$$

For $n = 0$, we have $0\mathbb{Z} = \{0\}$ and $\mathbb{Z}/0\mathbb{Z} \cong \mathbb{Z}$. For negative n, we have $n\mathbb{Z} = (-n)\mathbb{Z}$.

The Structure of Groups

Now we have structure theorems that enable us to think about groups not just as a list of axioms and subsequent theorems but as a crystalline concept that we can imagine in our minds. As we saw in theorem 13.20, a group G is precisely a group of permutations of a set of objects. In particular, it is a subgroup of the permutation group S_G permuting the underlying set G.

If we attempt to take a partition of a group G into subsets and define a group structure on the partition, then this can be done if, and only if, one of the subsets in the partition is a normal subgroup K of G and the other subsets in the partition are cosets of K.

If we have a homomorphism (a function preserving the group operation) $\phi : G \to H$ from a group G to another group H, then $\ker(\phi)$ (the elements in G mapping onto the identity in H) is a normal subgroup of G, the image $\text{im}(\phi)$ is a subgroup of H, and the quotient group $G/\ker(\phi)$ is isomorphic to $\text{im}(\phi)$.

In general, a group G formulated in terms of a set-theoretic definition may be seen as a group of permutations on a set X.

We saw earlier that the permutation group $\mathbb{S}_3$ can be considered as the group of symmetries of an equilateral triangle, where ρ is a rotation through an angle $2\pi/3$, μ is a mirror reflection in a line of symmetry through one of the vertices, and the permutations are of the form $\rho^p \mu^q$ where and $0 \leq p \leq 2$, $0 \leq q \leq 1$.

In the same way, other geometric figures have a group of symmetries. For instance, a square has eight symmetries: four rotations (one being the identity) and four reflections. A regular n-gon has $2n$ symmetries; a circle has infinitely many.

The theory of groups can be used to formulate properties of symmetries, particularly in geometry. Historically, however, the first developments in group theory arose in algebra in the early nineteenth century. In the next section we consider the broader evolution of mathematical ideas as part of an overall vision without being encumbered by the step-by-step detail of that development. This is intended to give you an overall picture of the use of group theory in courses you may encounter later in your studies.

Major Contributions of Group Theory throughout Mathematics

The abstract notion of a group developed from groups of permutations, which were first made explicit by Évariste Galois in connection with solutions of polynomial equations using algebraic formulas. By this we mean formulas that are constructed from the coefficients using addition, subtraction, multiplication, and division, but also pth roots for integers p. (We may assume p is prime here, since, for example, $\sqrt[15]{a} = \sqrt[3]{\sqrt[5]{a}}$ and so on.)

Niels Henrik Abel had proved that no such formula exists for the quintic equation, but Galois placed his result in a more general context: when can a polynomial equation be solved by a formula, and when not? His answer was that there is a group G of permutations associated with any such equation—basically, those permutations of its roots that preserve all algebraic relations between them—and there is a formula if and only if this group has a very special kind of structure. Namely, G has a sequence of subgroups

$$G = G_0 \supseteq G_1 \supseteq G_2 \supseteq G_3 \supseteq \ldots \supseteq G_k = \{1\},$$

where each G_{j+1} is a normal subgroup of G_j and the quotient group G_j/G_{j+1} has prime order. Roughly speaking, each piece corresponds to part of the formula that takes the pth root of something, where p is the prime concerned.

Galois observed in particular that when the equation is a general quintic, the group G consists of all permutations of the five roots, so it is $\mathbb{S}_5$. This has order 120, and it has only three normal subgroups: $\mathbb{S}_5$ itself, the trivial subgroup $\{1\}$, and a subgroup called $\mathbb{A}_5$ with order 60. Since 120 is not prime, we have to start the sequence with $G_1 = \mathbb{A}_5$. But Galois also observed that the only normal subgroups of $\mathbb{A}_5$ are $\mathbb{A}_5$ and $\{1\}$. (A normal subgroup of a normal subgroup of G need not be a normal subgroup of G, so this also needs proof.) Since neither 1 nor 60 is prime, the sequence gets stuck at $\mathbb{A}_5$; that is, there is no sequence of the required kind. Therefore the quintic can't be solved by a formula.

A similar algebraic technique resolved the classic problems of whether one could duplicate a cube or trisect an angle in Euclidean geometry. The answer is a resounding 'No!' Using algebra to interpret the intersections of lines and circles essentially involves finding the solution of successive quadratic equations which each have two solutions, and the permutation group of each successive quadratic equation involves groups of order 2, 2^2, and so on. But duplicating a cube with sides of length 1 involves constructing a cube side x whose volume satisfies $x^3 = 2$. This equation has 3 complex roots and the corresponding group of permutations is of order 3, not a power of 2. So it is not possible to duplicate the cube in Euclidean geometry.

If there were a technique for trisecting an angle, then we could apply it to trisect 30°, which would lead to the construction of the angle $\theta = 10°$ and, in particular, to the value of $x = \sin 10°$. But using the formulae for $\sin 3\theta$ it can be shown that $x = \sin \theta$ satisfies a cubic equation whose solutions have a permutation group of order 3, not a power of 2. Again, an algebraic proof using permutations solved a geometric problem that had puzzled mathematicians for 2000 years.

It took about 40 years after Galois' death in a duel for his ideas to be properly appreciated. For instance, Klein realised their implications in geometry. In 1872, he published a unified framework (called the Erlanger Programme) for classifying different forms of geometry. For two millennia, geometry had focused on Euclidean geometry in the plane and in three-dimensional space, using concepts such as congruent triangles and the parallel property, but over time, new forms of geometry had occurred.

During the Renaissance, painters had developed the idea of representing scenes on a canvas by imagining they were looking through a glass with their eye in a fixed position and painting the scene on the canvas to represent what they saw. This gave rise to *projective geometry*, projecting three dimensions onto a two-dimensional plane. The picture could be transformed by moving the position of the eye to look at the scene from a different viewpoint. Under such a transformation, points remained points, straight lines remained straight, but angles could change and circles could be transformed into ellipses. From Klein's viewpoint, points and straight lines were invariant concepts in projective geometry but angles and circles were not.

Klein realised that different forms of geometry could be described by generalising the algebraic language of permutations introduced by Galois, with each form of geometry operating on a set and focusing on properties that remained invariant under the transformations available in the theory.

The notion of symmetry could now be interpreted in a more general sense as a bijection on a set that preserves some specified kind of structure. It could be a shape (rigid motions), it could be an algebraic formula (Galois group), it could be a property like 'being a solution of a specific differential equation'. In this way, groups are 'really' about symmetry and offer powerful new principles in mathematics (See [33]).

For instance, we have already seen that the group of permutations $\mathbb{S}_3$ can be represented as the group of symmetries of an equilateral triangle, consisting of a rotation ρ through a third of a full turn, a flip μ over a line of symmetry through a vertex, together with combinations of these permutations where $\rho^3 = \mu^2$ is the identity and $\mu\rho = \rho^2\mu$.

We have also already seen that if we replace the equilateral triangle by other subsets of the plane, we obtain similar results: a square has eight symmetries (four rotations including the identity and four reflections), a regular n-gon has $2n$ symmetries, a circle has infinitely many.

These ideas can be further generalised to the entire plane. For instance, a tiling of the plane by congruent squares, like an infinite chessboard, has infinitely many translational, rotational, and reflectional symmetries.

This leads to a new way of viewing Euclidean geometry as the study of rigid motions of the plane that preserve the distance between points. These include *translations* that shift all points in a fixed direction, *rotations* through an angle about a fixed point, and *reflections*, which flip the plane over a fixed line to produce a mirror image. We can show that all rigid motions of the plane arise by combinations of translations, rotations, and reflections.

Formally, the rigid motion of the plane can be described as a map $f : \mathbb{R}^2 \to \mathbb{R}^2$ where the distance between two points x and y is the same as the distance between $f(x)$ and $f(y)$. This means that if we pick three non-collinear points A, B, C forming a triangle then the transformed triangle $A'B'C'$ will have the lengths of the sides maintained: $AB = A'B'$, $BC = B'C'$, $CA = C'A'$. We can show that any rigid motion can then be constructed by using a translation, a rotation, and a reflection, as follows.

First translate the plane so that A moves to A'. Then, because AB and $A'B'$ are the same length, rotate the plane around A' so that the rotated line coincides with $A'B'$. At this stage the rotated triangle may coincide with the triangle $A'B'C'$, or it may be a reflection in the line $A'B'$. The rigid motion of the plane is then seen as a successive application of a translation, then a rotation, and, if needed, a reflection.

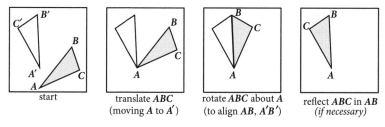

| start | translate ABC (moving A to A') | rotate ABC about A (to align AB, $A'B'$) | reflect ABC in AB (if necessary) |

Fig. 13.7 A rigid transformation as a translation, rotation, and perhaps a reflection

Formally, the successive transformations can be written as follows. The first translation takes any point (x, y) to $(x + a, y + b)$ and can be written as

$$T_{(a,b)}(x, y) = (x + a, y + b),$$

or more compactly as

$$T_u(z) = z + u, \tag{13.4}$$

where $z = (x, y)$ is a general point in the plane and $u = (a, b)$ is the specific vector representing the translation.

The same expression can also be interpreted as addition of complex numbers where $z = x + iy$ and $u = a + ib$.

The rotation $R_{v,\theta}$ around a fixed point $v = (c, d)$ through an angle α can be expressed in Cartesian coordinates. However, it has a more compact expression using complex numbers, where multiplying by $e^{i\alpha}$ turns a complex number through the angle α to give:

$$R_{v,\alpha}(z) = v + (z - v)e^{i\alpha}. \tag{13.5}$$

Finally, reflecting the plane in the horizontal axis can be expressed simply in complex numbers by taking $z = x + iy$ to $\bar{z} = x - iy$. To perform a general reflection of a point z in a line that is at an angle β to the horizontal through a point $v = (c, d)$ is more sophisticated, but it can be done in successive moves, first by moving z to $z - v$ (to move the point z to the origin), then turn the plane through an angle $-\beta$ to move the line horizontal to $z - v$ to $(z - v)e^{-i\beta}$, then flipping the plane over the horizontal axis to $\overline{(z - v)e^{-i\beta}} = \overline{(z - v)}\ \overline{e^{-i\beta}} = \overline{(z - v)}\,e^{i\beta}$, and finally, turning the plane back through an angle $+\beta$ to return the line to its original position. This gives the final mirror position of the original point z as

$$M_{v,\beta}(z) = \overline{(z - v)}\,e^{i\beta}e^{i\beta} = \overline{(z - v)}\,e^{2i\beta}. \tag{13.6}$$

The full transformation to shift the triangle ABC to the position $A'B'C'$ can therefore be performed by a translation T_v, followed by a rotation $R_{v,\alpha}$, and then, if it is necessary to flip the triangle over, to perform $M_{v,\beta}$. Any rigid motion of the plane can be written as a composite function taking z to

$$(M_{v,\beta})^k \circ R_{v,\alpha} \circ T_v(z) \text{ for } v \in \mathbb{R}^2, 0 \le \alpha < 2\pi, 0 \le \beta < 2\pi \text{ and } k = 0 \text{ or } 1.$$

This shows that translations, rotations, and mirror images generate the whole of the group of rigid transformations. To complete the description of the group we need to identify the relations between these rigid transformations. These include relationships such as:

$$T_0 = i$$

$$T_u \circ T_v = T_{u+v}$$

$$R_{v,0} = i$$

$$R_{v,\theta} \circ R_{v,\phi} = R_{v,\theta+\phi}$$

$$(M_v)^2 = i$$

and all possible pairwise combinations of T_u, $R_{v,\alpha}$, $M_{w,\beta}$ for different values of $u, v, w \in \mathbb{R}^2$, $\alpha, \beta \in \mathbb{R}$. The details include the possible combinations of translations, rotations, reflections in different directions, rotations round different points, or reflections in different lines. They could be generalised to rigid motions in $\mathbb{R}^3$, and even move on to higher dimensions in $\mathbb{R}^n$. Such activities take us beyond the goals of a focus on the foundations of mathematics and are postponed for possible study in later courses.

The Way Ahead

The theory of groups developed in this chapter can be extended to apply in many areas of mathematics using the notion of symmetry as a bijection of a set that preserves certain kinds of structure. A symmetry could relate to a shape (rigid motion), to an algebraic formula (Galois group), it could be a property like 'being a solution of a specific differential equation'.

In many applications of mathematics, the symmetries of a system tell us a lot about the system itself. For example, the symmetries of a drum impose constraints on its vibrational frequencies, and the symmetries of a growing organism affect the shapes it can take up. Deep areas of physics turned out to be governed by the symmetries of the basic equations of relativity and quantum mechanics. Modern particle physics, up to and including the recently discovered Higgs boson, builds on the study of such symmetries.

Pure mathematics also benefits from generalising the ideas of this chapter to more general algebraic structures. Algebraic structures may have several operations, some of which have a group structure relating to theorems proved here. For example, rings, fields, and vector spaces include a commutative operation of addition and this may interact with other operations such as multiplication in a ring or field or operations by scalar quantities in vector spaces.

In all these cases, the additive structure is commutative, so additive subgroups are normal and have additive quotient structures, such as $\mathbb{Z}/n\mathbb{Z} \cong \mathbb{Z}_n$. In this particular case, multiplication also works in $\mathbb{Z}_n$ to make it a ring (and also a field if p is prime). In this example, we not only have $n\mathbb{Z}$ closed under multiplication, we can also multiply any element $nk \in n\mathbb{Z}$ by any element $m \in \mathbb{Z}$ to get the product mnk, also in $n\mathbb{Z}$. This turns out to be the fundamental property for introducing quotient structures in ring theory.

Definition 13.38: If R is a ring and I is a subgroup under addition then I is an *ideal* if $x \in R$, $y \in I$ implies $xy \in I$.[3]

---

[3] Here, as elsewhere in this book, when we don't say otherwise, we speak of a ring with commutative multiplication. In a non-commutative ring, we would require both xy and yx to belong to I.

An ideal is not only closed under multiplication of elements within it, but also by multiplication by an element from the whole ring.

Combining the properties of addition and multiplication, we have

Equivalent Definition 13.39: If R is a ring and I is a non-empty subset of R, then I is an *ideal* if

(i) $x, y \in I$ implies $x - y \in I$
(ii) $x \in R, y \in I$ implies $xy \in I$.

Example 13.40: $n\mathbb{Z}$ is an ideal in Z.

In ring theory, it is possible to define quotient structures R/I for a ring R and an ideal I in R using the same techniques as for a group and a normal subgroup. For example, if I is an ideal in the ring R, then because addition is commutative, the cosets written additively as $x + I$ or $I + x$ are equal and, by the structure theorem for groups, addition may be defined on R/I by

$$(x + I) + (y + I) = (x + y) + I \qquad (13.7)$$

so that R/I is an additive group.

We can also define multiplication by

$$(x + I)(y + I) = xy + I. \qquad (13.8)$$

Theorem 13.41: If R is a ring and I is an ideal in R, then R/I is a ring where addition and multiplication are given by (13.7) and (13.8).

Proof: Because R is a commutative group under addition, R/I is already known to be a commutative group under addition. We need to check that multiplication is well defined and satisfies the associative, commutative, and distributive laws. These are all straightforward.

If $x + I = x' + I$ and $y + I = y' + I$, then

$$xy - x'y' = xy - x'y + x'y - x'y' = (x - x')y + x'(y - y').$$

By the definition of an ideal,

$$x - x' \in I, y' \in R \Rightarrow (x - x')y' \in I, \quad x' \in R, \ (y - y') \in I \Rightarrow x'(y - y') \in I.$$

Hence $xy - x'y' \in I$ and so $xy + I = x'y' + I$.

Thus multiplication is well defined. The multiplicative identity is $1 + I$ and the associative, commutative, and distributive laws follow from the corresponding properties in R. $\qquad \square$

Example 13.42: $\mathbb{Z}_n = \mathbb{Z}/n\mathbb{Z}$ is a ring and if p is prime, then it is a field. You should verify these properties for yourself.

This use of quotient rings will prove to be insightful in chapter 15.

Exercises

1. Express the following permutations in disjoint cycle form:

   (i) $\begin{pmatrix} 1 & 2 & 3 & 4 & 5 \\ 3 & 2 & 5 & 1 & 4 \end{pmatrix}$

   (ii) $\begin{pmatrix} 1 & 2 & 3 & 4 & 5 & 6 \\ 6 & 5 & 4 & 3 & 2 & 1 \end{pmatrix}$

   (iii) $\begin{pmatrix} 1 & 2 & 3 & 4 & 5 & 6 & 7 & 8 & 9 \\ 5 & 9 & 7 & 4 & 1 & 3 & 6 & 2 & 8 \end{pmatrix}$

   (iv) $(12)(12)(145)(23)$.

2. Express the following permutations in standard form:

   (i) $(1234)(23)(12)$

   (ii) $(1235)(43)$

   (iii) $(43)(34)(123)$

   (iv) $(12)(13)(12)(143)(2)$.

3. Find σ^{-1} in both standard and disjoint cycle form where σ is:

   (i) $\begin{pmatrix} 1 & 2 & 3 & 4 \\ 2 & 4 & 3 & 1 \end{pmatrix}$

   (ii) $(1234)(56)$

   (iii) $(12)(12)(12)(13)(14)$.

4. Calculate the product $\sigma\tau$ and $\tau\sigma$, using the convention that permutations are written on the left (i.e. $\sigma\tau$ is τ followed by σ):

   (i) $\sigma = \begin{pmatrix} 1 & 2 & 3 & 4 & 5 \\ 4 & 2 & 3 & 1 & 5 \end{pmatrix}, \tau = \begin{pmatrix} 1 & 2 & 3 & 4 & 5 \\ 3 & 4 & 1 & 5 & 2 \end{pmatrix}$

   (ii) $\sigma = (123), \tau = (23)$

   (iii) $\sigma = (1234), \tau = \begin{pmatrix} 1 & 2 & 3 & 4 \\ 1 & 4 & 3 & 2 \end{pmatrix}$.

5. Prove that if X is a finite set with n elements, then the number of bijections from X to itself is $n! = n(n-1)(n-2)\ldots 3.2.1$.

   Hint: Prove by induction the slightly more general theorem that if X, Y are finite sets, each having n elements, then the number of bijections from X to Y is $n!$.

6. Write out the axioms for a group. Which of the following sets is a group under the given operation? If it is, specify the identity element and the inverse of each element; if not, give one reason why it fails.

   (i) $\mathbb{Z}$ under addition
   (ii) $\mathbb{Z}$ under multiplication
   (iii) $\mathbb{R}$ under addition
   (iv) $\mathbb{R}$ under multiplication
   (v) $\mathbb{R}^+$, the positive real numbers, under multiplication
   (vi) $\mathbb{R}_5$, the integers modulo 5 under addition
   (vii) $\mathbb{R}_5$, the integers modulo 5 under multiplication.

7. Show that the set $\{1_5, 2_5, 3_5, 4_5\}$ forms a group under multiplication modulo 5. Show that $\{1_8, 3_8, 5_8, 7_8\}$ forms a group under multiplication modulo 8. Find the largest subset of the integers modulo 12 that is a group under multiplication modulo 12. In each case, write out the multiplication table.

8. The set $\{1_3, 2_3\}$ of non-zero integers modulo 3 form a group under multiplication modulo 3, but the set $\{1_4, 2_4, 3_4\}$ of non-zero integers modulo 4 do not form a group under multiplication modulo 4. Explain why, and investigate what happens to the set $\{1_n, 2_n, \ldots, (n-1)_n\}$ of non-zero integers modulo n.

9. Show that $\mathbb{Z}_7^*$, the set of non-zero elements modulo 7, is in the form $\{1_7, a, a^2, \ldots, a^5\}$ for $a = 3_7$. Deduce that, for any integer n, either $n^6 \equiv 0 \bmod 7$, or $n^6 \equiv 1 \bmod 7$.

10. Show that for any $1 \leq k < n$, the elements $1_n k_n, 2_n k_n, \ldots, (n-1)_n k_n$ are all different. Hence, or otherwise, prove that the non-zero elements in Z_n form a group if and only if n is a prime.

11. Find all subgroups of $\mathbb{S}_4$.

12. Prove that the complex nth roots of unity satisfying $\omega^n = 1$ form a cyclic group of order n under multiplication. Relate this fact to the rotational symmetries of a regular n-gon.

13. Let S be a non-empty set of permutations of a finite set X satisfying the *closure property*:

   $$\text{If } \sigma, \tau \in S \text{ then } \sigma \circ \tau \in S.$$

   Prove that if X is finite then the following properties also hold:

   $$\text{The identity } i_X \in S.$$
   $$\text{If } \sigma \in G \text{ then } \sigma^{-1} \in S.$$

   Hence deduce that a non-empty set S of permutations of a finite set X satisfying the closure property is a group.

14. Let A be a finite set and P be the set of subsets of A. Let the operation $\triangle$ on P be the symmetric difference: $X \triangle Y = \{x \in X \cup Y \mid x \notin X \cap Y\}$. Show that P is a group under the operation $\triangle$, with identity element $\varnothing$. What is the inverse of X? What happens if $A = \varnothing$?

15. If a, b are any two elements of a group, show that $a^{-1}b^{-1} = (ba)^{-1}$. Hence, or otherwise, show that if G is a group such that x^2 is the identity for every $x \in G$, then G is abelian (i.e. that $ab = ba$ for all a, $b \in G$).

16. Prove that H is a subgroup of G if and only if $H \neq \varnothing$ and $HH^{-1} = H$.

17. Find an example to show that if a subgroup H is not normal then the product of two cosets gH and kH of H need not be a coset of H.

18. Suppose that $M, N \triangleleft G$ and M is a subgroup of N. Prove that $M \triangleleft N$ and $(G/M)/(N/M) \cong G/N$.

    Hint: Prove that the composition of two homomorphisms is a homomorphism, and then consider the corresponding quotient groups.

19. Using complex numbers, define the rotation ρ_α around the origin turning through an angle α moving $z = x + iy$ to $\rho_\alpha(z) = e^{i\alpha}z$.

    Define the mirror image μ_β in a line through the origin making an angle β with the horizontal axis by $\mu_\beta(z) = e^{2i\beta}\bar{z}$.

    Prove that these definitions agree with the ones given in chapter 11, and prove the following properties:

$$R_0 = i$$
$$R_\theta \circ R_\phi = R_{\theta+\phi}$$
$$M_0(x, y) = (x, -y)$$
$$R_\theta \circ M_\phi = R_{\phi+\theta/2}$$
$$M_\phi \circ R_\theta = R_{\phi-\theta/2}.$$

CHAPTER 14

Cardinal Numbers

What is infinity?

When some first-year university students were asked this question, the consensus answer was 'something bigger than any natural number'. In a precise sense, this is correct; one of the triumphs of set theory is that the concept of infinity can be given a clear interpretation. However, there is a surprise: when we compare the sizes of sets, we find not one infinity, but many—a vast hierarchy of infinities. This discovery came about by reformulating the question. Instead of asking 'how many' elements there are in a given set and using counting, it is much more profitable to compare two sets, and ask if there are as many elements in one of them as there are in the other. This idea can be made precise by saying that sets A and B have 'the same number of elements' if there is a bijection $f : A \rightarrow B$.

Rather than beginning with the full hierarchy of infinities, let's begin with what turns out to be the smallest. Here the standard set, for comparison purposes, is the natural numbers $\mathbb{N}$. It is useful to consider $\mathbb{N}$ rather than $\mathbb{N}_0 = \mathbb{N} \cup \{0\}$ because a bijection $f : \mathbb{N} \rightarrow B$ organises the elements of B into a sequence; we can call $f(1)$ the *first* element of B using this bijection, $f(2)$ the *second*, and so on. Using this process we set up a method for *counting B*. Of course, if we actually say the elements one after another using this bijection, '$f(1), f(2), \ldots$', we never reach the end, but we do know that if $b \in B$ then $b = f(n)$ for *some* $n \in \mathbb{N}$, so we reach that particular element eventually.

Recall from chapter 8 that we defined $\mathbb{N}(0) = \varnothing$, and for $n \in \mathbb{N}$,

$$\mathbb{N}(n) = \{m \in \mathbb{N} \mid 1 \leq m \leq n\}.$$

Definition 14.1: A set X is *finite* if there exists a bijection $f : \mathbb{N}(n) \rightarrow X$ for some $n \in \mathbb{N}_0$. A set X is *countable* if either X is finite or there exists a bijection $f : \mathbb{N} \rightarrow X$. If there is a bijection $f : \mathbb{N}(n) \rightarrow X$, then we say that X has n elements.

Finding such a bijection for a finite set is just the usual process of counting. Why not generalise to infinite sets too? We can get started:

Definition 14.2: If there is a bijection $f : \mathbb{N} \to X$ then X has $\aleph_0$ elements. We say that X is *countably infinite*.

The symbol $\aleph$ is the first letter 'aleph' of the Hebrew alphabet, and $\aleph_0$ is our first example of a new concept of number, used to state how big an infinite set is. If there is a bijection between $\mathbb{N}$ and X, it makes sense to say that 'X has the same (cardinal) number of elements as $\mathbb{N}$.' That number is given a new symbol, $\aleph_0$.

Before discussing cardinal numbers in general, we take a closer look at the notion of countability.

Example 14.3: $\mathbb{N}_0$ is countable. Define $f : \mathbb{N} \to \mathbb{N}_0$ by $f(n) = n - 1$; then f is a bijection. This is the first fascinating property of this method of 'counting infinite sets'. $\mathbb{N}$ is a *proper* subset of $\mathbb{N}_0$, so intuitively it should have fewer elements, yet in the sense of a bijection between the sets, they have the same size.

Galileo gave an even more graphic example in 1638:

Example 14.4 (Galileo): There is a correspondence between the natural numbers and the perfect squares:

$$1 \ 2 \ 3 \ 4 \ \ldots \ n \ \ldots$$
$$\downarrow \downarrow \downarrow \downarrow \quad \ \downarrow$$
$$1 \ 4 \ 9 \ 16 \ldots n^2 \ldots$$

In modern set-theoretic terms, if $S = \{n^2 \in \mathbb{N} \mid n \in \mathbb{N}\}$, the map $f : \mathbb{N} \to S$ given by $f(n) = n^2$ is a bijection.

This result is very curious, because we get the squares from $\mathbb{N}$ by removing all of the numbers that are not square. There are infinitely many of these; moreover, squares get thinner on the ground as we progress to larger numbers. Intuitively, a 'random' natural number is probably not a perfect square.

For over two centuries, this seeming contradiction blighted any attempt to contemplate infinity in a precise sense. Leibniz went as far as to suggest that we should only ever consider finite sets—that the apparent contradiction arose because the natural numbers are infinite. His resolution of the conflict was that if we consider only finite sets of natural numbers—say the numbers less than 100—there is no correspondence between *these* natural numbers and those of them that are squares. Indeed, of the 100 numbers in this range exactly 11 are squares.

This is a bit retrictive. It rules out any sensible concept of 'number' for infinite sets. Georg Cantor realised that we can do better. His solution of the paradox in the 1870s was even more dramatic. He showed that if we interpret 'as many' to mean that there is a bijection between two sets, then any infinite set has 'as many' elements as a proper subset! Here 'infinite' is interpreted in the technical sense that B is infinite if there is no bijection $f : \mathbb{N}(n) \to B$ for any $n \in \mathbb{N}_0$. From Cantor's point of view, there is no paradox; just a counterintuitive theorem. As we have said many times: when you generalise a mathematical concept, some of its original properties may no longer be true.

Proposition 14.5 (Cantor): If a set B is infinite, then there exists a proper subset $A \underset{\neq}{\subseteq} B$ and a bijection $f : B \to A$.

Proof: First, choose a countably infinite subset X of B. Since no bijection exists between $\mathbb{N}(0)$ and B, B is non-empty and there exists some element in B which we call x_1. Define $g : \mathbb{N} \to B$ inductively by $g(1) = x_1$, and if distinct elements $x_1, x_2, \ldots, x_n$ have been found, then since g cannot give a bijection $g : \mathbb{N}(n) \to B$, there must be another element, which we name $x_{n+1} \in B$, that is distinct from $x_1, \ldots, x_n$. Define $g(n + 1) = x_{n+1}$. Let

$$X = \{x_n \in B \mid n \in \mathbb{N}\}.$$

Let $A = B \backslash \{x_1\}$, define $f : B \to A$ by

$$f(x_n) = x_{n+1} \quad \text{for } x_n \in X$$

and

$$f(b) = b \quad \text{for } b \notin X.$$

Then f is a bijection. $\qquad\square$

We can do better than this. We can start with an infinite set B and remove an *infinite* subset to leave a subset C with a bijection from B to C. For example, if we take the set $\mathbb{N}$ of natural numbers, then the sets E of even numbers and O of odd numbers allow us to define bijections

$$f : \mathbb{N} \to E \text{ where } f(n) = 2n$$
$$g : \mathbb{N} \to O \text{ where } g(n) = 2n - 1.$$

If we start with the infinite set $\mathbb{N}$ and remove the infinite subset O then this leaves the infinite subset E and a bijection $f : \mathbb{N} \to E$.

More generally, for any infinite set B, we can remove an infinite subset and still be left with a subset A with a bijection $f : B \to A$. To do this, choose a countably infinite subset X of B as in the proof of proposition 14.5. Let Y be the subset $\{x_n | n \text{ is odd}\}$ and let A be the subset of B with the elements of Y removed:

$$A = B \backslash Y.$$

Define $f : B \to A$ by

$$f(x_n) = x_{2n} \text{ for } x_n \in Y, \text{ and } f(x) = x \text{ for } x \notin Y.$$

Then f is a bijection which maps all of B onto A.

We can therefore start with any infinite set B, remove a (countably) infinite subset Y and still be left with a subset A which has 'as many elements' as B!

Cantor's Cardinal Numbers

Cantor's solution to the problem 'how many elements?' for infinite sets was to introduce the concept of a cardinal number. For the moment we assume that for every set X, there is a concept, more briefly called a *cardinal*, with the property that if there is a bijection $f : X \to Y$, then X and Y have the same cardinal, and if there is no bijection, then the cardinals concerned are different. We denote the cardinal of X by $|X|$.

We haven't yet said what cardinals are, just what they do. To place them on a firm basis, we have to construct them set-theoretically. Cantor didn't get that far, and neither will we. However, it can be done.

In the case of finite sets, a convenient candidate for the cardinal number is close at hand. If there is a bijection $f : \mathbb{N}(n) \to X$, the cardinal of X is n. Likewise, given a bijection $f : \mathbb{N} \to X$, the cardinal of X is $\aleph_0$. For other infinite sets, we may have to invent new symbols for their cardinals. In general we denote the cardinal of X by $|X|$, on the understanding that if there is a bijection $f : X \to Y$, then $|X| = |Y|$. If there exists an *injection* $f : X \to Y$, we say that $|X| \le |Y|$. As usual, we define $|X| < |Y|$ to mean $|X| \le |Y|$ and $|X| \ne |Y|$.

In general, if X is a subset of Y, then the inclusion $i : X \rightarrow Y, i(x) = x$, is an injection, so we have

$$X \subseteq Y \Rightarrow |X| \leq |Y|.$$

Proposition 14.5 says that for any infinite set B there exists a proper subset A such that $|A| = |B|$. Thus for infinite sets,

$$X \underset{\neq}{\subset} Y \nRightarrow |X| < |Y|.$$

The dilemma posed by Galileo's example is not so much mathematical as psychological. When we extend the system of natural numbers and counting to embrace infinite cardinals, the larger system need not have all of the properties of the smaller one. However, familiarity with the smaller system leads us to expect certain properties, and we can become confused when the pieces don't seem to fit. Insecurity arose when the square of a complex number violated the real number principle that all squares are positive. This was resolved when we realised that the complex numbers cannot be ordered in the same way as their subset of reals. Likewise we resolve the seeming contradiction that Galileo discovered by realising that when we interpret 'same cardinal' in terms of a bijection between sets, proper inclusion of A in B does not prevent A and B from having the same *infinite* cardinal.

We return to the notion of countability. Given any infinite set B, as in the proof of proposition 14.5, we can select a countably infinite subset $X \subseteq B$. This means that $\aleph_0 = |X| \leq |B|$, so $\aleph_0$ is the smallest infinite cardinal. Surprisingly, many familiar sets that seem much bigger than $\mathbb{N}$ also have cardinality $\aleph_0$.

Example 14.6: The integers are countable. Define $f : \mathbb{N} \rightarrow \mathbb{Z}$ by:

$$f(2n) = n, f(2n - 1) = 1 - n \text{ for } n \in \mathbb{N},$$

then we get the bijection

$$
\begin{array}{ccccccc}
1 & 2 & 3 & 4 & 5 & 6 & 7 & \ldots \\
\downarrow & \downarrow & \downarrow & \downarrow & \downarrow & \downarrow & \downarrow & \\
0 & 1 & -1 & 2 & -2 & 3 & -3 & \ldots.
\end{array}
$$

Although f is a bijection, it doesn't preserve the order (in the sense that $m < n$ does not imply $f(m) < f(n)$; for instance $f(2) > f(3)$). When we set up bijections between sets with an order on them, we may have to do it in a very higgledy-piggledy way, as the next example shows.

Example 14.7: The rationals are countable.

We'll prove this in stages, first by counting the positive rationals. A positive rational is p/q, where p and q are natural numbers. One way of counting the rationals is to think of them written out as an array:

$$\begin{array}{llll}
1/1 & 1/2 & 1/3 & 1/4 \ldots \\
2/1 & 2/2 & 2/3 & 2/4 \ldots \\
3/1 & 3/2 & 3/3 & 3/4 \ldots \\
4/1 & 4/2 & 4/3 & 4/4 \ldots \\
\cdot & \cdot & \cdot & \cdot \quad \ldots \\
\cdot & \cdot & \cdot & \cdot \quad \ldots \\
\cdot & \cdot & \cdot & \cdot \quad \ldots
\end{array}$$

Now read them off along the 'cross diagonals', first 1/1, next 1/2, 2/1, then 1/3, 2/2, 3/1, and so on:

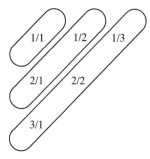

Fig. 14.1 Counting the positive rationals

This process strings the positive rationals out as a list 1/1, 1/2, 2/1, 1/3, 2/2, 3/1, ... However, this list includes repeats, because 1/1 = 2/2 and later on we get 3/3, 4/4, and so on. Similarly 1/2 = 2/4 = 3/6 = ... That prevents the construction of a bijection. So we consider each element in the list in turn, and delete it if it has occurred before. That leaves 1/1, 1/2, 2/1, 1/3, 3/1, ... Suppose that the nth rational in the remaining sequence is a_n. Then the function f from the natural numbers to the positive rationals for which $f(n) = a_n$ is a bijection. Now we include negative rationals as well: the list 0, $a_1, -a_1, a_2, -a_2, \ldots, a_n, -a_n, \ldots$ includes every rational precisely once. So the map $g : \mathbb{N} \to \mathbb{Q}$ given by

$$g(1) = 0, g(2n) = a_n, \ldots, g(2n + 1) = -a_n \quad \text{for } n \in \mathbb{N}$$

is a bijection, as required.

Although we have not given an explicit formula for $g(n)$, we *have* given an explicit prescription for it. The first few terms are

$$1 \quad 2 \quad 3 \quad 4 \quad 5 \quad 6 \quad 7 \quad 8 \quad 9 \quad 10 \quad 11 \quad \ldots$$
$$\downarrow \downarrow \downarrow \downarrow \downarrow \downarrow \downarrow \downarrow \downarrow \downarrow \downarrow$$
$$0 \quad 1 \quad -1 \quad \tfrac{1}{2} \quad -\tfrac{1}{2} \quad 2 \quad -2 \quad \tfrac{1}{3} \quad -\tfrac{1}{3} \quad 3 \quad -3 \quad \ldots$$

and you should be able to continue as far as you wish. Later we develop a more powerful result, the Schröder–Bernstein theorem, which lets us prove that two sets have the same cardinal without constructing an explicit bijection. By invoking the theorem, we can deal with the rationals more cleanly.

The reason why we allowed 'countable' to include 'finite' as well as 'countably infinite' is the next result:

Proposition 14.8: A subset of a countable set is countable.

Proof: Given a bijection $f : \mathbb{N} \to A$ and $B \subseteq A$, either B is finite, or we can define $g : \mathbb{N} \to B$ by

$$g(1) \text{ is the least } m \text{ such that } f(m) \in B,$$

having found $g(1), \ldots, g(n)$, then

$$g(n + 1) \text{ is the least } m \text{ such that } f(m) \in B \backslash \{g(1), \ldots, g(n)\}.$$

Informally, this just amounts to writing out the elements of A as a list

$$f(1), f(2), f(3), \ldots, f(n), \ldots,$$

deleting those terms not in B, and leaving the terms in B listed in the same order. □

The remarkable fact about countable sets is that we can build up sets from them that seem a lot bigger, but once more are countable, in the following precise sense:

Proposition 14.9: A countable union of countable sets is countable.

Proof: Given a countable collection of sets, we can use $\mathbb{N}$ as the index set and write the sets as $\{A_n\}_{n \in \mathbb{N}}$. (If there is only a finite number of sets, $A_1, \ldots, A_k$, put $A_n = \varnothing$ for $n > k$.) Since each A_n is countable, we can write the elements of A_n as a list $a_{n_1}, a_{n_2}, \ldots, a_{n_m}, \ldots$, which terminates if A_n is finite but is an infinite sequence if A_n is countably infinite. Now tabulate the elements of $\bigcup_{n \in \mathbb{N}} A_n$ as a rectangular array, and read them off along the cross diagonals as in the previous example:

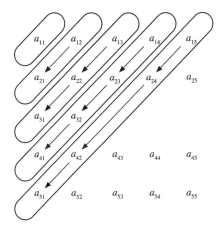

Fig. 14.2 Counting down successive cross diagonals

There may be gaps in the array because some of the sets are finite (as in row three of the above illustration) or because there is only a finite number of sets. There may be repeats when two sets A_n, A_m have elements in common, so an element in row n is repeated in row m. We just pass over the gaps and delete elements that have occurred earlier. The list is then either finite, or an infinite sequence with no repeats. This shows that $\bigcup_{n \in \mathbb{N}} A_n$ is countable. ☐

Proposition 14.10: The cartesian product of two countable sets is countable.

Proof: If A and B are countable, write the elements of A as a sequence $a_1, a_2, \ldots, a_n, \ldots$ (which terminates if A is finite). Similarly, write the elements of B as $b_1, b_2, \ldots, b_m, \ldots$. Now write the elements of $A \times B$ as a rectangular array and read them off along the cross diagonals:

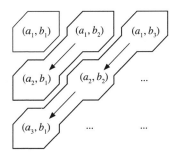

Fig. 14.3 Counting ordered pairs

If either A or B is finite, there are gaps that should be passed over; if both are finite, then $A \times B$ is finite. If both are infinite, an explicit bijection $f : \mathbb{N} \to A \times B$ is not hard to write down. There is one element in the first cross diagonal, two in the second, and, in general, n in the nth. So there are $1 + 2 + \cdots + n = \frac{1}{2}n(n + 1)$ elements in the first n. The rth element in the next cross diagonal is (a_r, b_{n+2-r}), so an explicit formula for the bijection $f : \mathbb{N} \to A \times B$ is

$$f(m) = (a_r, b_{n+2-r}) \text{ for } m = \tfrac{1}{2}n(n + 1) + r \ (1 \leq r \leq n + 1). \qquad \square$$

An instance of proposition 14.10 in action is:

Example 14.11: The set of points in the plane with rational coordinates is countable.

At this stage of the game, the reader may be forgiven for thinking that every infinite set is countable, but that is not so, as we see by looking at the real numbers.

Example 14.12: The real numbers are not countable. We prove this by contradiction, by showing that no map $f : \mathbb{N} \to \mathbb{R}$ can be surjective, so there cannot be a bijection $f : \mathbb{N} \to \mathbb{R}$. Given a map $f : \mathbb{N} \to \mathbb{R}$, express each $f(m) \in \mathbb{R}$ as a decimal expansion,

$$f(m) = a_m \cdot a_{m_1} a_{m_2} \ldots a_{m_n} \ldots (a_m \in \mathbb{Z}, a_{m_r} \in \mathbb{N}_0, 0 \leq a_{m_r} \leq 9)$$

where, for definiteness, if the decimal terminates, we write it that way, ending in a sequence of zeros, not a sequence of nines. Now we write down a real number, different from all the $f(m)$. Let

$$\beta = 0 \cdot b_1 b_2 ... b_n ...$$

where

$$b_n = \begin{cases} 1 \text{ if } a_n = 0 \\ 0 \text{ if } a_n \neq 0. \end{cases}$$

Then β is different from $f(n)$ because it differs in the nth place. We have avoided the possible ambiguity that might arise from an infinite sequence of nines in the expansion, by making sure that the expansion of β doesn't have any.

Let $\aleph$ be the cardinal number of $\mathbb{R}$. Since $\mathbb{N} \subseteq \mathbb{R}$ we have $\aleph_0 \leq \aleph$, and the last example shows that $\aleph_0 \neq \aleph$. So at last we have found a cardinal strictly bigger than $\aleph_0$.

In fact, for any cardinal we can find a strictly bigger one. The cardinal must be associated with some set A. We show that the power set of A *always* has strictly bigger cardinality:

Proposition 14.13: If A is a set then $|\mathbb{P}(A)| > |A|$.

Proof: Evidently the map $f : A \to \mathbb{P}(A)$ given by $f(a) = \{a\}$ is an injection, so $|A| \leq |\mathbb{P}(A)|$. It remains to show that $|A| \neq |\mathbb{P}(A)|$. To do so, we prove that no map $f : A \to \mathbb{P}(A)$ can be a surjection. For such a map, $f(a) \in \mathbb{P}(A)$ for each $a \in A$, so $f(a)$ is a subset of A. We ask 'does a belong to the subset $f(a)$?' The answer is always 'yes' or 'no'. We select those elements for which the answer is 'no' to get the subset

$$B = \big\{ a \in A \,|\, a \notin f(a) \big\}.$$

We claim that B is not mapped onto by any element of A under the function f. For if B were equal to $f(a)$ for some $a \in A$, the question 'does a belong to B?' leads to a contradiction:

$$a \in B \Rightarrow a \notin f(a) = B,$$
$$a \notin B \Rightarrow a \in f(a) = B.$$

So B is not mapped onto by f and f is not surjective. Even more so, it cannot be a bijection. $\qquad\square$

Proposition 14.13 leads us to a hierarchy of infinities. We begin with $\aleph_0 = |\mathbb{N}|$. Then $|\mathbb{P}(\mathbb{N})|$ is strictly bigger, then $|\mathbb{P}(\mathbb{P}(\mathbb{N}))|$, and so on.

The Schröder–Bernstein Theorem

An obvious question to ask concerning the relation $\leq$ between cardinals is

if $|A| \leq |B|$ and $|B| \leq |A|$, can we conclude that $|A| = |B|$?

The answer to this question is in the affirmative, and the content of this statement is the Schröder–Bernstein theorem. The proof is trickier than might seem necessary for such a simple-looking proposition. The main problem is that $|A| \leq |B|$ tells us that there is *some* injection $f : A \to B$, and $|B| \leq |A|$ tells us that there is *some* injection $g : B \to A$, but these injections need not be related in any useful way. Nevertheless, somehow we must use them to construct a bijection between A and B. This requires some ingenuity, and it took a while to find the proof.

Theorem 14.14 (Schröder–Bernstein): Given sets A, B, then $|A| \leq |B|$ and $|B| \leq |A|$ implies $|A| = |B|$.

Proof: We have injections $f : A \to B, g : B \to A$. We can use f to pass from A to B or g to pass from B to A. Repeating the process, we can pass to and fro obtaining $f(a), g(f(a)), f(g(f(a))), \ldots.$

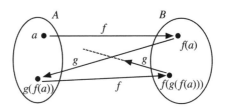

Fig. 14.4 Tracing a chain forwards

The key to the proof is to try to trace such a chain *backwards*. Start with $b \in B$ and see if there exists $a \in A$ such that $f(a) = b$; *if* such an a exists, it is unique. Then see if there is a $b_1 \in B$ such that $g(b_1) = a$, then $a_1 \in A$ such that $f(a_1) = b_1$, attempting to build up a chain, $b, a, b_1, a_1, \ldots, b_n, a_n$, where $f(r_r) = b_r, g(b_r) = a_{r+1}$. In tracing back a chain of elements in this fashion, three things can happen:

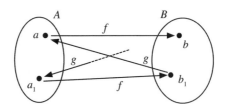

Fig. 14.5 Tracing a chain backwards

   (i) we reach $a_N \in A$ and stop because there is no $b^* \in B$ with $g(b^*) = a_N$;
  (ii) we reach $b_N \in B$ and stop because no $a^* \in A$ satisfies $f(a^*) = b_N$;
 (iii) the process goes on forever.

This partitions B into three sets:

   (1) B_A, the subset of elements in B whose ancestry originates in A, as in (i).

(2) B_B, the subset of elements in B whose ancestry originates in B, as in (ii).

(3) B_∞, the subset of elements in B whose ancestry can be traced back forever, as in (iii).

Note that B_A, B_B, B_∞ are disjoint and their union is B, so they do indeed give a partition. Similarly we can partition A into A_A, A_B, A_∞ whose ancestry originates in A, B, or goes back forever, respectively.

It is easily seen that the restriction of f to A_A gives a bijection $f : A_A \to B_A$, the restriction of g to B_B gives a bijection $g : B_B \to A_B$, and the restrictions of f, g both give bijections $f : A_\infty \to B_\infty, g : B_\infty \to A_\infty$. Using the first two and one of the third, we concoct a bijection $F : A \to B$ by setting

$$F(a) = \begin{cases} f(a) & \text{if } a \in A_A \\ g^{-1}(a) & \text{if } a \in A_B \\ f(a) & \text{if } a \in A_\infty \end{cases}$$

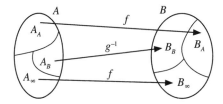

Fig. 14.6 Where does tracing back end?

This completes the proof. $\qquad\qquad\qquad\qquad\qquad\qquad\square$

As an example of this theorem, we give an alternative proof that the rationals are countable. The inclusion $i : \mathbb{N} \to \mathbb{Q}$ shows that $|\mathbb{N}| \leq |\mathbb{Q}|$, and since any rational can be written uniquely in its lowest terms as $(-1)^n p/q$ where $n, p, q \in \mathbb{N}$, by unique factorisation the function $f : \mathbb{Q} \to \mathbb{N}, f\big((-1)^n p/q\big) = 2^n 3^p 5^q$, is an injection, so $|\mathbb{Q}| \leq |\mathbb{N}|$.

A more interesting example shows that $|\mathbb{P}(\mathbb{N})| = \aleph$. An injection $f : \mathbb{P}(\mathbb{N}) \to \mathbb{R}$ can be obtained by

$$f(A) = 0 \cdot a_1 a_2 \ldots a_n \ldots$$

where

$$a_n = \begin{cases} 0 & \text{if } n \notin A \\ 1 & \text{if } n \in A. \end{cases}$$

For each subset $A \subseteq \mathbb{N}$, this gives a unique decimal expansion and f is an injection.

To get an injection $g : \mathbb{R} \to \mathbb{P}(\mathbb{N})$ requires a little more cunning. Instead of writing a real number as a decimal expansion, we express it as a *bicimal*,[1] which means that we write it as the limit of fractions of the form

$$a_0 + a_1/2 + a_2/4 + \cdots + a_n/2^n$$

where a_0 is an integer and a_n is 0 or 1 for $n \geq 1$. If we exclude such expressions concluding with an infinite sequence of 1s (the bicimal equivalent of the decimal problem involving an infinite sequence of 9s), then such a bicimal expansion is unique. Now express the integer a_0 in binary notation as

$$a_0 = (-1)^m \, b_k \ldots b_2 b_1$$

where m and the digits $b_1, b_2, \ldots, b_k$ are all 0 or 1, then we have a unique bicimal expansion for each real number x in the form

$$x = (-1)^m \, b_k \ldots b_2 b_1 \cdot a_1 a_2 \ldots a_n \ldots$$

where m and each digit $b_1, \ldots, b_k, a_1, \ldots, a_n, \ldots$ is 0 or 1. For convenience, in this case write $b_n = 0$ for $n > k$. Now write the terms out as a sequence in the order $m, a_1, b_2, a_2, b_2, \ldots, a_n, b_n, \ldots$. This is a sequence of 0s and 1s and defines a unique subset A of $\mathbb{N}$ according to the rule

$$r \in A \text{ if and only if the } r\text{th term of the sequence is } 1.$$

In this way we obtain a function $g : \mathbb{R} \to \mathbb{P}(\mathbb{N})$ by defining $g(x)$ to be the subset A determined in this manner. This is an injection, and the Schröder–Bernstein theorem shows that $|\mathbb{R}| = |\mathbb{P}(\mathbb{N})|$.

Cardinal Arithmetic

Just as we can add, multiply, and take powers of finite cardinals, we can mimic the set-theoretic procedures involved and define corresponding operations on infinite cardinals. Some, but not all, of the properties of ordinary arithmetic carry over to all cardinals, and it is most instructive to see which ones. First of all the definitions:

Definition 14.15: The operations on cardinal numbers are as follows:
Addition: Given two cardinals α, β (finite or infinite), select disjoint sets A, B such that $|A| = \alpha, |B| = \beta$. (This can always be done. If A and B are not disjoint, replace them by $A' = A \times \{0\}$ and $B' = B \times \{1\}$. Obvious bijections

---

[1] Classical scholars will be horrified, but the word seems unavoidable because of its connotations with 'binary' and 'decimal'.

show that $|A'| = |B'|$, and it is clear that A' and B' are disjoint.) Define $\alpha + \beta$ to be the cardinal of $A \cup B$.

Multiplication: If $\alpha = |A|, \beta = |B|$, then $\alpha\beta = |A \times B|$.

Powers: If $\alpha = |A|, \beta = |B|$, then $\alpha^\beta = |A^B|$ where A^B is the set of all functions from B to A.

You should check that when the sets concerned are finite, these definitions correspond to standard arithmetic. In particular, when $|A| = m$ and $|B| = n$, then on defining a function $f : B \to A$, each element $b \in B$ has m possible choices of image, giving m^n functions in all. Addition and multiplication are quite easy in the finite case.

Notice that the sets in the definition of addition have to be disjoint, but this is not necessary for the other two operations. For addition, the reason is that if $|A| = m, |B| = n$, and $A \cap B \neq \varnothing$, then $|A \cup B| < m + n$. The most important fact to check about these definitions is that they are well defined. Starting with cardinals α, β, we must choose sets A, B with $|A| = \alpha, |B| = \beta$: it is essential to check that if different sets A', B' were used, then the cardinal found in each case would be the same as before. In the case of multiplication, for instance, if $|A| = |A'|, |B| = |B'|$, then there are bijections $f : A \to A', g : B \to B'$, which induce a bijection

$$h : A \times B \to A' \times B'$$

given by

$$h(a, b) = (f(a), g(b)).$$

Thus $|A \times B| = |A' \times B'|$, and the product cardinal is well defined. There are corresponding proofs for addition and powers of cardinals.

If we investigate the properties of these arithmetic operations, we find that many properties of finite numbers continue to hold for cardinals:

Proposition 14.16: If α, β, γ are cardinals (finite or infinite), then

    (i) $\alpha + \beta = \beta + \alpha$,

   (ii) $(\alpha + \beta) + \gamma = \alpha + (\beta + \gamma)$,

  (iii) $\alpha + 0 = \alpha$,

  (iv) $\alpha\beta = \beta\alpha$,

   (v) $(\alpha\beta)\gamma = \alpha(\beta\gamma)$

  (vi) $1\alpha = \alpha$,

 (vii) $\alpha(\beta + \gamma) = \alpha\beta + \alpha\gamma$,

(viii) $\alpha^{\beta+\gamma} = \alpha^\beta \alpha^\gamma$,

  (ix) $\alpha^{\beta\gamma} = (\alpha^\beta)^\gamma$,

  (x) $(\alpha\beta)^\gamma = \alpha^\gamma \beta^\gamma$.

Proof: Let A, B, C be (disjoint) sets with cardinals α, β, γ, respectively. 0 is the cardinal of $\varnothing$ and 1 is the cardinal of any one-element set, say $\{0\}$.

(i)–(iii) follow trivially because $A \cup B = B \cup A$, $(A \cup B) \cup C = A \cup (B \cup C)$, and $A \cup \varnothing = A$.

(iv)–(vi) follow because there are obvious bijections $f : A \times B \to B \times A$ given by $f((a, b)) = (b, a)$, $g : (A \times B) \times C \to A \times (B \times C)$ given by $g(((a, b), c)) = (a, (b, c))$, and $h : \{0\} \times A \to A$ given by $h((0, a)) = a$.

(vii) results from the equality $A \times (B \cup C) = (A \times B) \cup (A \times C)$.

If the last three seem harder, it is because we are less familiar with the set of functions A^B from B to A. It is enough to set up the appropriate bijections.

(viii) Define $f : A^{B \cup C} \to A^B \times A^C$ by starting with a map $\phi : B \cup C \to A$, defining $\phi_1 : B \to A$ to be the restriction of ϕ to B, $\phi_2 : C \to A$ to be the restriction of ϕ to B, then put $f(\phi) = (\phi_1, \phi_2)$. This function f is a bijection.

(ix) Define $g : A^{B \times C} \to (A^B)^C$ by starting with a function $\phi : B \times C \to A$, then defining the function $g(\phi) : C \to A^B$ by $[g(\phi)](c) : B \to A$ as the function that takes $b \in B$ to

$$([g(\phi)](c))(b) = \phi((b, c)).$$

As this is less familiar, it is worth demonstrating that g is a bijection. It is injective, for if $g(\phi) = g(\psi)$ for two maps ϕ, ψ from $B \times C$ to A, then

$$([g(\phi)](c))(b) = ([g(\psi)](c))(b) \quad \text{for all } b \in B, c \in C$$

so, by definition,

$$\phi((b, c)) = \psi((b, c)) \quad \text{for all } b \in B, c \in C,$$

which means that $\phi = \psi$.

To show that g is surjective, start with a function $\theta \in (A^B)^C$. That is, $\theta : C \to A^B$. Then define $\phi : B \times C \to A$ by

$$\phi(b, c) = [\theta(c)](b) \quad \text{for all } b \in B, c \in C.$$

We have $g(\phi) = \theta$, as required.

(x) The final equality between cardinals follows from the bijection $h : (A \times B)^C \to A^C \times B^C$ given by writing any $\phi : C \to A \times B$ in terms of

$$\phi(c) = (\phi_1(c), \phi_2(c)) \quad \text{for } c \in C,$$

and then setting $h(\phi) = (\phi_1, \phi_2)$. Checking the details is left to you. $\qquad\square$

Now we perform some explicit calculations with cardinals. As a corollary of proposition 14.9, we find that

$$n + \aleph_0 = \aleph_0 + n = \aleph_0 \text{ for any finite cardinal } n,$$

$$\aleph_0 + \aleph_0 = \aleph_0.$$

This shows that there is no possibility of defining subtraction of cardinals where infinite cardinals are involved, for what would $\aleph_0 - \aleph_0$ be? According to the above results it could be any finite cardinal or $\aleph_0$ itself, so subtraction cannot be defined to ensure that

$$\aleph_0 - \aleph_0 = \alpha \Leftrightarrow \aleph_0 = \aleph_0 + \alpha.$$

From proposition 14.10 it is easy to deduce that

$$n\aleph_0 = \aleph_0 n = \aleph_0 \text{ for } n \in \mathbb{N},$$

$$\aleph_0 \aleph_0 = \aleph_0.$$

It is interesting to calculate $0\aleph_0$. This turns out to be zero. In fact

$$0\beta = 0 \text{ for each cardinal number } \beta.$$

This is because

$$A = \varnothing \Rightarrow A \times B = \varnothing \text{ for any other set } B,$$

for if A has no elements, then there are no ordered pairs (a, b) for $a \in A, b \in B$. This means that, in terms of cardinal numbers, zero times infinity is zero, no matter how big the infinite cardinal is.

Likewise, it is instructive to calculate α^0 and α^1 for any cardinal α. By definition, if $|A| = \alpha$, then α^0 is the cardinal number of the set of functions from $\varnothing$ to A. You might be forgiven for thinking that there are *no* functions from $\varnothing$ to A, but the set-theoretic definition of a function $f : \varnothing \to A$ as a subset of $\varnothing \times A$ exhibits just one such function, the empty subset of $\varnothing \times A$. So $\alpha^0 = 1$. Since $|\{0\}| = 1, \alpha^1$ is the cardinal number of the set of functions from $\{0\}$ to A. A function $f : \{0\} \to A$ is uniquely determined by the element $f(0) \in A$, so there is a bijection $g : A^{\{0\}} \to A$ given by $g(f) = f(0)$, showing $\left|A^{\{0\}}\right| = |A|$, or $\alpha^1 = \alpha$. By induction using proposition 14.16(viii), we get

$$(\aleph_0)^0 = 1, (\aleph_0)^n = \aleph_0 \text{ for } n \in \mathbb{N}.$$

If we calculate 2^α for any cardinal α, we get an interesting result in terms of the power set. Suppose that $|A| = \alpha$, then, since $|\{0, 1\}| = 2$, we have

$$\left|\{0, 1\}^A\right| = 2^\alpha.$$

But a function $\phi : A \to \{0, 1\}$ corresponds precisely to a subset of A, namely

$$\{a \in A \mid \phi(a) = 1\}.$$

Define $f : \{0, 1\}^A \to \mathbb{P}(A)$ by $f(\phi) = \{a \in A \mid \phi(a) = 1\}$, then f is a bijection, so $|\mathbb{P}(A)| = 2^{\alpha}$. From proposition 14.13 we see that

$$2^{\alpha} > \alpha \text{ for all cardinal numbers } \alpha.$$

Order Relations on Cardinals

We have already proved a number of results concerning the order of cardinals at various points in this chapter. It is now an opportune moment to collect these together and make the list more comprehensive by filling in the gaps:

Proposition 14.17: If $\alpha, \beta, \gamma, \delta$ are cardinals (finite or infinite) then

(i) $\alpha \le \beta, \beta \le \gamma \Rightarrow \alpha \le \gamma$,
(ii) $\alpha \le \beta, \beta \le \alpha \Rightarrow \alpha = \beta$,
(iii) $\alpha \le \beta, \gamma \le \delta \Rightarrow \alpha + \gamma \le \beta + \delta$,
(iv) $\alpha \le \beta, \gamma \le \delta \Rightarrow \alpha\gamma \le \beta\delta$,
(v) $\alpha \le \beta, \gamma \le \delta \Rightarrow \alpha^{\gamma} \le \beta^{\delta}$.

Proof: Select sets A, B, C, D with cardinals $\alpha, \beta, \gamma, \delta$.

(i) If $f : A \to B, g : B \to C$ are injections, then $gf : A \to C$ is an injection.
(ii) This is the Schröder–Bernstein theorem.
(iii) Given injections $f : A \to B, g : C \to D$ where $A \cap C = \varnothing, B \cap D = \varnothing$, define $h : A \cup C \to B \cup D$ by

$$h(x) = \begin{cases} f(x) & \text{for } x \in A \\ g(x) & \text{for } x \in C. \end{cases}$$

Since $A \cap B = \varnothing$, this is well defined, and since $B \cap D = \varnothing$, the fact that f, g are injections implies h is an injection.

(iv) Given injections $f : A \to B, g : C \to D$, define $p : A \times C \to B \times D$ by

$$p((a, c)) = (f(a), g(c)) \text{ for all } a \in A, c \in C.$$

Clearly p is an injection (for if $p((a_1, c_1)) = p((a_2, c_2))$, then $\big(f(a_1), g(c_1)\big) = \big(f(a_2), g(c_2)\big)$, so $f(a_1) = f(a_2), g(c_1) = f(c_2)$, and the injectivity of f, g implies $a_1 = a_2, c_1 = c_2$).

(v) This is best visualised by considering $A \subseteq B, C \subseteq D$. (If we are given injections $f : A \to B, g : C \to D$, replace A by $f(A) \subseteq B$, and C by $g(C) \subseteq D$ in the argument that follows.)

For $A \subseteq B$, $C \subseteq D$, to define a map $\mu : A^C \to B^D$, all we need to do is to show how to extend a function $\varnothing : C \to A$ to a function $\mu(\phi) : D \to B$. (The function $\mu(\phi) : D \to B$ isn't usually an injection; don't confuse this with the function $\mu : A^C \to B^D$.) The easiest way to do this is to select an element $b \in B$, (any one will do, the exceptional case $B = \varnothing$ easily implies (v) by a separate argument); then define $\mu(\phi) \in B^D$ by

$$[\mu(\phi)](d) = \begin{cases} \phi(d) & \text{for } d \in C, \\ b & \text{for } d \in D \backslash C. \end{cases}$$

Then $\mu : A^C \to B^D$ is an injection because $\mu(\phi_1) = \mu(\phi_2)$ implies

$$[\mu(\phi_1)](d) = [\mu(\phi_2)][(d)] \text{ for all } d \in D;$$

in particular, this means that

$$\phi_1(d) = \phi_2(d) \text{ for all } d \in C,$$

so $\phi_1 = \phi_2$. $\qquad\qquad\qquad\qquad\qquad\qquad\qquad\qquad\qquad\qquad\quad$ $\square$

Looking at this last proposition, there is a notable omission from the list of properties we might expect of an order relation. We have not asserted that any two cardinal numbers are comparable; that is, given cardinals α, β then either $\alpha \leq \beta$ or $\beta \leq \alpha$. What this would amount to is selecting sets A, B with cardinals α, β respectively and showing that there is either an injection $f : A \to B$, or $g : B \to A$, (or both). To be able to construct such an injection, we would either have to know something about the sets A and B, or we would need some general method of proceeding with the construction of a suitable injection. Given specific sets, we can proceed in an ad hoc fashion and use our ingenuity to try to set up an injection from one to the other. A general method that works for *all* sets requires us to be much more precise about what we mean by a set. It strains the bounds of set theory. Until we put specific restrictions on what we mean by the word 'set' we cannot say how to compare two of them. The theory of sets has grown into a large and living plant; to nourish it we must put down stronger roots into the foundations.

Exercises

1. Let X be the set of points $(x, y, z) \in \mathbb{R}^3$ such that $x, y, z \in \mathbb{Q}$. Is X countable?

2. Let S be the set of spheres in $\mathbb{R}^3$ whose centres have rational coordinates and whose radii are rational. Show that S is countable.

3. Let $[0, 1[$ be the set of real numbers x such that $0 \leq x < 1$. By writing each one as a decimal expansion, prove that $[0, 1[$ is uncountable.

4. Which of the following sets are countable? (Prove or disprove each case.)
   (a) $\{n \in \mathbb{N} \mid n \text{ is prime}\}$
   (b) $\{r \in \mathbb{Q} \mid r > 0\}$
   (c) $\{x \in \mathbb{R} \mid 1 < x < 10^{-1,000,000}\}$
   (d) $\mathbb{C}$
   (e) $\{x \in \mathbb{R} \mid x^2 = 2^a 3^b \text{ for some } a, b \in \mathbb{N}\}$.

5. If $a, b \in \mathbb{R}$ and $a < b$, the *closed interval* $[a, b]$ is

$$[a, b] = \{x \in \mathbb{R} \mid a \leq x \leq b\},$$

   the *open interval* is

$$]a, b[= \{x \in \mathbb{R} \mid a < x < b\},$$

   and the *half-open intervals* are

$$[a, b[= \{x \in \mathbb{R} \mid a \leq x < b\},$$
$$]a, b] = \{x \in \mathbb{R} \mid a < x \leq b\}.$$

   Prove for $a < b, c < d$, that $f : [a, b] \to [c, d]$ given by

$$f(x) = \frac{(b - x) c}{b - a} + \frac{(x - a) d}{b - a}$$

   is a bijection. Deduce that any two closed intervals have the same cardinal number.

   Prove also that $[a, b],]a, b[, [a, b[,]a, b]$ all have the same cardinal number. (*Hint*: Show that $[a, b]$ has the same cardinal number as any one of the other three by choosing c, d such that $a < c < d < b$, and then using the Schröder–Bernstein theorem.)

6. Prove that the cardinal number of a closed interval, an open interval, and a half-open interval is $\aleph$.

7. Prove that between any two distinct real numbers there are a countable number of rationals and an uncountable number of irrationals.

8. Construct an explicit bijection from $[0, 1]$ to $[0, 1[$. (If all else fails, try using the Schröder–Bernstein construction on the injections $f : [0, 1] \to [0, 1[, f(x) = \frac{1}{2}x$, and $g : [0, 1[\to [0, 1], g(x) = x$.)

9. If A_1, A_2 are arbitrary sets, prove

$$|A_1| + |A_2| = |A_1 \cup A_2| + |A_1 \cap A_2|.$$

Generalise to n sets $A_1, \ldots, A_n$.

10. Find counterexamples which demonstrate that the following general statements are *false* for cardinal numbers α, β, γ:
    (a) $\alpha < \beta \Rightarrow \alpha + \gamma < \beta + \gamma$
    (b) $\alpha < \beta \Rightarrow \alpha\gamma < \beta\gamma$
    (c) $\alpha < \beta \Rightarrow \alpha^\gamma < \beta^\gamma$
    (d) $\alpha < \beta \Rightarrow \gamma^\alpha < \gamma^\beta$.

11. (a) Define $f : [0, 1[\times [0, 1[\to [0, 1[$ by

$$f(0 \cdot a_1 a_2 \ldots a_n \ldots, 0 \cdot b_1 b_2 \ldots b_n \ldots) = 0 \cdot a_1 b_1 a_2 b_2 \ldots a_n b_n \ldots$$

Deduce that $\aleph^2 = \aleph$.

(b) Prove the result of (a) more elegantly by using $2^{\aleph_0} = \aleph$, and the properties of cardinal arithmetic.

(c) Using $1\aleph \leq \aleph_0\aleph \leq \aleph\aleph$, or otherwise, find $\aleph_0\aleph$.

(d) What is $n\aleph$ for $n \in \mathbb{N}$?

(e) Prove $\aleph^{\aleph_0} = \aleph$ and $\aleph^\aleph = 2^\aleph$.

(f) Find $\aleph_0^\aleph$.

12. Given an infinite cardinal α, it may be shown that there exists a cardinal number β such that $\alpha = \aleph_0\beta$. Use this to show that $\aleph_0\alpha = \alpha$.

13. (The proof by which Cantor showed that there exist transcendental numbers without actually specifying any!)

    A real number is *algebraic* if it is a solution of a polynomial equation

$$a_n x^n + \cdots + a_1 x + a_0 = 0$$

with integer coefficients. If not, it is *transcendental*.

    (a) Show that the set of polynomials with integer coefficients is countable.
    (b) Show that the set of algebraic numbers is countable.
    (c) Show that some real numbers must be transcendental.
    (d) How many transcendental numbers are there?

CHAPTER 15

Infinitesimals

As an ongoing theme, we have built a formal model of the real numbers $\mathbb{R}$ as a complete ordered field. This construction reveals the real numbers as the unique structure, up to isomorphism, that satisfies the given axioms. Visually, the real numbers fill the geometric real line, so it seems impossible to fit any more points between them. For instance, an element $x \in \mathbb{R}$ cannot be arbitrarily small, in the sense that $0 < x < r$ for all positive $r \in \mathbb{R}$. If we try to find such an $x \in \mathbb{R}$, then $r = \frac{1}{2}x$ would be smaller—a contradiction.

Yet, in the historical development of the calculus, the idea arose of quantities x, y that can vary by 'arbitrarily small' quantities dx and dy, and such ideas remain today in practical applications. Given a relation such as $y = x^2$, when x changes to $x + dx$, then y changes to $y + dy = (x + dx)^2$, and Leibniz calculated

$$\frac{dy}{dx} = \frac{(x + dx)^2 - x^2}{dx} = 2x + dx.$$

He went on to argue that if dx is infinitesimally small, it does not change the value of $2x$ significantly, so the rate of change dy/dx can be taken to be $2x$ *exactly*. Newton used a physical image of 'flowing' quantities to justify a similar calculation in different notation.

This proposal led to centuries of dispute over the legitimacy of such arguments, focusing on the problem that if dx is not zero, then $2x + dx$ is not *precisely* equal to $2x$, but if dx is equal to zero then it cannot be used as the denominator of the quotient dy/dx.

The problem was eventually resolved to the satisfaction of pure mathematicians by introducing the idea of limit and the modern definitions of analysis. Replace 'arbitrarily close' by reformulating the question of *how close* as a finite challenge. The limiting value is specified as a real number L that satisfies:

Tell me how accurate you want the result to be by specifying an error $\varepsilon > 0$, and then I will specify $\delta > 0$ so that when dx is non-zero and smaller in size than δ then the difference between $(f(x + dx) - f(x))/dx$ and L is less than the error ε you require.

This approach led to the modern formulation of analysis, but an accidental by-product was the elimination of infinitesimals. First, when Cantor introduced infinite cardinals he showed that they can be added and multiplied, but not subtracted or divided. As a consequence, an infinitesimal cannot exist as the reciprocal of an infinite cardinal.

Second, he was also the first person to construct the real numbers using Cauchy sequences of rationals and he proved the completeness axiom for the real numbers. Introducing irrational numbers to 'fill in the gaps' between the rationals did not leave any room on the number line for even smaller infinitesimals.

Richard Dedekind formulated an alternative construction of the real numbers using 'Dedekind cuts'. These divide the rational numbers $\mathbb{Q}$ into two disjoint subsets, one subset to the left L and one to the right R, where every element of the left subset is to the left of every element in the right subset. There are two kinds of cut. The first occurs when there is a rational number a such that all rationals less than a are in L and all those larger than a are in R; then a can be placed in either. The other is typified by the case where R consists of all positive rationals r satisfying $r^2 > 2$ and L is everything else. This 'cut' does not occur at a rational number; it corresponds to a new phenomenon on the number line the square root of two.

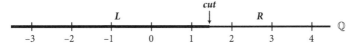

Fig. 15.1 An irrational cut

Dedekind cuts are an alternative method to construct a system of real numbers containing the rational and irrational numbers.

The theories of Cantor and Dedekind supported the idea that the real number line is complete—not only in the axiomatic mathematical sense that all Cauchy sequences of rational numbers converge to a unique real number, but in the intuitive sense that the extra irrational numbers fill up the number line. This was later formulated as:

The Cantor–Dedekind Axiom: The real numbers are order isomorphic to the linear continuum of geometry.

According to this proposed axiom, the geometric real line corresponds precisely with the arithmetic decimal number line up to order isomorphism: the unique complete ordered field. Completeness of the real line was taken to mean that there is no room on it to fit in infinitesimals.

This view was widely accepted in the early twentieth century, and infinitesimals were usually excluded from mathematical analysis. Yet applied mathematicians continued to use 'arbitrarily small' quantities as a meaningful way of thinking about the calculus. Infinitesimals seemed to be useful in practice but problematic in theory.

We have seen that such a phenomenon is common when mathematical systems have been successively generalised over centuries. Intuitive assumptions sometimes acquire iconic status, and can't be questioned even in a new context. To avoid this error we ask: if there are no infinitesimals in the real numbers, could such quantities exist in a *larger* system than the reals? We already know that we can place the real number line in the broader complex plane. Is there any possible way that we could introduce an extension of the real number line that incorporates infinitesimals?

For instance, can we imagine an ordered field K that contains a subfield isomorphic to $\mathbb{R}$, but in which there are elements $x \in K$ such that $0 < x < r$ for all positive $r \in \mathbb{R}$? If we can, the earlier contradiction can no longer be obtained by taking $r = \frac{1}{2}x$, because $\frac{1}{2}x$ is not in $\mathbb{R}$; only in K.

We therefore give a formal definition:

Definition 15.1: If K is a field with $\mathbb{R}$ as an ordered subfield, then $x \in K$ is said to be *infinitesimal* in K if $x \neq 0$ and $-r < x < r$ for all positive $x \in \mathbb{R}$.

Such a possibility does not contradict Cantor's theory of real numbers, nor his theory of infinite cardinals. The infinitesimals in a field are not the reciprocals of infinite cardinal numbers nor are they real numbers. They are elements in the ordered field K.

Ordered Fields Larger than the Real Numbers

Many fields contain the real numbers as a subfield. An easy example is the field $\mathbb{R}(x)$ of rational expressions with elements

$$\frac{a_n x^n + \cdots + a_0}{b_m x^m + \cdots + b_0} \text{ (where } a_r, b_r \in \mathbb{R}, \ b_m \neq 0).$$

This forms a field in which elements of the form $a_0/1$ correspond to the real numbers. In this way we can consider $\mathbb{R}$ as a subfield of $\mathbb{R}(x)$.

The field $\mathbb{R}(x)$ can be given an order in various ways. For example, on page 214, example 10.3, we showed how to order $\mathbb{R}(t)$ by saying that $f(t) > g(t)$ if the graph of f is higher than the graph of g for *large* real values of t. In this sense t is larger than any real number k, so t is infinite in this order and $1/t$ is infinitesimal.

For convenience we now consider the order using $x = 1/t$ and speak of the ordering on $\mathbb{R}(x)$ where x is to be infinitesimal. This involves comparing graphs for *small* positive real values of x saying that $f(x) < g(x)$ if the graph of f is below the graph of f in a sufficiently small interval to the right of the origin.

For instance, the following picture shows three such graphs, $y = x$, $y = x^2$, and $y = 2$.

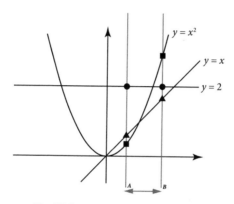

Fig. 15.2 Graphs of rational functions

At different values of x a vertical line meets these graphs at various points and the comparative order may be different. For instance, marking the point where the graph $y = 2$ meets a vertical line with a circle ●, $y = x$ with a triangle ▲ and $y = x^2$ with a square, ■, we can see that in position A we have the order by height as ■ < ▲ < ●, but in position B, we have ▲ < ● < ■. As the vertical line varies, the constant elements in $\mathbb{R}$ (such as $y = 2$) remain in the same place, but the others vary.

However, if we consider what happens as the line A moves to the left, getting closer and closer to the vertical y-axis, the order settles down to ■ < ▲ < ●, which suggests the order $x^2 < x < 2$. This also happens if the constant 2 is replaced by any real number $r > 0$: for all x in $0 < x < r$ we have $0 < x^2 < x < r$.

This suggests a possible way to order the rational functions so that x is positive and satisfies $x < r$ for all positive real numbers r. To give the field $\mathbb{R}(x)$ the structure of an ordered field, we define a subset $\mathbb{R}(x)^+$ that satisfies axioms (O1)–(O3) on page 189.

Definition 15.2: A rational function $f(x)$ is in $\mathbb{R}(x)^+$ if it is either zero or strictly positive on some interval $0 < x < k$.

To prove that this makes $\mathbb{R}(x)$ into an ordered field in which x is infinitesimal, note that a non-zero rational function $a(x) = p(x)/q(x)$ has only a finite number of places where the polynomial $p(x) = 0$, and only a finite number of places where $q(x) = 0$ and the rational function is undefined. Let k be the smallest positive value among these points. Then $a(x)$ is non-zero on the interval $0 < x < k$, and it cannot be positive in one place and negative in another because it would then be zero somewhere in between. (Here we are assuming the intermediate value theorem, proved in any course on analysis.)

Proposition 15.3: The field $\mathbb{R}(x)$ is an ordered field, with $\mathbb{R}(x)^+$ as the set of elements that are zero or positive.

Proof: (O1) If $a(x)$, $b(x) \in \mathbb{R}(x)^+$, then each is a rational function that is either zero or strictly positive in some interval to the right of the origin, so their sum and product is either zero or strictly positive in the smaller of the two intervals concerned.

(O2) If $a(x) \in \mathbb{R}(x)$ then either $a(x) = 0$ or $a(x)$ is strictly positive or strictly negative in an interval to the right of the origin, so either $a(x) \in \mathbb{R}(x)^+$ or $-a(x) \in \mathbb{R}(x)^+$.

(O3) If $a(x) \in \mathbb{R}(x)^+$ and $-a(x) \in \mathbb{R}(x)^+$ then $a(x)$ cannot be both strictly positive and strictly negative, so it must be zero. $\qquad\square$

The order on $\mathbb{R}(x)$ is defined in a technical manner, but it satisfies the required axioms of an ordered field. When we try to imagine the elements of this field, there are several possibilities. The first is to consider the field as a purely symbolic set of quotients of polynomials in a single unknown x with the usual algebraic operations on the elements. Another is to visualise the elements as graphs of rational functions.

A third possibility is to imagine points where the graphs meet a vertical line $y = v$ as v is a variable real number that becomes smaller. This represents the elements of $\mathbb{R}(v)$ as points on the vertical line where x is replaced by v. Now we can think of the terms symbolically as rational functions in v where v is a variable quantity.

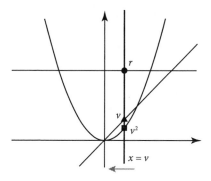

Fig. 15.3 Elements of $\mathbb{R}(v)$ as variable quantities

The figure shows the vertical line with three points on it corresponding to the values v, v^2, and r, where r is a fixed real number. In an interval to the right of the origin where $x < r$, the points are in ascending order v^2, v, and r, representing the order $v < v^2 < r$.

This is an interesting idea as it reveals two kinds of quantity: *constant quantities* which correspond to real numbers in $\mathbb{R}$ that remain in a fixed position, and *variable quantities* corresponding to non-constant rational functions, represented as points that vary as v becomes small.

In particular, a variable quantity like v, that becomes smaller than any fixed real number as the line $x = v$ moves towards the vertical axis $x = 0$, is an *infinitesimal* in this ordered field. The point marked v satisfies $0 < v < r$ as v gets smaller than r. And $0 < v^2 < v$ for $v < 1$, which shows that v^2 is an even smaller infinitesimal than v.

Can we take v to be so small that it is infinitesimal? No. Mathematically a complete ordered field cannot contain an infinitesimal. Furthermore, the way in which Dedekind and Cantor completed the real line by introducing the irrational numbers suggests that there is simply *no room* on the number line to fit in infinitesimals. But is there another way of visualising infinitesimals? The answer is 'Yes!' We can achieve this, but not by restricting ourselves to the real numbers. We simply work in a larger ordered field.

Super Ordered Fields

Formal mathematics lets us define new concepts with useful properties, and then to use these new concepts as a basis for further proofs. In our search for infinitesimals, we define a new concept:

Definition 15.4: A *super ordered field* is an ordered field K that contains the real numbers $\mathbb{R}$ as a proper ordered subfield.

You will not find this definition in any other texts at the moment. We have taken the opportunity to formulate a new definition to show how mathematical theory evolves into the future. This definition proves to be *precisely* what is needed to provide the precise formal structure for an infinitesimal in an ordered field.

The Structure Theorem for Super Ordered Fields

Structure Theorem 15.5: Let K be super ordered field. Then an element $k \in K$ where $k \notin \mathbb{R}$ satisfies precisely one of the following:

- (a) $k > r$ for all real numbers r,
- (b) $k < r$ for all real numbers r,
- (c) there is a unique real number c so that $k = c+e$ where e is infinitesimal.

Proof: Either k satisfies (a) or (b), or there exist $a, b \in \mathbb{R}$ with $a < c < b$. Consider the set $S = \{x \in \mathbb{R} \mid x < k\}$. This is non-empty (because $a \in S$) and bounded above by b, so it has a least upper bound c where $a \leq c \leq b$. Let $e = k - c$, then $k = c + e$. The element e cannot be zero because $k \notin \mathbb{R}$. If e is positive, either e is infinitesimal or there is $r \in \mathbb{R}$ such that $0 < r < e$. Adding c leads to $c < c + r < c + e = k$. This gives a real number $c + r$ less than k and therefore in S, contradicting c being an upper bound of S. Hence e is infinitesimal.

On the other hand, if e is negative and not infinitesimal, then $c < -r < 0$ for some positive $r \in \mathbb{R}$ and $k = c + e < c - r < c$, giving a real number $c - r$ exceeding k. This is an upper bound, but less than the purported *least* upper bound. So again, e is infinitesimal. □

This theorem gives information about the structure of *any* super ordered field K. Such a field has properties that resonate with historical ideas of finite and infinitesimal quantities. We choose to name the elements of K *quantities*. They are either

> *constant quantities*: elements in $\mathbb{R}$,
> *positive infinite quantities*: elements $k > r$ for all $r \in \mathbb{R}$,
> *negative infinite quantities*: elements $k < r$ for all $r \in \mathbb{R}$,

or

> *finite quantities* of the form $k = c + e$ where $c \in \mathbb{R}$ and e is infinitesimal.

This resonates strongly with the historical view of constant and variable quantities. A super ordered field consists precisely of *quantities* that are either *constant*, *infinite* (positive or negative), or a *constant plus an infinitesimal*.

In particular, a finite quantity k is either a constant real number or $k = c + e$ where c is a unique real number and e is an infinitesimal.

Definition 15.6: For any finite number x in a super ordered field, the unique real number c such that $x = c + e$ where e is zero or infinitesimal is called the *standard part* of x and is denoted by

$$c = \text{st}(x).$$

This allows us to specify the unique real number that differs from a finite quantity by an infinitesimal. There are no infinitesimals in $\mathbb{R}$, just as Cantor asserted. However, they do occur in *every* super ordered field that extends the real numbers. So formal mathematics *guarantees* the existence of infinitesimals. We now have a choice: to restrict the study of calculus only to real numbers, which leads to the standard and perfectly viable formulation of analysis using epsilon–delta definitions, or to use infinitesimals in an extended system.

In the applications of mathematics, infinitesimal quantities are often considered as variable points on the real number line. Cauchy took this viewpoint by defining an infinitesimal to be a *variable* quantity that becomes arbitrarily small. In modern notation, this idea can be represented as a *null sequence*, which is simply a sequence that tends to zero.

Cauchy considered such a sequence to be a *variable quantity*—an infinitesimal. From this he developed continuous functions and calculus. For instance, he operated symbolically with a quantity $\alpha = (a_n)$ by defining $f(x + \alpha)$ to be the sequence of values $f(x + a_n)$. He defined a function f to be continuous at x if $f(x + \alpha) - f(x)$ is infinitesimal whenever α is infinitesimal. He then developed a theory of calculus using infinitesimals, even imagining a number line with infinitesimal quantities upon it.

However, at his time in history, the notion of completeness of the real numbers had yet to be formalised and there was no obvious way to represent infinitesimals on a number line. The structure theorem for super ordered fields offers a solution.

Visualising Infinitesimals on a Geometric Number Line

To visualise an infinitesimal in a super ordered field, we use the structure theorem to *see* infinitesimal quantities. In chapter 1, when we attempted to draw a physical picture of the real number line, we realised that, on a given scale, two distinct points can be so close together that to the human eye they are indistinguishable. A ruler marked with centimetres and millimetres lets us distinguish between the mark at 1·4 cm and the next mark at 1·5 cm. If we tried to mark $\sqrt{2}$ as accurately as possible, we could mark it at approximately 1·41 cm, between 1·4 and 1·5. But the difference between 1·414 cm and 1·4142 cm with ordinary implements would be impossible to see. Our response was to *magnify* the line, to distinguish between 1·414 and 1·415, with 1·4142 nestling between them. When performing the magnification, we redrew the lines without making them thicker.

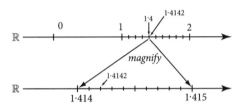

Fig. 15.4 Magnifying $\mathbb{R}$

If we wish to see the difference between two extremely close numbers, say 1 and $1 + 1/10^{100}$, we magnify the difference by a factor of 10^{100}. The map $m : \mathbb{R} \to \mathbb{R}$ with $m(x) = 10^{100}(x - 1)$ gives $m(1) = 0$, $m(1 + 1/10^{100}) = 1$. Under this map the very close numbers 1 and $1 + 1/10^{100}$ are mapped to 0 and 1.

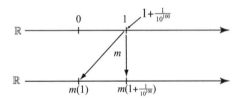

Fig. 15.5 Seeing two extremely close points on $\mathbb{R}$

More generally, we can magnify part of the real line by a huge scale factor so that two very close real numbers can be seen as two separate points. The same technique can be used in a super ordered field K by introducing the map

$$m : K \to K \text{ where } m(x) = \frac{x - a}{e} \text{ for any } a, e \in K, e \neq 0.$$

Now $m(a) = 0$, $m(a + e) = 1$. Thus whatever the non-zero value of e, be it finite, infinite, or infinitesimal, we can define the map m which maps a and $a + e$ onto the distinct points 0, 1.

Usually we take $e > 0$, so that $a + e > a$, because this maintains the direction on the line so that for $a < b$ we have $m(a) < m(b)$.

Definition 15.7: The *e-lens pointed at a* is the map $m : K \to K$ where

$$m(x) = \frac{x - a}{e}.$$

This map makes sense for *any* non-zero e; in particular, for infinitesimals. If we take e to be a specific infinitesimal $\varepsilon > 0$, then an ε-lens can be used to *see* infinitesimal detail on an extended number line. For instance, we may imagine an extended number line K as a geometric number line, with the origin, the natural numbers, the rationals, and reals all in their usual places. Infinite quantities $\alpha < 0$ and $\beta > 0$ are too far off to the left and right to see on a normal scale, while the two points a, $a + \varepsilon$ for $a \in \mathbb{R}$ and ε infinitesimal are too close together to be marked separately. In figure 15.6 we have drawn a to the right of 1, but it could be anywhere else on the extended number line K.

Fig. 15.6 The line to a normal scale

Now use $m(x) = (x - a)/\varepsilon$ to map the whole extended number line K onto a second number line K.

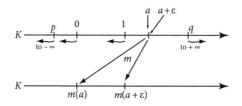

Fig. 15.7 Magnifying the whole extended line

This map sends a to $m(a) = 0$ and $a + \varepsilon$ to the distinct point $m(a + \varepsilon) = 1$. Meanwhile, the image of a general point x is $(x - a)/\varepsilon$, which may be finite or infinite.

Definition 15.8: The *field of view* of the map $m(x) = (x - a)/e$ where a, $a + e \in K$ and $e \neq 0$ is the set $\{x \in K \mid (x - a)/e \text{ is finite}\}$.

The field of view of m is precisely the set of elements that map onto the finite part of K. Points outside the field of view map to infinite elements of K, which are too far to the left or right to be seen in a finite picture.

Definition 15.9: If $u, v \in K$ are both non-zero, then u is said to be of *higher order than* v if u/v is infinitesimal. It is of *lower order than* v if u/v is infinite. The two are *the same order* if u/v is finite but not infinitesimal.

Example 15.10: If ε is infinitesimal, then ε^2 is higher order than ε, and $1/\varepsilon$ is lower order than any finite element. The element $17\varepsilon + 1066\varepsilon^2$ is the same order as $5\varepsilon + \pi \varepsilon^2 + 10^{100}\varepsilon^5$.

In general, when using the map $m(x) = (x - a)/e$, points that differ from a by a quantity of order greater than e are mapped to infinite quantities, points differing by a quantity of the same order as e are mapped onto finite points and points differing by a quantity of lower order are mapped onto points that differ by an infinitesimal.

Because the human eye cannot see infinitesimal quantities, we can strip away infinitesimal differences by taking the standard part of the image of the map m.

Definition 15.11: The *optical lens* $o : K \to \mathbb{R}$ based on the e-lens m is given by

$$o(x) = \text{st}(m(x)).$$

This maps the field of view onto the real line $\mathbb{R}$.

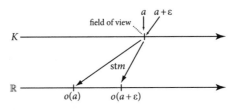

Fig. 15.8 An optical lens

Here we are interested in the case of an optical lens pointing at $a \in \mathbb{R}$ where e is an infinitesimal ε. The field of view consists of points that differ from a by infinitesimals of the same or higher order than ε. For any $r \in \mathbb{R}$, $o(x + r\varepsilon) = r$, so the optical lens maps onto the whole of $\mathbb{R}$. But lower-order infinitesimal detail is lost because if δ is of lower order than ε, then the two elements $x, x + \delta$ map to the same element of $\mathbb{R}$.

One further technical convention makes the visual picture even simpler to grasp. When we make geographical maps, we draw a representation of a particular geographical region R on a map M. We can think of this as a function $s : R \rightarrow M$ from the original region R to the physical map M. But when we mark the position of a specific place, such as the position of London on a map of the United Kingdom, we do not write s(London) on the map, we write the original name 'London'.

Using this convention, we modify the picture by naming the image points in $\mathbb{R}$ with the same original names in K, on the understanding that what we see in $\mathbb{R}$ is simply the standard part of the image of original. Now we are able to 'see' points that are infinitely close in K by using an optical lens to move them apart.

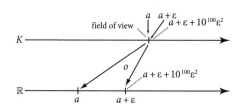

Fig. 15.9 Seeing infinitesimal detail

The field of view is magnified to fill the whole real line. The image of a is distinct from the image of $a + \varepsilon$, yet the latter has the same image as $a + \varepsilon + 10^{100}\varepsilon^2$ even though the number 10^{100} is immense in human terms. Despite its vast size, it is still *finite* and the quantity $10^{100}\varepsilon^2$ is of smaller order than ε.

In this representation we are again 'abusing notation' by denoting the image of an element x in the field of view by the same name x. However, by using this notation while being fully aware that the picture represents not only the physical image drawn on paper but the full meaning of the formal theory, we are offering a natural view of a formal concept.

We can go even further. When we to attempt to draw a super ordered field K as a line to an appropriate scale that allows us to distinguish between real numbers, then all we can draw is (part of) the line L consisting of the finite elements. Let L be the subset of finite elements of K and I the subset of infinitesimals.

Theorem 15.12: The standard part map st : $L \to \mathbb{R}$ maps L to the whole of $\mathbb{R}$ and is a ring homomorphism satisfying

$$\mathrm{st}(x \pm y) = \mathrm{st}(x) \pm \mathrm{st}(y), \ \mathrm{st}(xy) = \mathrm{st}(x)\mathrm{st}(y)$$
$$\mathrm{st}(x/y) = \mathrm{st}(x)/\mathrm{st}(y) \text{(for } \mathrm{st}(y) \neq 0).$$

Proof: This is left as an exercise. □

As a homomorphism of the additive group, st : $L \to \mathbb{R}$ is a group homomorphism with kernel I, and maps onto the whole of $\mathbb{R}$. By the structure theorem for group homomorphisms (chapter 13, theorem 13.36) L/I is isomorphic to $\mathbb{R}$ as an additive group whose elements are cosets of the form $x + I$ for $x \in L$.

By defining the relation $x \sim y$ if and only if $x - y \in I$ the equivalence class $x + I$ contains precisely one real number $\mathrm{st}(x)$. The equivalence class containing a real number a is called the *monad*[1] or 'halo' around a and will be denoted by M_a. This is the cluster of points around a including a and any other element that differs from it by an infinitesimal.

The order on L satisfies $x < y$ where $x = a + \varepsilon$ and $y = b + \delta$ if and only if either $a < b$ or $a = b$ and $\varepsilon < \delta$. This allows us to see why a super ordered field cannot be complete. In a negative sense we already know that a complete ordered field cannot contain an infinitesimal. However, the notion of a monad offers positive proof that completeness fails.

Theorem 15.13: A super ordered field K is not complete.

Proof: Every monad M_a for $a \in \mathbb{R}$ is non-empty (because $a \in M_a$) and bounded above by any $b \in \mathbb{R}$ where $b > a$. However it cannot have a least upper bound $c \in \mathbb{R}$. For if c is a least upper bound of M_a, then either $c \in M_a$ or $c \in M_b$ where $b > a$. It cannot lie in M_a because there will be elements in M_a that are bigger than c so it is not an upper bound. It cannot lie in M_b for $b > a$, for then there would be elements in M_b that are upper bounds for M_a that are smaller than c. Hence a super ordered field K contains subsets that are bounded above but have no upper bound in K. □

In our mind's eye, we can imagine a super ordered field as a number line in which the finite part is the field of real numbers with a halo around each

---

[1] Leibniz used the term 'monad' in his philosophical theory to specify indivisible entities that make up the entire universe of thought. While these equivalence classes consist of tiny elements too small for the human eye to perceive, they are different from the notion of Leibniz as each one consists of an infinite set of elements.

number consisting of elements differing from it by an infinitesimal quantity. The standard part map allows us to collapse the monads into single real numbers to visualise the real number line in the usual mathematical representation. Optical microscopes allow us to see infinitesimal detail magnified to an appropriate visible level.

We can now see how the notion of infinitesimal quantities that evolved over many centuries can be re-evaluated using the notion of super ordered field. In developing the calculus, Leibniz conceived of the idea of an infinitesimal quantity that is arbitrarily small in size. Then Euler produced remarkable results by thinking of an infinitesimal as a symbol that he could manipulate using algebraic rules. The first example given in this chapter begins with the field $\mathbb{R}(x)$ where elements are manipulated purely symbolically and x is an infinitesimal, as in figure 15.10(1).

expressions including $k \in \mathbb{R}$, x, x^2,
with general term:
$$\frac{a_n x^n + \ldots + a_0}{b_m x^m + \ldots + b_0} \quad (b_0 \neq 0)$$

1. elements of $\mathbb{R}(x)$ as algebraic expressions

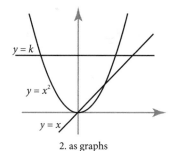

2. as graphs

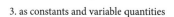

3. as constants and variable quantities   4. as points on an extended number line

Fig. 15.10 Four isomorphic representations

We moved on from the algebraic manipulation of symbols in $\mathbb{R}(x)$ to visualise the corresponding rational functions as graphs as in figure 15.10(2). Now an infinitesimal is a whole graph and we compare the order of items by how the graphs are ordered a little to the right of the origin.

Next we considered where the graphs of rational functions meet the vertical line $x = v$. As v gets small, constant functions meet the line in a fixed point but variable functions meet the vertical line in a variable point where the order is determined by what happens as v gets small as in figure 15.10(3). This example is consistent with the idea of a number line including constant quantities that are real numbers and variable quantities that can include infinitesimals.

More generally, the structure theorem for any super ordered field reveals how we can imagine an infinitesimal as a point on a number line that can be visualised horizontally or vertically as a number line. Figure 15.10(4) shows a vertical presentation as the ultimate form of the super ordered field $\mathbb{R}(\varepsilon)$ where ε is an infinitesimal.

However, this visualisation of an infinitesimal now works not just in the example of $\mathbb{R}(\varepsilon)$, but in *any* super ordered field K.

Magnification in Higher Dimensions

The idea of infinite magnification on the extended number line K can be easily used in two or more dimensions by using e-lenses on each axis separately.

Definition 15.14: The *ε-δ-lens pointed at* $(a, b) \in K^2$ is the map $m : K^2 \to K^2$ given by

$$m(x, y) = \left(\frac{x - a}{\varepsilon}, \frac{y - b}{\delta} \right).$$

The *optical ε-δ-lens pointed at* $(a, b) \in K^2$ is the map $o : K^2 \to \mathbb{R}^2$ given by

$$o(x, y) = (\text{st}((x - a)/\varepsilon, \text{st}((y - b)/\delta).$$

The *field of view* of an optical ε-δ-lens pointed at $(a, b) \in K^2$ is the set

$$\{(x, y) \in K^2 \mid (x - a)/\varepsilon, (y - b)/\delta \text{ are both finite}.$$

The elements a, b, ε, δ may be any elements in K provided that ε and δ are non-zero. For example, we can choose a or b to be infinite to view the situation 'at infinity', or we can choose ε and δ to be infinitesimal to look at 'infinitesimal detail'.

Example 15.15: Let $f(x) = x^2$ and suppose that $x \in \mathbb{R}$ and ε is infinitesimal. Then an optical ε-δ-lens with $\varepsilon = \delta$, pointed at (x, x^2), sees a nearby point $(x + h, (x + h)^2)$.

$$o(x + h, (x + h)^2) = \left(\text{st}\left(\frac{x + h - x}{\varepsilon} \right), \text{st}\left(\frac{(x + h)^2 - x^2}{\varepsilon} \right) \right)$$
$$= \left(\text{st}\left(\frac{h}{\varepsilon} \right), \text{st}\left(\frac{2xh + h^2}{\varepsilon} \right) \right)$$
$$= \left(\text{st}\left(\frac{h}{\varepsilon} \right), \text{st}(2x + h)\text{st}\left(\frac{h}{\varepsilon} \right) \right).$$

If this is in the field of view, then $\text{st}(x/\varepsilon)$ must be finite, so h is of the same order as ε or less, so h is also infinitesimal and $\text{st}(2x + h) = 2x$. Writing $\lambda = \text{st}(x/\varepsilon)$, this gives

$$o(x + h, (x + h)^2) = (\lambda, 2x\lambda).$$

So under the optical lens, the field of view is mapped precisely to the whole real line, represented parametrically by $\lambda(1, 2x)$ for any real number λ. In representing the picture of the map o from the field of view as a subset of K^2 to the real plane $\mathbb{R}^2$, we again use the convention that the image $o(x + h, f(x + h))$ is also denoted by $(x + h, f(x + h))$ and the image $o(x + h, f(x + h))$ is denoted by $(x + h, f(x + h))$. The optical lens magnifies an infinitesimal part of the graph, centred on $(x, f(x))$, to an infinite straight line in $\mathbb{R}^2$ passing through $(x, f(x))$.

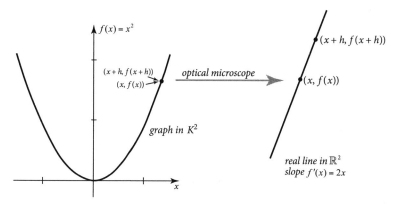

Fig. 15.11 Magnifying a locally straight graph to see a full straight line

Calculus with Infinitesimals

This experience suggests that we may be able to do calculus logically with infinitesimals. However, one more step is needed to make this fully operational. When calculating the derivative of a function $f(x)$, we form the ratio

$$\frac{f(x + h) - f(x)}{h}$$

for infinitesimal h and take the standard part. To do this, we must be able to calculate f not only for elements in $\mathbb{R}$, but also for elements in the extension field K.

For instance, if $f(x) = x^2$ then, for infinitesimal h,

$$\frac{f(x+h) - f(x)}{h} = \frac{(x+h)^2 - x^2}{h} = 3x^2 + 3xh + h^2$$

with standard part $3x^2$.

This calculation can be performed in the extension field $\mathbb{R}(\varepsilon)$ with the infinitesimal $h = \varepsilon$ because $((f(x+\varepsilon) - f(x))/\varepsilon$ is a rational function in ε. However, if we are to consider functions other than rational functions, then we need more powerful theory.

The standard functions in calculus such as $\sin x$ and $\cos x$ can be represented as power series:

$$\sin x = x - \frac{x^3}{3!} + \frac{x^5}{5!} - \cdots$$
$$\cos x = 1 - \frac{x^2}{2!} + \frac{x^4}{4!} - \cdots$$

These can be handled in the field $\mathbb{R}((x))$ consisting of power series in ε with a finite number of negative powers:

$$a_k \varepsilon^{-k} + \cdots + a_1 \varepsilon^{-1} + b_0 + b_1 \varepsilon + \cdots + b_n \varepsilon^n + \cdots$$

for an integer $k \geq 0$. This extension field serves for functions given by rational functions or power series given as combinations of polynomials, trigonometric functions, exponentials, logarithms, and so on, as encountered in school calculus.

However, this still does not cope with all possible functions. For instance, a sequence $a_1, a_2, a_n, \ldots$ is a function $a : \mathbb{N} \to \mathbb{R}$ where $a(n) = a_n$. How do we extend this sequence to work in an appropriate extension field?

In calculus, if we wish to calculate the derivative of a general function $f : D \to \mathbb{R}$, we form the quotient

$$\frac{f(x+h) - f(x)}{h}$$

where $x \in D$ and h is infinitesimal.

This was no problem for Leibniz as his functions were given by a formula and he assumed that the same formula would work for infinitesimals. But modern mathematical analysis works with general functions defined set-theoretically that may not have a simple formula.

Now we need to extend a set-theoretic function $f : D \to \mathbb{R}$ to a larger domain $f : {}^*D \to K$ where the extended domain ${}^*D$ contains not only real numbers but also elements $x + h$ where h is infinitesimal. This is not all that is required. For example, a sequence (s_n) is a function $s : \mathbb{N} \to \mathbb{R}$ where $s_n = s(n)$ and we need to consider how to extend such functions in an appropriate way.

Non-standard Analysis

Abraham Robinson introduced a theory in 1966 [29], called *non-standard analysis*. Whereas standard mathematical analysis only uses the real numbers, non-standard analysis works in a super ordered field called the *hyperreals*, denoted by the symbol ${}^*\mathbb{R}$.

The techniques for constructing the extension from $\mathbb{R}$ to ${}^*\mathbb{R}$ are essentially the same as those used in chapter 9 to construct the extension from $\mathbb{Q}$ to $\mathbb{R}$. This began with Cauchy sequences in $\mathbb{Q}$ and putting an equivalence relation on them so that the equivalence classes became elements of $\mathbb{R}$. To construct ${}^*\mathbb{R}$ from $\mathbb{R}$, we begin with the set S of all sequences (a_n) for $a_n \in \mathbb{R}$. Such a sequence is a function $s : \mathbb{N} \to \mathbb{R}$ where $a_n = s(n)$, so the full set of such sequences is $S = \mathbb{R}^{\mathbb{N}}$.

We introduce an equivalence relation on S so that the equivalence classes become the elements of ${}^*\mathbb{R}$. The equivalence class containing (a_n) is written as $[a_n]$ or as $[a_1, a_2, \ldots, a_n, \ldots]$ and we embed $\mathbb{R}$ in ${}^*\mathbb{R}$ by identifying $a \in \mathbb{R}$ with the element $[a, a, \ldots, a, \ldots]$.

The construction requires us to define a relation $(a_n) \sim (b_n)$ on S satisfying the usual properties of an equivalence relation:

(E1) $(a_n) \sim (a_n)$ for all $(a_n) \in S$
(E2) If $(a_n) \sim (b_n)$ then $(b_n) \sim (a_n)$
(E3) If $(a_n) \sim (b_n), (b_n) \sim (c_n)$ then $(a_n) \sim (c_n)$.

Then we need to define the usual operations of addition, multiplication, and order on the equivalence classes as elements of ${}^*\mathbb{R}$ to make it into an ordered field extension of $\mathbb{R}$.

We could begin by suggesting that $(a_n) \sim (b_n)$ if $(a_n), (b_n)$ agree at all but a finite number of places.

Definition 15.16: A subset $T \subseteq \mathbb{N}$ is said to be *cofinite* if its complement $T^c = \mathbb{N} \backslash T$ is finite.

As a first step, we require that $(a_n) \sim (b_n)$ if $a_n = b_n$ for n in a cofinite set. For example, if we took a sequence (a_n) and changed a finite number of terms to get a sequence (b_n), then $[a_n] = [b_n]$. In particular, if N is the largest element in $\mathbb{N}$ such that $(a_n) \neq (b_n)$, then $(a_n) = (b_n)$ for all $n > N$, meaning that the terms of the two sequences are identical from some point on.

However, we would need to decide what to do with sequences such as

$$(a_n) = (1, 0, 1, 0, \ldots) \text{ and } (b_n) = (1, 1, 1, 1, \ldots).$$

Do we claim that $(a_n) \sim (b_n)$ or not? In this case they are equal for $n \in O$ (the odd numbers) and they are different for $n \in E$ (the even numbers). To make a decision requires us to make a *choice*. If we choose to focus only on O, they are equal, but if we focus only on E, they are different.

The clever idea that Robinson conceived can be expressed in a simple way. For every subset $T \subseteq \mathbb{N}$, he decided that he must make the choice between what happens on T and what happens on its complement $T^c = \mathbb{N} \backslash T$. His approach amounts to assuming that it is possible to select a subset U of subsets of $\mathbb{N}$ so that precisely one of T and its complement T^c is in U. Then a statement such as $[a_n] = [b_n]$ would be declared to be true if $T \in U$, and false if $T^c \in U$. This leads to the following definition:

Definition 15.17: $(a_n) \sim (b_n)$ if and only if $\{n \in \mathbb{N} \mid a_n = b_n\} \in U$.

The choice of U may not be unique, for we may have one choice of U in which the odd numbers $O \in U$, in which case

$$[1, 0, 1, 0, \ldots] = [1, 1, 1, 1, \ldots],$$

and another in which the even numbers $E \in U$, in which case

$$[1, 0, 1, 0, \ldots] \neq [1, 1, 1, 1, \ldots].$$

This means that we may have different ways of constructing an appropriate extension field and the choice may not be unique. However, what matters is that the choice is fit for purpose. So we continue by asking what kind of properties are required.

First, U is a set of subsets of $\mathbb{N}$ so $U \subseteq \mathbb{P}(\mathbb{N})$ and for every $T \subseteq \mathbb{N}$ we require:

(U₁) If $T \subseteq \mathbb{N}$ then either $T \in U$ or $\mathbb{N} \backslash T \in U$, but not both.

We also require:

(U₂) If T is cofinite, then $T \in U$.

It is clear that if a statement is true on a set $T \subseteq \mathbb{N}$, then it is also true on any subset of T, which requires:

(U3) If $T \in U$ and $S \subseteq T$ then $S \in U$.

We must then check to confirm that we have an equivalence relation satisfying (E1)–(E3). In the proof that follows later, we will find that we require a further condition:

(U4) $T_1, T_2 \in U \Rightarrow T_1 \cap T_2 \in U$.

We will shortly see that these conditions are all that are required, so we make the definition:

Definition 15.18: An *ultrafilter* on $\mathbb{N}$ is a collection U of subsets of $\mathbb{N}$ satisfying:

(U1) if $T \subseteq \mathbb{N}$ then either $T \in U$ or $T^c \in U$, but not both
(U2) if T is cofinite, then $T \in U$
(U3) If $T \in U$ and $S \subseteq T$ then $S \in U$
(U4) $T_1, T_2 \in U \Rightarrow T_1 \cap T_2 \in U$.

We postpone the discussion of how to construct such an ultrafilter until the next chapter where we discuss more sophisticated methods appropriate for the task. For the rest of this chapter, we assume that we have an ultrafilter satisfying (U1)–(U4) to consider how the theory works. We begin with a lemma:

Lemma 15.19: If U is an ultrafilter on $\mathbb{N}$, then the equivalence relation

$$(a_n) \sim (b_n) \text{ if and only if } \{n \in \mathbb{N} \mid a_n = b_n\} \in U$$

given in definition 15.17 is an equivalence relation on the set S of all real sequences.

Proof: To prove (E1), let $(a_n) \in S$, then

$$T = \{n \in \mathbb{N} \mid a_n = a_n\} = \mathbb{N}$$

so T is cofinite and, by (U3), $T \in U$ and $(a_n) \sim (a_n)$ for all $(a_n) \in S$.
(E2) If $(a_n) \sim (b_n)$, then $a_n = b_n$ for all n in some set $T \in U$, so $b_n = a_n$ for all $n \in T$ and $(b_n) \sim (a_n)$.
(E3) If $(a_n) \sim (b_n)$ and $(b_n) \sim (c_n)$ then

$$a_n = b_n \text{ for all } n \text{ in some set } T_1 \in U$$

and

$$b_n = c_n \text{ for all } n \text{ in some set } T_2 \in U.$$

So $a_n = c_n$ for all n in $T_1 \cap T_2$ and, by (U4), $T_1 \cap T_2 \in U$. $\qquad\square$

Having proved that definition 15.17 gives an equivalence relation on S, we define $^*\mathbb{R}$ to be the set of equivalence classes. In particular, if the equivalence class containing (a_n) is denoted by $[a_n]$, then we have

$$[a_n] = [b_n] \text{ if and only if } \{n \in \mathbb{N} \mid a_n = b_n\} \in U.$$

Now we need to define the field operations on $^*\mathbb{R}$, check that they are well defined, and prove that they satisfy the axioms for a field.

Proposition 15.20: The set $^*\mathbb{R}$ with operations on equivalence classes given by

$$[a_n] + [b_n] = [a_n + b_n], \ [a_n][b_n] = [a_n b_n]$$

is a field containing $\mathbb{R}$ as a subfield.

Proof: First the operations are well defined, because if $[a_n] = [a'_n]$ and $[b_n] = [b'_n]$ then the sets $T_1 = \{n \in \mathbb{N} \mid a_n = a'_n\}$ and $T_2 = \{n \in \mathbb{N} \mid b_n = b'_n\}$ satisfy $T_1, T_2 \in U$, so that $a_n + b_n = a'_n + b'_n$ for $n \in T_1 \cap T_2$. By (U4), $T_1 \cap T_2 \in U$, so $[a_n + b_n] = [a'_n + b'_n]$.

The proof for the product is similar.

The proofs of commutativity, associativity, and distributivity of addition and multiplication are straightforward. (You should explain them to yourself.) The zero of $^*\mathbb{R}$ is $[0, 0, \ldots, 0, \ldots]$, the unit is $[1, 1, \ldots, 1, \ldots]$, and $\mathbb{R}$ can be embedded in $^*\mathbb{R}$ by identifying $a \in \mathbb{R}$ with $[a, a, \ldots, a, \ldots]$. The additive inverse of $[a_n]$ is $[-a_n]$.

The only difficult part is to define the multiplicative inverse $1/[a_n]$ of $[a_n]$ because the simple solution defining $1/[a_n]$ to be $[1/a_n]$ will not work if any of the a_n are zero. To cope with this, we note that $[a_n] = [0]$ if and only if the set

$$T = \{n \in \mathbb{N} \mid a_n = 0\} \in U.$$

If $[a_n] \neq [0]$, then $T \notin U$ and, by (U1), the set $T^c = \{n \in \mathbb{N} \mid a_n \neq 0\} \in U$. Let

$$b_n = \begin{cases} a_n & \text{if } a_n \neq 0 \\ 1 & \text{if } a_n = 0. \end{cases}$$

Then $b_n \neq 0$ for all n. Because $\{n \in \mathbb{N} \mid a_n \neq 0\} \in U$, by definition $[a_n] = [b_n]$.

Now define

$$1/[a_n] = [1/b_n].$$

This completes the proof that $^*\mathbb{R}$ is a field and $\mathbb{R}$ may be embedded as a subfield by identifying $a \in \mathbb{R}$ with $[a, a, \ldots, a, \ldots] \in {}^*\mathbb{R}$. $\qquad\square$

It is now only necessary to extend the order from $\mathbb{R}$ to make $^*\mathbb{R}$ an ordered field.

Definition 15.21: $^*\mathbb{R}^+ = \{[a_n] \in {}^*\mathbb{R} \mid [a_n] \geq [0]\}$ or, equivalently,

$$[a_n] \geq [b_n] \text{ if and only if } \{n \in \mathbb{N} \mid a_n \geq b_n\} \in U.$$

Theorem 15.22: $^*\mathbb{R}$ is a super ordered field.

Proof: We first need to check that the order is well defined and then that it satisfies the standard properties of order.

To check that it is well defined, we must show that if $[a_n] = [a'_n]$ and $[b_n] = [b'_n]$ then $[a_n] \geq [b_n]$ is the same as $[a'_n] \geq [b'_n]$.

If $[a_n] = [a'_n]$ and $[b_n] = [b'_n]$ then $T_1 = \{n \in \mathbb{N} \mid a_n = a'_n\}$, $T_2 = \{n \in \mathbb{N} \mid b_n = b'_n\}$ satisfy $T_1, T_2 \in U$. So

$$a_n = a'_n \text{ and } b_n = b'_n \text{ for } n \in T_1 \cap T_2.$$

By (U4), $T_1 \cap T_2 \in U$.

If $[a_n] \geq [b_n]$ and $T_3 = \{n \in \mathbb{N} \mid a_n \geq b_n\}$, then $T_3 \in U$ and by (U4) again

$$T = (T_1 \cap T_2) \cap T_3 \in U.$$

For $n \in T$ we have $a_n = a'_n$, $b_n = b'_n$ and $a_n \geq b_n$ which gives $a'_n \geq b'_n$ for $n \in T$ and $T \in U$ as required.

Now consider the standard properties of order:

(O1) $[a_n], [b_n] \in {}^*\mathbb{R}^+ \Rightarrow [a_n] + [b_n], [a_n][b_n] \in {}^*\mathbb{R}^+$
(O2) $[a_n] \in {}^*\mathbb{R} \Rightarrow [a_n] \in {}^*\mathbb{R}^+$ or $-[a_n] \in {}^*\mathbb{R}^+$
(O3) If $[a_n] \in {}^*\mathbb{R}$ and $-[a_n] \in {}^*\mathbb{R}^+$ then $[a_n] = [0]$.

To prove (O1), suppose that $[a_n], [b_n] \in {}^*\mathbb{R}^+$; then, by definition 15.21,

$$T_1 = \{n \in \mathbb{N} \mid a_n \geq 0\} \in U, \ T_2 = \{n \in \mathbb{N} \mid b_n \geq 0\} \in U$$

and, by (U4),

$$T = T_1 \cap T_2 \in U$$

so, for $n \in T$ we have

$$a_n + b_n \geq 0 \text{ and } a_n b_n \geq 0 \text{ where } T \in U.$$

Using definition 15.21 again, this gives

$$[a_n] + [b_n] = [a_n + b_n] \in {}^*\mathbb{R}^+, \ [a_n][b_n] = [a_n b_n] \in {}^*\mathbb{R}^+$$

as required.

To prove (O2), suppose $[a_n] \in \mathbb{R}$, and let

$$T = \{n \in \mathbb{N} \,|\, a_n \geq 0\}.$$

By (U1), either $T \in U$, in which case, by definition 15.21, $[a_n] \in {}^*\mathbb{R}^+$, or

$$T^c = \{n \in \mathbb{N} \,|\, a_n < 0\} \in U$$

in which case

$$\{n \in \mathbb{N} \,|\, -a_n \geq 0\} \in U$$

and so

$$-[a_n] = [-a_n] \in {}^*\mathbb{R}^+.$$

To prove (O3), suppose that $[a_n] \in {}^*\mathbb{R}$ and $-[a_n] \in {}^*\mathbb{R}^+$; then by definition 15.21,

$$T_1 = \{n \in \mathbb{N} \,|\, a_n \geq 0\} \in U, \ T_2 = \{n \in \mathbb{N} \,|\, -a_n \geq 0\} \in U.$$

Again, by (U4), we have

$$T = T_1 \cap T_2 \in U$$

so we have

$$a_n \geq 0 \text{ and } -a_n \geq 0 \text{ for } n \in T,$$

which gives

$$a_n = 0 \text{ for } n \in T \text{ where } T \in U.$$

This completes the proof. $\qquad\qquad\qquad\qquad\qquad\qquad\qquad\qquad\square$

Once ${}^*\mathbb{R}$ has been shown to be a super ordered field, the floodgates open. For example, we can define $\omega = [1, 2, 3, \ldots, n, \ldots]$; then clearly ω is infinite because for any real number k, $n \geq k$ for all $n \geq k$. Furthermore, $1/\omega = [1, \frac{1}{2}, \ldots, \frac{1}{n}, \ldots]$ is an infinitesimal and $\omega + 1 = [1, 2, 3, \ldots, n, \ldots]$ satisfies $\omega + 1 > \omega$ where $\omega + 1 \neq \omega$ because the nth terms $n + 1$ and n

are always different. Now these elements have a full arithmetic and you may check to see that

$$\omega - 2 < \omega - 1 < \omega < \omega + 1 < \ldots < \ldots 2\omega - 1 < 2\omega < \ldots < \omega^2 \ldots,$$

and so on.

You will be able to show that there are elements of different orders; for example, $\omega^2 = [1, 4, 9, \ldots, n^2, \ldots]$ is a higher-order infinite element than ω, and if $\varepsilon = 1/\omega$ then $\varepsilon^2 = [1, 1/2^2, \ldots, 1/n^2, \ldots]$ is of lower order than ε.

The hyperreals $^*\mathbb{R}$ are much more powerful than the original conception of Leibniz, who imagined that infinitesimal elements were first order, second order, and so on. This is true in the field of rational functions $\mathbb{R}(\varepsilon)$ where ε is infinitesimal. If we take the order of ε to be 1, then ε^n is of order n. But there is no element in $\mathbb{R}(\varepsilon)$ whose square is ε. However, in $^*\mathbb{R}$, the element $\varepsilon = [1, 1/2, \ldots, 1/n, \ldots]$ has square root

$$\sqrt{\varepsilon} = [1, 1/\sqrt{2}, \ldots, 1/\sqrt{n}, \ldots]$$

and every non-negative element in $^*\mathbb{R}$ has a square root.

Furthermore, any real function $f : D \to \mathbb{R}$ can be extended very naturally to a function on $^*\mathbb{R}$. The method is astonishingly simple. First, let $^*D$ be the elements of the form $[x_n]$ where all the x_n are in D, giving

$$^*D = \{[x_n] \in {}^*\mathbb{R} \mid x_n \in D\}.$$

Then extend f to $^*D$ by defining

$$f([x_n]) = [f(x_n)].$$

How breathtakingly beautiful this is! The extension $^*D$ of D is made up from equivalence classes whose elements are sequences in D and the extended function $f : {}^*D \to {}^*\mathbb{R}$ is defined in a natural way using these sequences whose elements are already in D and so the definition is mind-bogglingly simple!

Amazing Possibilities in Non-standard Analysis

Once we have the ideas of the hyperreals, we can do amazing things. Consider the extension $^*\mathbb{N}$ of the natural numbers $\mathbb{N}$. By definition, $^*\mathbb{N}$ includes all equivalence classes of sequences of natural numbers, so $\omega = [1, 2, 3, \ldots, n, \ldots] \in {}^*\mathbb{N}$. This shows us that $^*\mathbb{N}$ contains infinite elements. To calculate a limit of a sequence (x_n), we consider the function $f : \mathbb{N} \to \mathbb{R}$ given by $f(n) = x_n$, extend it to $f : {}^*\mathbb{N} \to \mathbb{R}$ and consider $f(N) = x_N$ for infinite $N \in {}^{*}\mathbb{N}$. If x_N is finite then we calculate st(x_N) and, if we get the same value for all infinite N, then this is the limit of the sequence. For instance, if

$$x_n = \frac{6n^2 + n}{2n^2 - 1}$$

then

$$x_N = \frac{6N^2 + N}{2N^2 - 1} = \frac{6 + 1/N}{2 - 1/N^2}$$

and, as $1/N$ is infinitesimal,

$$\text{st}(x_N) = \frac{6 + 0}{2 - 0} = 3.$$

Other definitions such as continuity or uniform continuity can be expressed simply in terms of infinitesimals:

Definition 15.23: $f : D \to \mathbb{R}$ is continuous at $x \in D$ if

$\forall y \in {}^{*}D : x - y$ infinitesimal implies $f(x) - f(x)$ is infinitesimal.

Definition 15.24: $f : D \to \mathbb{R}$ is uniformly continuous in D if

$\forall x \in {}^{*}D, \forall y \in {}^{*}D : x - y$ infinitesimal implies $f(x) - f(x)$ is infinitesimal.

Essentially, the difference between the two is that continuity involves $x \in D, y \in {}^{*}D$ and uniform continuity involves $x, y \in {}^{*}D$.

These ideas extend to more general functions $f : D \to \mathbb{R}^n$ where $D \subseteq \mathbb{R}^m$, and all relationships involving such functions remain true when extended. For instance, if D is the inside of a unit sphere $x^2 + y^2 + z^2 < 1$ in $\mathbb{R}^3$ then D generalises to the unit sphere ${}^{*}D$ in ${}^{*}\mathbb{R}^3$ with the same formula.

Predicates $P(x_1, x_2, \ldots, x_1)$ in n variables such as the commutative law or associative law for elements in $\mathbb{R}$

$$x + y = y + x, \ x(y + z) = xy + xz,$$

extend to the same relationships in ${}^{*}\mathbb{R}$.

If we quantify relationships such as

$$\forall x \in \mathbb{R} \, \forall y \in \mathbb{R} : x + y = y + x$$
$$\exists 0 \in \mathbb{R} \, \forall x \in \mathbb{R} : x + 0 = 0$$
$$\forall x \in \mathbb{R} \, \exists y \in \mathbb{R} : x + y = 0$$

then all these relationships generalise to

$$\forall x \in {}^*\mathbb{R} \; \forall y \in {}^*\mathbb{R} : x + y = y + x$$
$$\exists 0 \in {}^*\mathbb{R} \; \forall x \in {}^*\mathbb{R} : x + 0 = 0$$
$$\forall x \in {}^*\mathbb{R} \; \exists y \in {}^*\mathbb{R} : x + y = 0.$$

But, some properties do not generalise, for instance the completeness axiom. If we look at the completeness axiom, it says

$\forall S \subseteq \mathbb{R} :$ S non-empty and bounded above implies S has a least upper bound.

This axiom quantifies a *set* S. All the other axioms for a complete ordered field only quantify *elements* of a set. It is this observation that makes non-standard analysis work.

Definition 15.25: A quantified predicate is said to be a *first-order* logical statement if it only quantifies *elements* of sets.

All the axioms for a complete ordered field except the completeness axiom are first-order statements. All the first-order axioms extend from $\mathbb{R}$ to $^*\mathbb{R}$. The completeness axiom does not.

The axioms for the natural numbers $\mathbb{N}$ also exhibit the same phenomenon. (N1) and (N2) are first-order statements. However, the induction axiom (N3), which says

$$\forall S \subseteq \mathbb{N} : \text{if} (1 \in S \text{ and } n \in S \Rightarrow n + 1 \in S) \text{ then } S = \mathbb{N},$$

is not. The extension $^*\mathbb{N}$ satisfies (N1) and (N2), but not (N3). For example, the set $S = \mathbb{N}$ is a subset of $^*\mathbb{N}$ and satisfies $1 \in S$ and $n \in S \Rightarrow n + 1 \in S$ but S does not equal $^*\mathbb{N}$ because $\omega \in {}^*\mathbb{N}$ and $\omega \notin \mathbb{N}$.

Non-standard analysis can be shown to satisfy:

The Transfer Principle: Any true first-order logical statements involving elements in $\mathbb{R}$ remain true when extended to $^*\mathbb{R}$.

If this principle is taken as an axiom, then it can be used as a basis for developing the theory of non-standard analysis. However, it is not our intention to take these matters further: this is a book on foundations of mathematics, not non-standard analysis. Our main reason for including material on infinitesimal ideas is to show that, as mathematics evolves, new theories are developed that change the way that we think about mathematics.

At this stage in history, analysis is studied using standard epsilon–delta techniques and there is a very good reason for this. To set up calculus using infinitesimals can be pictured using super ordered fields that give us a natural sense of ideas that have arisen in various forms in earlier generations.

To do non-standard analysis properly requires the construction of an ultrafilter U on the natural numbers. This requires deciding for every subset $T \subseteq \mathbb{N}$ whether T or its complement is in U in a manner that fits the definition of an ultrafilter. It involves making an infinite, even uncountable number of choices. As human beings, we certainly can't do this unaided in our finite lifetime.

The definition of the natural numbers $\mathbb{N}$ requires a potential infinity of elements but we can at least imagine that theoretically we can reach any given element in the sequence, even if this may be utterly impracticable for very large numbers. But to contemplate making an uncountable number of choices to define an ultrafilter seems to demand more than the human brain can take.

If we look back to the different strands of mathematics that emerged at the beginning of the twentieth century in terms of intuitionism, logicism, and formalism, we have a number of different options. An intuitionist would reject non-standard analysis because the construction of an ultrafilter is beyond our human capacity to accomplish in a finite sequence of steps. Errett Bishop took this position in his book on constructive analysis [14]. On the other hand, a logicist may be happy to use first-order logic to formulate the theory, and that is how Abraham Robinson developed the idea [29]. A formalist mathematician, who may use natural ideas to get initial inspiration, subsequently requires theories formulated using set-theoretic definitions and mathematical proof.

In today's mathematical world, pure mathematics broadly follows the formalist approach because the logical foundation required as a basis for non-standard analysis has a high initial cost in terms of the logic required. In this chapter we have shown how ideas about infinitesimals may be visualised in a natural way on a number line using the idea of magnification based on algebraic operations. We have also shown how this leads to a way of defining the system formally in terms of an ultrafilter. This requires a further stretch of imagination that some may be willing to accept as part of a more sophisticated form of mathematics but others may consider to be unattainable.

In the final chapter of this book we will contemplate the next step, strengthening the foundations of mathematics by axiomatising set theory itself. This allows us to include a further axiom—the axiom of choice—that, if taken as an additional axiom for set theory, may be used to prove more powerful results, including the construction of the logical development of infinitesimal calculus.

Exercises

1. In the field $\mathbb{R}(x)$ using the order specified in definition 15.2, which of the following are positive:
   (a) $x^3 - 2x$
   (b) $1/(x^3 - 2x)$
   (c) $x - 1000x^2$
   (d) $1000x^2 - x$
   (e) $a + bx + cx^2$ for real values of a, b, c (taking all possible cases into account)
   (f) $1/(a + bx + cx^2)$ for various real values of a, b, c.

2. Place the following elements of $\mathbb{R}(x)$ in order:

$$x, 0, 2x^2, -x^3, 1/(1 - x), 25, -x, -x^3/(1 - 3x).$$

3. How would you test whether a general rational function

$$\frac{a_n x^n + \cdots + a_0}{b_m x^m + \cdots + b_0} \quad (\text{where } a_r, b_r \in \mathbb{R}, \ b_m \neq 0)$$

is
   (a) infinite
   (b) infinitesimal
   (c) finite.
   Write out a full explanation that makes sense to you in a way that you can explain to someone else, taking care of every possible case.

4. Let F be any ordered field, which must contain the rational numbers. An element $k \in F$ is said to be positive infinite if $k > x$ for all $x \in \mathbb{Q}$. Make similar definitions to say when an element k is
   (a) negative infinite
   (b) finite
   (c) positive infinitesimal
   (d) negative infinitesimal.
   Prove the following:
   (e) k is positive infinite if and only if $1/k$ is positive infinitesimal.
   (f) If k is infinitesimal, then k^2 is infinitesimal.
   (g) if k is infinite and h is finite, then $k - h$ is infinite.
   (h) If $k \in \mathbb{Q}$ and positive and $h > k$, then h cannot be infinitesimal.

5. In proposition 15.20, write out in full detail the proof that the commutative, associative, and distributive properties of the operations of addition and multiplication on $*\mathbb{R}$ all hold.

6. Let K be a super ordered field. Write out a proof that the set I of infinitesimals is bounded above but does not have a least upper bound.

7. Let K be the super ordered field $\mathbb{R}(\varepsilon)$ where ε is infinitesimal and $\varepsilon > 0$. Show that for any infinitesimal $\delta \in \mathbb{R}(\varepsilon)$, the element δ/ε is finite. Let F be the set of finite elements in $\mathbb{R}(\varepsilon)$. Show that the function $\tau : F \to \mathbb{R} \times F$ given by

$$\tau(a + \delta) = (a, \delta/\varepsilon)$$

is an order-preserving bijection in which

$$a + \delta < b + \gamma \Leftrightarrow a < b \text{ or } (a = b \text{ and } \delta < \gamma).$$

In this bijection, show that the monad

$$M_a = \{a + \delta \in K \mid \delta \text{ is } \textit{infinitesimal}\} \text{ in } F$$

corresponds to the vertical line through $a \in \mathbb{R}$. This representation should give you a better sense of why the monads are bounded above but do not have upper bounds. Explain this in your own words.

8. Use the transfer principle with the statement

$$\forall x \in \mathbb{R}, x > 0 \, \exists y > 0 : y^2 = x$$

to deduce that every positive $x \in {}^*\mathbb{R}$ has a square root in $*\mathbb{R}$ and that if $\varepsilon > 0$ is infinitesimal, then its square root $\delta = \sqrt{\varepsilon}$ is a higher-order infinitesimal.
Show that the function $\tau(a + \delta) = (a, \delta/\varepsilon)$ maps the finite elements $a + \delta$ to $\mathbb{R} \times {}^*\mathbb{R}$ in which the image of the monad M_a lies in the vertical line $*\mathbb{R}$ through the point a on the horizontal real line and that $a + \delta < b + \gamma$ for real a, b and infinitesimal δ, γ if and only if $a < b$ or $a = b$ and the elements are in the same monad with $\delta < \gamma$.

9. Using the notation $[a_n]$ to represent the equivalence class of the sequence (a_n) of real numbers, write down an element $[a_n]$ which is:
   (a) the sum of an infinite number and an infinitesimal
   (b) the cube root of $[a_n]$
   (c) a number equal to $\omega = [1, 2, \ldots, n, \ldots]$ where $(a_n) \neq (1, 2, \ldots, n, \ldots)$
   (d) a number of higher order than ω.

10. Reflect on the chapter as a whole, reading it through to explain the ideas to yourself and to discuss these ideas with others. At this point you may not be fluent in operating with the ideas, but it is important to gain a sense of how infinitesimal and infinite quantities can be imagined visually and manipulated algebraically to lead to more sophisticated possibilities in later developments.

PART V
Strengthening the Foundations

Part IV showed how the material developed so far can lead into the main body of mathematics, into ever higher realms. But this final part will lead in the opposite direction: down into the depths.

There is a reason for this.

Having constructed such a fine building, it becomes prudent to re-examine the ground on which it rests. We have replaced a very complicated system of intuitions about numbers by a rather simpler system of intuitions about *sets*. But our set-theoretic basis is still intuitive and informal. If we had built a bungalow, this might not have been important; but we have built a skyscraper—and one that can be extended to much greater heights. It is time to dig a little deeper into the foundations, to see whether they really can support all that weight. Or, in a horticultural analogy, we must make sure that the roots of our plant will support the fully grown organism, which may require us to improve the soil, use higher quality fertiliser, and sow better seed.

In mathematical terms, as expressed by Klein's quote at the end of the opening chapter, the power of mathematics depends not only on building ever more sophisticated branches of the mathematical tree, but also in growing deeper roots to support the ever-growing branches that reach up to the sky.

Our aim here is to indicate what *can* be done, but not actually to do it. So we talk in an informal way about the possibility of a formal system of axioms for set theory itself. It may seem that the argument has come full circle: here we are, right back at the beginning, worrying about the same things as before. In fact this is not so: we have come more in a *spiral*, returning to the same point but at a higher level. We now understand the problems involved, and their solutions, much better than before. The material we have covered so far is quite adequate for almost all of a university course in mathematics. But we should not imagine that we have reached a complete and final solution, or that total perfection has now been attained.

CHAPTER 16

Axioms for Set Theory

U p to this point, we have concentrated on deriving a formal structure for arithmetic based on set theory. This analysis has provided
a deeper understanding of the various number systems, how they
work, and their place in the scheme of things. It should also have sharpened your critical faculties and your appreciation for logical rigour. It may
have sharpened them sufficiently to see that one fundamental ingredient is
still lacking. We have axiomatised everything we can lay hands on, with one
notable exception: set theory itself.

Having taken such pains with the structural detail of the number systems,
it would be a great pity if the basis on which we worked should turn out to be
defective—unable to support the weight of the superstructure erected on it.
In the ultimate analysis, it is hardly more satisfactory to base a formal theory
of numbers on an informal, intuitive, and naive theory of sets than it is to
start with an informal, intuitive, and naive theory of numbers themselves.

However, we may yet escape this criticism by returning to our starting
point and axiomatising set theory as well. (It would, indeed, have been pleasant to have started off from an axiomatic basis for set theory, except that
there are enormous psychological barriers involved in doing something so
far removed from reality with no idea why it is needed.) We will not go into
the details very deeply (see Mendelson [27] if you want to do this), nor shall
we adopt an overly formal style in discussing them. Our aim is merely to
make clear the unconscious assumptions that have been made about sets, to
discard some over-optimistic ones that lead to paradoxes, and to list a system
of axioms that offers a stronger basis for formal mathematical theory.

Historically, some mathematicians hoped for more than this. At the turn
of the century a number of them, led by David Hilbert, embarked upon a
kind of Arthurian Quest for Truth: a firm and immutable basis for mathematics and a guarantee that the truths of mathematics can be rendered
absolute. In this impermanent and uncertain universe, it is hardly surprising
that the Holy Grail turned out, in the end, to be a mare's nest.

Some Difficulties

The problems with naive set theory are of two kinds. First, there are the *paradoxes*: apparently contradictory results obtained by apparently impeccable logic. Then there are purely technical difficulties: are infinite cardinals always comparable? Is there a cardinal between $\aleph_0$ and $2^{\aleph_0}$?

By way of motivation, we consider two paradoxes. The first, due to Bertrand Russell, was alluded to in chapter 3. If

$$S = \{x \,|\, x \notin x\}$$

then is $S \in S$ or $S \notin S$? Either answer directly implies the other!

For the second, let U be the set of *all* things, defined (say) by

$$U = \{x \,|\, x = x\}.$$

Now $X \subseteq U$ for every set X. In particular, the power set $\mathbb{P}(U) \subseteq U$. Taking cardinals,

$$|U| \geq |\mathbb{P}(U)|,$$

but by proposition 12.5 of chapter 12,

$$|U| < |\mathbb{P}(U)|.$$

This is a contradiction: what's wrong?

Many responses are possible, among them:

The Ostrich. Ignore the difficulties and maybe they'll go away.

The Drop-out. The paradoxes point to unavoidable defects in mathematics. Give up, and take up something more profitable such as knitting or sociology.

The Optimist. Re-examine the reasoning, isolate the source of the difficulties, and try to salvage what is worth saving while disposing of the paradoxes.

If you agree with the Ostrich, stop reading here. If with the Drop-out, burn this book. If with the Optimist, read on . . .

Sets and Classes

In the next few sections we discuss one possible solution to the problems, known as *von Neumann–Bernays–Gödel set theory*. This starts from the observation that a plausible source of trouble is the freedom to form weird and very large sets (for example, the two sets S and U defined above). All of the known paradoxes seem to 'cheat' in this way.

We therefore distinguish two things: *classes*, which may be thought of as arbitrary collections (what we hitherto have naively called 'sets'), and *sets*, which are respectable classes. Then we restrict our ability to define weird or large creatures to classes only. This is the idea: the details are roughly as follows.

Classes are introduced as a primitive, undefined term, along with a relation $\in$ (corresponding to the intuitive idea of membership) and its negation $\notin$. If X and Y are classes, then one or other of

$$X \in Y, X \notin Y$$

is required to hold. We define equality of classes $X = Y$ by

$$(\forall Z)(Z \in X \Leftrightarrow Z \in Y).$$

We say that a class X is a *set* if $X \in Y$ for some class Y. This is the crucial definition: sets are those classes that can be *members* of other classes.

This is quite different from the intuitive feeing that sets are things of which other things are *members*. The difference is what makes it hard to define weird and large sets. To make this work, we agree that an expression like

$$\{x \mid P(x)\}$$

means 'the *class* of all *sets* x for which $P(x)$ is true'. This restriction is forced upon us, because only sets can be members of classes anyway. It has the beneficial effect of blocking paradoxes. For example, consider Russell's class

$$S = \{X \mid X \notin X\}.$$

In the new interpretation, this is the class of all *sets* X such that $X \notin X$. Let us run through the usual argument for a contradiction, and see what happens. Suppose that $S \in S$. Then S is a member of something, so it is a set, so $S \notin S$, a contradiction. Now suppose $S \notin S$. If S is a set, then it satisfies the defining property $X \in X$, so by the definition, $S \in S$. This is a contradiction too. There remains, however, the possibility that S is *not* a set. In this case we *cannot* deduce that $S \in S$; elements of S have to be sets as well as not being members of themselves.

The upshot is that we don't get a paradox. All we get is a proof that S is not a set. Classes that are not sets are called *proper classes*; we have just proved that they exist. Similarly U may be proved to be a proper class, and again there is no paradox.

The Axioms Themselves

The majority of the axioms required come as an anticlimax, because all they do is state that things we obviously want to be sets *are* sets. For convenience, we assume that the usual notation of set theory also applies to classes, in the obvious way. For instance we define

$$\varnothing = \{x \mid x \neq x\},$$
$$\{x, y\} = \{x \mid x = u \text{ or } x = y\},$$

and so on.

From now on we make the convention that small letters x, y, z, ... stand for sets, whereas capitals X, Y, Z, ... stand for classes—which may or may not be sets.

(S1) *Extensionality.* $X = Y \Leftrightarrow (\forall Z)(X \in Z \Leftrightarrow Y \in Z)$.

We have defined equality of classes as 'having the same members'. This purely technical axiom says that equal classes belong to the same things.

(S2) *Null set.* $\varnothing$ is a set.

(S3) *Pairs.* $\{x, y\}$ is a set for all sets x, y.

We now define *singletons* by $\{x\} = \{x, x\}$, then *ordered pairs* using the Kuratowski definition $(x, y) = \{\{x\}, \{x, y\}\}$, then functions, relations, as before.

(S4) *Membership.* $\in$ is a relation, that is, there exists a class M of ordered pairs (x, y) such that $(x, y) \in M \Leftrightarrow x \in y$.

(S5) *Intersection.* If X, Y are classes, there is a class $X \cap Y$.

(S6) *Complement.* If X is a class, its complement X^c exists and is a class.

(S7) *Domain.* If X is a class of ordered pairs, there exists a class Z such that $u \in Z \Leftrightarrow (u, v) \in X$ for some v.

Much more interesting is an axiom for defining a class by a property of its elements, analogous to $\{x \mid P(x)\}$. We state here a general axiom: it can be derived if desired from a small number of more specialised axioms of the same type.

(S8) *Class existence.* Let $\phi(X_1, \ldots, X_n, Y_1, \ldots, Y_m)$ be a compound predicate statement in which only set variables are quantified. Then there exists a class Z such that

$$(x_1, \ldots, x_n) \in Z \Leftrightarrow \phi(x_1, \ldots, x_n, Y_1, \ldots, Y_m).$$

We write

$$Z = \{(x_1, \ldots x_n) \mid \phi(x_1, \ldots x_n), (Y_1, \ldots Y_m)\} .$$

Notice that the xs here are *sets*. In particular, the class

$$Z = \{x \mid P(x)\}$$

contains as members only those *sets* x for which $P(x)$ is true. This, as we saw above, allows us to avoid paradoxes.

(S9) *Union.* The union of a set of sets is a set.

(S10) *Power set.* If x is a set, so is $\mathbb{P}(x)$.

(S11) *Subset.* If x is a set and X a class, then $x \cap X$ is a set.

There is also an axiom that asserts a slight generalisation of the following:

(S12) *Replacement.* If f is a function whose domain is a set, then its image is a set.

These axioms suffice for almost all of the constructions we have made using set theory. However, they all hold good even if we restrict ourselves only to *finite* sets. We therefore need an axiom to say that infinite sets exist, otherwise we cannot construct any of our beloved number systems. We therefore add an axiom introduced in chapter 8 (von Neumann's brainwave):

(S13) *Axiom of infinity.* There exists a set x such that $\varnothing \in x$, and whenever $y \in x$ it follows that $y \cup \{y\} \in x$.

Using von Neumann's definition of natural numbers, this axiom boils down to the assertion that the natural numbers form a *set*. It is pretty clear that without some such assertion, set theory would not be much use.

The thirteen axioms listed so far suffice for almost all of our previous work, though a detailed proof is (as usual) somewhat involved and tedious. However, some of the problems in the chapters on cardinals and infinitesimals require more delicate axioms yet.

The Axiom of Choice

Proposition 12.5 of chapter 12 used an argument that involved selecting an element x_1 from a set B, then x_2 from $B \backslash \{x_1\}, \ldots$, and in general an element x_{n+1} from $B \backslash \{x_1, \ldots, x_n\}$. Although this looks like a recursion argument, it is not covered by the recursion theorem (theorem 8.3 of chapter 8), since x_{n+1}

is found by an arbitrary choice and not in terms of a previously specified function. Roughly speaking, the method asks us to make 'infinitely many arbitrary choices'. It turns out (though not easily!) that the list of axioms we have so far produced is insufficient to justify this. We therefore state an additional axiom:

(S14) *Axiom of choice.* If $\{x_a\}_{a\in a}$ is an indexed family of sets (with an index set a) then there exists a function f such that

$$f : a \to \bigcup_{\alpha \in a} x_\alpha$$

and

$$f(\alpha) \in x_\alpha \text{ for each } \alpha \in a.$$

In other words, f 'chooses' for each $\alpha \in a$ an element of x_α. This seems quite reasonable. After all, it is essentially saying that if we have a family of sets, we can choose an element from each one of them all at the same time. But its logical status proves to be difficult to grasp, though it is now well understood by mathematical logicians.

Neither its truth nor its falsity contradicts axioms (S1)–(S13) (in the same way that neither the truth nor the falsity of the commutative law contradicts the axioms for a group: there exist both commutative and non-commutative groups). The first fact was proved by Kurt Gödel in 1940, the second (a long-unsolved problem) by Paul Cohen in 1963. For this reason it is customary in mathematics to point out whenever the axiom of choice is being used, whereas the ordinary axioms (S1)–(S13) are not normally mentioned.

Assuming the axiom of choice allows us to tidy up two loose ends that arose in the chapters on infinite cardinals and infinitesimals. It implies that for any sets x, y, either $|x| \geq |y|$ or $|y| \geq |x|$, so that any two infinite cardinals can be compared. (For a proof, see Mendelson [27] p. 198.) It also gives a proof that an ultrafilter can be defined on the natural numbers, hence providing a proof of the existence of the hyperreal number system. This requires considering each subset $T \subseteq \mathbb{N}$ and placing it into the set U of subsets so that the conditions (U1)–(U4) are satisfied. We can start by placing every cofinite set in U and every finite set into its complement U^c. Then we consider other sets that have not yet been assigned and decide whether they should be placed in U or not, while still maintaining the conditions (U1)–(U4). Since the sets concerned are in the power set $\mathbb{P}(\mathbb{N})$, which has cardinal number greater than that of $\mathbb{N}$, it turns out that we cannot prove this by a regular induction proof but we can prove it using the axiom of choice (see, for example, [9] on the internet).

As mathematics grows more sophisticated, it turns out that new possibilities occur that need additional axioms. For example, Cantor formulated the *Continuum Hypothesis* that

there is no infinite cardinal lying properly between $\aleph_0$ and $2^{\aleph_0}$.

It happens that neither its truth nor its falsity contradicts (S1)–(S13), or even (S1)–(S14). The proofs are again due to Gödel and Cohen. It is perhaps surprising that so specific a problem should have such an unspecific answer; but it shows how delicate the problems are.

Other, different, axioms have also been proposed at various times, and many of the relations between them are now understood quite well. We refer the reader to more specialised texts.

Consistency

However, there is one final problem. Having got our set of axioms, how do we *know* that no paradoxes arise? We certainly seem to have avoided them (for instance, no one has ever been able to find any), but how can we be *certain* there are no hidden contradictions? A firm, final answer to this question is now known. Unfortunately, this is it: we can *never* be certain.

To explain this, we must go back to the time of Hilbert. Call a system of axioms *consistent* if it does not lead to logical contradictions. Hilbert wanted to prove that the axioms for set theory are consistent.

For some axiom systems this is easy. If we can find a *model* for the axioms, that is, a structure that satisfies them, they must be consistent—or else the model could not exist. The trouble is, what materials do we allow for the construction of the model? It is generally agreed that a *finite* model is unexceptionable, because any assertion about it can be checked, in principle, in a finite time. But the axiom of infinity, for example, means that we cannot find a finite model for set theory.

Hilbert's idea was that something less restricted should suffice: what he called a *decision procedure*. This is, so to speak, a program consisting of a finite sequence of decisions which, when fed a formula in set theory, can decide whether it is true (like the truth-table method for propositions). If we can find such a program, and *prove* that it always works, then we can feed it the equation

$$0 \neq 0$$

and see what it says. If it says 'true' then our axioms must be inconsistent, since any contradiction implies the above proposition (use a vacuous argument by contradiction: *anything* is true in an inconsistent system!).

For a while it looked as if Hilbert's idea might work.

Then Gödel dashed all hopes by proving two theorems. The first is that there exist, in set theory, theorems that are true, but for which there neither exists a proof nor a disproof.[1] The second: that if set theory is consistent, then there does not exist any decision procedure that will prove it to be.

The proofs of Gödel's theorems are quite technical: they are sketched in Stewart [32] pp. 294–5. But they demolish Hilbert's hope of a complete consistency proof.

Does this mean that it is, after all, pointless to seek greater logical rigour in mathematics? After all, if at the end the whole thing hovers in limbo, it hardly seems worth bothering in the first place. This is emphatically *not* the moral to be drawn. Without a proper search for rigour, we would never have reached Gödel's theorems. What they do is pin down certain problems inherent in the axiomatic approach itself.

They do not demonstrate it to be futile: on the contrary, it provides an adequate framework for the whole of modern mathematics, and an inspiration for the development of new ideas. But with Gödel's theorems we can avoid deluding ourselves that everything is perfect, and understand the limitations of the axiomatic method as well as its strengths.

Exercises

1. Show that the axiom of choice implies that if $f : A \to B$ is a surjection, then $|A| \geq |B|$. Conversely, in the context of the other axioms of set theory, prove that the latter fact implies the axiom of choice.

2. Given a collection of sets $\{X_\alpha\}_{\alpha \in A}$ indexed by a set A, the *cartesian product* is defined to be the set of all functions $f : A \to \bigcup_{\alpha \in A} X_\alpha$ such that $f(\alpha) \in X_\alpha$. Show that for $A = \{1, 2, \ldots, n\}$ this corresponds to the usual definition of $X_1 \times X_2 \times \cdots \times X_n$.

   Prove that the axiom of choice is equivalent to the assertion that every cartesian product of non-empty sets is itself non-empty.

3. Show that there is a choice involved in the proof of proposition 12.1 of chapter 12. Express it in terms of a function from a set of subsets of B to B. Is it necessary in this case to include all the subsets of B in the choice?

4. Reconsider Goldbach's conjecture (exercise 13 at the end of chapter 8), which postulates that every positive even integer is the sum of two

[1] People always put it this way, but, interestingly, the *negation* of such a statement is *also* 'true'. Both the statement and its negation are consistent with the other axioms.

primes. Look at as many cases of this as you wish to see if there is any pattern to the primes which occur. Convince yourself that Goldbach's conjecture might be true but there may be no single proof which will work for every case. On the other hand, there is always the possibility that the conjecture is false for some very large integer which we have not yet found.

5. Given a predicate $P(n)$ valid for all $n \in \mathbb{N}$, such that a proof for each $P(n)$ exists in a finite number of lines as explained in chapter 6, is it reasonable to expect that there is a proof of

$$\forall n \in \mathbb{N} : P(n)$$

in this sense?

6. Read chapter 1 again and the introductions to each of the five parts into which the book is divided. Now review the exercises at the end of chapter 1. If you still have the solutions that you wrote out at the time you first read chapter 1, so much the better. If the book has achieved its purpose, your view on many of these topics will have matured and changed. You should now be in a position to appreciate the kind of thinking used in more advanced mathematics, together with an idea of the sort of problems in the foundations of the subject which are worthy of further study.

How to Read Proofs: The 'Self-Explanation' Strategy

Prepared by Lara Alcock, Mark Hodds, Matthew Inglis,
Mathematics Education Centre, Loughborough University

The 'self-explanation' strategy has been found to enhance problem solving and comprehension in learners across a wide variety of academic subjects. It can help you to better understand mathematical proofs: in one recent research study students who had worked through these materials before reading a proof scored 30% higher than a control group on a subsequent proof comprehension test (see [3]).

How to Self-Explain

To improve your understanding of a proof, there is a series of techniques you should apply. After reading each line:

- Try to identify and elaborate the main ideas in the proof.
- Attempt to explain each line in terms of previous ideas. These may be ideas from the information in the proof, ideas from previous theorems/proofs, or ideas from your own prior knowledge of the topic area.
- Consider any questions that arise if new information contradicts your current understanding.

Before proceeding to the next line of the proof you should ask yourself the following:

- Do I understand the ideas used in that line?
- Do I understand why those ideas have been used?
- How do those ideas link to other ideas in the proof, other theorems, or prior knowledge that I may have?
- Does the self-explanation I have generated help to answer the questions that I am asking?

On the next page you will find an example showing possible self-explanations generated by students when trying to understand a proof (the labels '(L1)' etc. in the proof indicate line numbers). Please read the example carefully in order to understand how to use this strategy in your own learning.

Example Self-Explanations

Theorem: No odd integer can be expressed as the sum of three even integers.

Proof:

(L1) Assume, to the contrary, that there is an odd integer x, such that $x = a + b + c$, where a, b, and c are even integers.

(L2) Then $a = 2k$, $b = 2l$, and $c = 2p$, for some integers k, l, and p.

(L3) Thus $x = a + b + c = 2k + 2l + 2p = 2(k + l + p)$.

(L4) It follows that x is even; a contradiction.

(L5) Thus no odd integer can be expressed as the sum of three even integers. □

After reading this proof, one reader made the following self-explanations:

- 'This proof uses the technique of proof by contradiction.'
- 'Since a, b, and c are even integers, we have to use the definition of an even integer, which is used in L2.'
- 'The proof then replaces a, b, and c with their respective definitions in the formula for x.'
- 'The formula for x is then simplified and is shown to satisfy the definition of an even integer also; a contradiction.'
- 'Therefore, no odd integer can be expressed as the sum of three even integers.'

Self-Explanation Compared with Other Comments

You must also be aware that the self-explanation strategy is not the same as *monitoring* or *paraphrasing*. These two methods will not help your learning to the same extent as self-explanation.

Paraphrasing

'a, b, and c have to be positive or negative, even whole numbers.'

There is no self-explanation in this statement. No additional information is added or linked. The reader merely uses different words to describe what is already represented in the text by the words 'even integers'. You should avoid using such paraphrasing during your own proof comprehension. Paraphrasing will not improve your understanding of the text as much as self-explanation will.

Monitoring

'OK, I understand that $2(k + l + p)$ is an even integer.'

This statement simply shows the reader's thought process. It is not the same as self-explanation, because the student does not relate the sentence to additional information in the text or to prior knowledge. Please concentrate on self-explanation rather than monitoring.

A possible self-explanation of the same sentence would be:

'OK, $2(k + l + p)$ is an even integer because the sum of 3 integers is an integer and 2 times an integer is an even integer.'

In this example the reader identifies and elaborates the main ideas in the text. They use information that has already been presented to understand the logic of the proof.

This is the approach you should take after reading every line of a proof in order to improve your understanding of the material.

Practice Proof I

Now read this short theorem and proof and self-explain each line, either in your head or by making notes on a piece of paper, using the advice from the preceding pages.

Theorem: There is no smallest positive real number.

Proof: Assume, to the contrary, that there exists a smallest positive real number.

Therefore, by assumption, there exists a real number r such that for every positive number s, $0 < r < s$.

Consider $m = r/2$.

Clearly, $0 < m < r$.

This is a contradiction because m is a positive real number that is smaller than r.

Thus there is no smallest positive real number. □

Practice Proof 2

Here's another more complicated proof for practice. This time, a definition is provided too. Remember: use the self-explanation training after *every* line you read, either in your head or by writing on paper.

Definition: An *abundant* number is a positive integer n whose divisors add up to more than $2n$. For example, 12 is abundant because $1 + 2 + 3 + 4 + 6 + 12 > 24$.

Theorem: The product of two distinct primes is not abundant.

Proof: Let $n = p_1p_2$, where p_1 and p_2 are distinct primes. Assume that $2 \leq p_1$ and $3 \leq p_2$.

The divisors of n are 1, p_1, p_2, and p_1p_2.

Note that $\frac{p_1+1}{p_1-1}$ is a decreasing function of p_1.

So $\max\left(\frac{p_1+1}{p_1-1}\right) = \frac{2+1}{2-1} = 3$.

Hence $\frac{p_1+1}{p_1-1} \leq p_2$.

So $p_1 + 1 \leq p_1p_2 - p_2$.

So $p_1 + 1 + p_2 \leq p_1p_2$.

So $1 + p_1 + p_2 + p_1p_2 \leq 2p_1p_2$. $\qquad\qquad\square$

Remember . . .

Using the self-explanation strategy has been shown to substantially improve students' comprehension of mathematical proofs. Try to use it every time you read a proof in lectures, in your notes or in a book.

REFERENCES
AND FURTHER READING

References

1. E. Bills and D. O. Tall. Operable definitions in advanced mathematics: the case of the least upper bound, *Proceedings of PME 22*, Stellenbosch, South Africa **2** (1998) 104–111.

2. A. W. F. Edwards. *Cogwheels of the Mind*, Johns Hopkins University Press, Baltimore 2004.

3. M. Hodds, L. Alcock, and M. Inglis. Self-explanation training improves proof comprehension, *Journal for Research in Mathematics Education* **45** (2014) 98–137.

4. F. Klein. *Vorträge uber den Mathematischen Unterricht an Höheren Schulen* (ed. R. Schimmack) 1907. The quote here comes from the English translation *Elementary Mathematics from an Advanced Standpoint*, Dover, New York 2004 p.15.

5. L. Li and D. O. Tall. Constructing different concept images of sequences and limits by programming, *Proceedings of the 17th Conference of the International Group for the Psychology of Mathematics Education, Japan* **2** (1993) 41–48.

6. M. M. F. Pinto. *Students' Understanding of Real Analysis*, PhD Thesis, University of Warwick 1998. See [36] chapter 10 for a fuller discussion including the work of other researchers.

7. C. Reid. *Hilbert*. Springer, New York 1996 p.57.

8. I. N. Stewart. Secret narratives of mathematics, in *Mission to Abisko* (eds J. Casti and A. Karlqvist), Perseus, Reading 1999, 157–185.

9. Ask Dr. Math, <http://mathforum.org/dr.math/faq/analysis_hyperreals.html>.

Further Reading

Books requiring extra mathematical background are marked with an asterisk.

10. L. Alcock. *How to Study for a Mathematics Degree*, Oxford University Press, Oxford 2012.

11. M. Anthony and M. Harvey. *Linear Algebra: Concepts and Methods*, Cambridge University Press, Cambridge 2012.

12. A. H. Basson and D. J. O'Connor. *Introduction to Symbolic Logic*, University Tutorial Press, London 1953.

13. D. W. Barnes and J. M. Mack. *An Algebraic Introduction to Mathematical Logic*, Springer, New York 1975.

14. *E. Bishop. *Foundations of Constructive Analysis*, Academic Press, New York 1967.

15. R. P. Burn. *Groups: A Path to Geometry*, Cambridge University Press, Cambridge 1985.

16. K. Ciesielski. *Set Theory for the Working Mathematician*, London Mathematical Society Student Texts **39**, Cambridge University Press, Cambridge 1997.

17. *K. Gödel. *On Formally Undecidable Propositions of Principia Mathematica and Related Systems* (translated by B. Meltzer), Oliver and Boyd, Edinburgh 1962.

18. P. R. Halmos. *Naive Set Theory*, Van Nostrand, New York 1960; reprinted by Martino Fine Books, Eastford 2011.

19. N. T. Hamilton and J. Landin. *Set Theory*, Prentice-Hall, London 1961.

20. S. Hedman. *A First Course in Logic*, Oxford University Press, Oxford 2008.

21. J. F. Humphreys. *A Course in Group Theory*, Oxford University Press, Oxford 1996.

22. *J. Keisler. *Foundations of Infinitesimal Calculus*, <http://www.math.wisc.edu/~keisler/foundations.html>.

23. F. Klein. *Elementary Mathematics from an Advanced Standpoint*, Dover, New York 2004.

24. M. Kline. *Mathematics in the Modern World: Readings from Scientific American*, Freeman, San Francisco 1969.

25. K. Kunen. *Set Theory*, College Publications, London 2011.

26. S. K. Langer. *An Introduction to Symbolic Logic*, Dover, Mineola 2003.

27. *E. Mendelson. *Introduction to Mathematical Logic*, Van Nostrand, Princeton 1964.

28. P. M. Neumann, G. A. Stoy, and E. C. Thompson. *Groups and Geometry*, Oxford University Press, Oxford 1994.

29. *A. Robinson. *Non-standard Analysis,* Princeton University Press, Princeton 1966 (2nd ed. 1996).

30. R. R. Skemp. *The Psychology of Learning Mathematics*, Penguin, Harmondsworth 1971.

31. M. Spivak. *Calculus*, Benjamin, Reading 1967.

32. I. N. Stewart. *Concepts of Modern Mathematics*, Penguin, Harmondsworth 1975.

33. *I. N. Stewart. *Galois Theory* (3rd edition), CRC Press, Boca Raton 2003.
34. I. N. Stewart. *Symmetry: A Very Short Introduction*, Oxford University Press, Oxford 2013.
35. *I. N. Stewart and D. O. Tall. *Complex Analysis*, Cambridge University Press, Cambridge 1983.
36. D. O. Tall. *How Humans Learn to Think Mathematically*, Cambridge University Press, New York 2013.

Online Reading

<http://en.wikipedia.org/wiki/Foundations_of_mathematics>
<http://en.wikipedia.org/wiki/Set_theory>
<http://mathworld.wolfram.com/SetTheory.html>
<http://www-groups.dcs.st-and.ac.uk/history/HistTopics/Beginnings_of_set_theory.html>
<http://en.wikipedia.org/wiki/Mathematical_logic>
<http://en.wikipedia.org/wiki/First-order_logic>
<http://mathworld.wolfram.com/First-OrderLogic.html>
<http://en.wikipedia.org/wiki/Formalism_%28mathematics%29>
<http://en.wikipedia.org/wiki/G%C3%B6del%27s_incompleteness_theorems>
<http://en.wikipedia.org/wiki/Abstract_algebra>
<http://mathworld.wolfram.com/AbstractAlgebra.html>
<http://www.extension.harvard.edu/open-learning-initiative/abstract-algebra>
<http://abstract.ups.edu/>
<http://mathworld.wolfram.com/AxiomaticSetTheory.html>
<http://en.wikipedia.org/wiki/Real_number>
<http://mathworld.wolfram.com/RealNumber.html>
<http://en.wikipedia.org/wiki/Complex_number>
<http://mathworld.wolfram.com/ComplexNumber.html>
<http://en.wikipedia.org/wiki/Cardinal_number>
<http://mathworld.wolfram.com/CardinalNumber.html>
<http://mathforum.org/dr.math/faq/analysis_hyperreals.html>

INDEX

A

Abel, N. H. 287
abelian group 287
absolute value 32, 235
Alcock, L. vii, 377, 383
all and some 125
Anthony, M. 384
Archimedes' condition 27
Argand, J. R. 231
arithmetic modulo n 84
associative 108, 116, 286
  general law 178, 288
automorphism 216
axiom of choice 371
axiom of infinity 176, 371
axiomatic systems 151, 157, 255, 257
axioms for set theory 367–375

B

Barnes, D. W. 384
Basson, A. H. 384
bijection 102
Bills, E. 383
binary operations 115
Bishop, E. 361, 384
bounded 36
  above 38, 41
  below 38, 42
  sets 41
Burn, R. P. 384

C

calculus with infinitesimals 349
Cantor, R. 202, 316, 317, 335, 339, 341, 373
Cantor–Dedekind Axiom 335
cardinal arithmetic 326
cardinal numbers vii, viii, 12, 275, 314–323
Cartesian product 75–76, 374
Cauchy, A. L. 202, 259, 341

Cauchy sequence 201
  in a general ordered field 227
Ciesielski, K. 384
class 368
codomain 96
cofinite set 351
Cohen, P. J. 372, 373
commutative 116
  ring 191
  group 267
  law 287
complement 65, 66, 370
complete ordered field 188, 190, 208–228,
  262, 334
completeness 36, 189, 207
complex numbers 7, 229–247
  construction 232
  conjugation 234
  modulus 235
composition of functions 107–109, 116, 280
compound predicate 135
compound statement 135
concept formation 4
congruence modulo n 84
conjugate
  of a complex number 235
  of a quaternion 250
connective 130
contextual technique 149, 180
consistency 373
continuum hypothesis 373
contradiction 137
  proof 139
contrapositive 139
convergence 34–36, 40, 125, 128, 129, 201,
  264–265
convergent sequence 36, 265
coprime 180
corollary 151
coset 296